Day's Veterinary Immunology

Michael Day's *Veterinary Immunology: Principles and Practice* is the adopted text in numerous veterinary schools throughout the world. Updated and revised by Brian Catchpole and Harm HogenEsch with advances in knowledge since 2014, this third edition reflects the rapid developments in the field internationally, while preserving the strengths of Day's original writing. It adds numerous case studies demonstrating the clinical context across companion and farm animals.

The textbook presents information on commonly used diagnostic test procedures and includes learning objectives at the start and key points at the end of each chapter, standard symbols in diagrams throughout the text to provide continuity, clinical examples and clinicopathological figures throughout, and a glossary of terms and list of commonly used abbreviations. Short animations are viewable via the Support Materials tab on the Routledge webpage, adding a new element of practical application.

Exploring the immunological principles of both large and small animals, the book emphasizes immunological principles while applying them to disease processes and to clinical practice. It provides a practical textbook for veterinary students and a handy reference for practitioners.

Day's Veterinary Immunology

Principles and Practice

3rd Edition

BRIAN CATCHPOLE AND HARM HOGENESCH

CRC Press
Taylor & Francis Group
Boca Raton London New York

CRC Press is an imprint of the
Taylor & Francis Group, an **informa** business

Third edition published 2023
by CRC Press
6000 Broken Sound Parkway NW, Suite 300, Boca Raton, FL 33487–2742

and by CRC Press
4 Park Square, Milton Park, Abingdon, Oxon, OX14 4RN

CRC Press is an imprint of Taylor & Francis Group, LLC

ISBN: 978-1-032-31717-5 (hbk)
ISBN: 978-1-032-31716-8 (pbk)
ISBN: 978-1-003-31096-9 (ebk)

DOI: 10.1201/9781003310969

Typeset in Janson
by Apex CoVantage, LLC

Printed in the UK by Severn, Gloucester on responsibly sourced paper

CONTENTS

Veterinary Immunology—Principles and Practice has the simple aim of providing undergraduate veterinary students with core knowledge of veterinary immunology, while emphasizing the clinical relevance of this subject area. Veterinary students often struggle with the complexity of the immune system and find it difficult to relate immunological concepts to veterinary practice. In many universities, immunology is taught to veterinary undergraduates at a superficial level and the subject is also often delivered by basic scientists with an inappropriate focus on murine or human systems. Despite this, immunology is a key subject in the veterinary curriculum; it links together subject areas such as biochemistry, microbiology, parasitology and pathology with clinical medicine. One of the main clinical applications of immunology is the practice of vaccination, and all graduating veterinarians must have a solid understanding of the principles of vaccinology.

There are relatively few textbooks of veterinary immunology to support the delivery of undergraduate courses in this subject. I hope that *Veterinary Immunology—Principles and Practice* will fill a key niche in providing students with an affordable and practical reference that will also have relevance as they progress through a practice career. The key features of this book include:

- A sufficient level of detail of core knowledge without this being diluted by minutiae.
- Clear definition of learning objectives and bullet-point summaries of key points, with key words highlighted throughout the text in bold font and listed in a glossary.
- An affordable price for a full colour text supported by numerous diagrams and photographic images.
- An emphasis on clinical examples with a series of 15 clinical case studies that present core information related to clinically significant immune-mediated diseases. This aspect is truly unique to this text and is found in no other reference book on this subject. These case studies should very clearly indicate to the student the major clinical relevance of immunology. They provide a bridge between this book and the more practitioner-orientated *Clinical Immunology of the Dog and Cat*. Both of these texts share fundamental aspects of presentation (e.g. diagrammatic symbols), which should make it straightforward to progress from one to the other.

This book is born from the lecture course in veterinary immunology that I have delivered over the past 20 years to students at the University of Bristol. I am delighted to have had the opportunity of working with Professor Ron Schultz on this project. Ron has made insightful suggestions on the text and ensured that the content also covers the curriculum needs of students in North America. I am very grateful to Ron for sharing his vast experience in veterinary immunology and helping shape this major new resource. Both of us share a passion for this subject and the teaching of it to veterinary undergraduates. However, it is our belief that *Veterinary Immunology—Principles and Practice* will not only have relevance to the student market, but should serve as a simple reference for veterinarians already in practice.

I would like to acknowledge the support and enthusiasm for this project from Michael Manson and Commissioning Editor Jill Northcott.

This is my fourth book for Manson Publishing, but the first truly undergraduate textbook. My thanks go to the production team of Kate Nardoni (project manager), Susan Tyler (illustrator) and Peter Beynon (copy editor) for their professional input into the finished product.

M. J. Day
January 2011

PREFACE (SECOND EDITION)

I was delighted with the response to publication of the first edition of *Veterinary Immunology—Principles and Practice* in 2011. The book was very well received by veterinary students and has now become the adopted text in numerous veterinary schools throughout the world. The text achieved its goal of providing a concise and affordable book supported by high-quality colour illustration. Many colleagues who teach veterinary immunology were highly complimentary about this new resource and offered constructive suggestions for changes to a second edition. Teachers of the subject were also very pleased with the opportunity, provided by the publishers, to obtain teaching sets of the images and diagrams used in the book.

Immunology is a rapidly developing subject area and in order for this textbook to remain current the text of the second edition has been widely updated with advances in knowledge since 2011. The format and style of the text remains the same, but the second edition contains around 20 new and updated figures, one new table and two new clinical case studies. New developments in fundamental and clinical veterinary immunology are reported, with expanded information on commonly used diagnostic test procedures and the inclusion of newly arising diseases such as bovine neonatal pancytopenia.

I have been delighted to have the opportunity once more of collaborating with my friend and colleague, Professor Ron Schultz, on this second edition. I would also like to acknowledge my Commissioning Editor, Jill Northcott, and the production team of CRC Press.

It is now over 15 years since I first began working with Manson Publishing on the production of my first book, *Clinical Immunology of the Dog and Cat*. This second edition of *Veterinary Immunology—Principles and Practice* would have been my sixth Manson publication, but as the final manuscript for this edition was submitted, Michael announced his retirement and the transfer of Manson Publishing to CRC Press. I would like to acknowledge my long and fruitful collaboration with Michael and wish him the very best for the future.

M. J. Day
May 2014

We are publishing the third edition of *Day's Veterinary Immunology: Principles and Practice* with mixed emotions. Work on the new edition commenced in 2019, when Professor Michael Day approached Professor Brian Catchpole to collaborate with this commission from the publishers. Despite making good headway with the revisions during the early months of 2020, progress was interrupted by the COVID-19 pandemic and subsequently Michael's untimely death in May of that year, following his battle with cancer. After a period of relatively slow progress, the project was reinvigorated when Professor Harm HogenEsch joined the team and we resumed the work on the revisions in 2021. Our goal was to provide an updated edition of the textbook that was faithful to the remit of the previous editions and which would honor Michael's legacy. He was a great scientist, an internationally renowned veterinary immunologist, a prolific author and superb communicator. Furthermore, Michael was one of the most encouraging, altruistic and (above all) kind people to have worked in academia, and he played a substantial part in the career development of many undergraduate and postgraduate students. It is satisfying to see the new edition of his textbook finally come to fruition and we hope that Michael would be pleased with the result.

The format and style of this new edition remains largely similar to the previous editions, but with some reorganization of chapters, updated figures and revised text. We continue to strive to maintain a balance between the increasing depth of detailed information that is published daily in our field and the need for understanding immunological concepts and providing practical information for veterinary students and practitioners. Thus, we have focussed on a principle-based approach to veterinary immunology, with a clinical context wherever possible. In doing so, we hope that this textbook will continue to be useful for veterinary students, general practitioners and veterinary clinical specialists.

This edition was largely written during the COVID-19 pandemic, which has made immunology more relevant than ever. We have witnessed the devastating effects of an infectious disease; the power of science in rapid development of novel immunoassays, vaccines and antiviral drugs; alongside the challenges surrounding testing and vaccination of large populations. This is also an exciting time in veterinary immunology; the tools and techniques to study immune responses and immune-mediated diseases in domestic animals are rapidly improving and catching up with human and mouse immunology. These advances not only demonstrate broad similarities in the immune system across various species but also highlight more subtle differences, emphasizing the importance of comparative immunology and not making assumptions based on human medicine when dealing with veterinary patients. Over the past decade, we have seen monoclonal antibodies for treatment of atopic dermatitis and pain in dogs and cats becoming commercially available, while novel forms of immunotherapy for cancer in companion animals are currently undergoing clinical trials. New vaccine technologies are being applied to protect food animals from infectious diseases, with improved safety and efficacy. Veterinary immunology has never been more interesting and relevant to clinicians and the general public.

Finally, we would like to thank our editor, Alice Oven, for her patience and support during the writing process; the production team of Taylor & Francis, CRC Press for their efforts with the draft manuscript; as well as the reviewers for their time and constructive comments.

Brian Catchpole and Harm HogenEsch,
September 2022

Brian Catchpole is Professor of Companion Animal Immunology at the Royal Veterinary College, University of London. He graduated from the RVC in 1992 and, after a period of time in veterinary clinical practice, returned to London to undertake a Wellcome Trust veterinary research training fellowship in the Infection & Immunity Group at King's College London. After gaining his PhD, he returned to the RVC, initially as a postdoctoral research fellow and subsequently as a member of the academic staff. Brian's research is focussed on canine immunology, including studies into infectious disease (leptospirosis), immunosenescence, cancer immunotherapy and immune-mediated disease in dogs. Brian has been teaching immunology to veterinary undergraduates for the past 20 years.

Harm HogenEsch is Distinguished Professor of Immunopathology in the Department of Comparative Pathobiology at the Purdue University College of Veterinary Medicine. He received his DVM degree from the University of Utrecht, the Netherlands, and his PhD in immunology from the University of Illinois at Urbana-Champaign. He is a diplomate of the American College of Veterinary Pathologists. He worked for three years at the Institute for Aging and Vascular Research in Leiden, the Netherlands, before joining Purdue University. Harm's research interests include vaccine development and the immunopathology of allergic skin disease. He has published more than 150 peer-reviewed articles and 22 book chapters. Harm teaches an immunology course for first-year veterinary students and graduate students.

"Information is clearly presented and well written, layering immunological knowledge with each chapter. I found the chapter lengths perfect—providing comprehensive information without being overwhelming. The objectives of each chapter are defined at the beginning and key points concisely reviewed at the end, allowing easy navigation and assisting comprehension. The excellent illustrations are a boon for visual learners. Immunological systems, diagnostic test procedures, clinical disease, therapy, and recent developments in the field are explored. Finally a series of case studies helps bring together and test your understanding. . . . A practical textbook for students and handy reference for clinicians. . . . the authors kept a challenging subject alive with analogy, illustration and clear explanations, providing a basic knowledge of immune function and clearly relating this to clinical practice. I would recommend this book to anyone challenged, intrigued or inspired by veterinary immunology."

—Caroline Blundell, BVetMed CertSAM RCVS, in the *Journal of Feline Medicine and Surgery*

". . . an excellent tool to understand the basic principles of immunology and veterinary immunology . . . updated with the most recent advances in veterinary immunology developments . . . the basic knowledge provided is always practice-oriented, with practical examples on disease pathogenesis and diagnosis. This is why this book is not only a useful tool for undergraduates but also for postgraduate veterinarians either doing a PhD, internship or residency in any area of veterinary medicine. But it is also a useful tool for veterinary clinicians who are keen to have a solid and deeper understanding of the immunology of diseases and routine procedures (such as vaccination or immunotherapy). . . . The format of this book is very helpful, with a great number of illustrations and graphics that I found extremely valuable to gain understanding of difficult concepts or mechanisms, as well as the learning objectives and key points boxes in each chapter."

—Albert Lloret, DVM, in *European Journal of Companion Animal Practice*

". . . an excellent companion to veterinary science students in their study of immunology and for the practising veterinarian interested in gaining a greater understanding of the principles behind clinical immunology."

—*Australian Veterinary Journal*

"This is a wonderful tool, very clearly designed and beautifully illustrated . . . the reader is guided very logically in a journey of discovery of the immune system . . . and the authors are keen on explaining clearly the latest knowledge in their field. . . ."

—*Tomorrow's Vets*

"The second edition of **Veterinary Immunology: Principles and Practice** is exactly the type of textbook necessary for teaching immunology to veterinary students. It is also an excellent reference for practitioners and residents who need to review immunology in preparation for board examinations. The format of the book is inviting, with diagrams or photographs on most pages. Tables with pertinent information are well placed and easily understood. Having taught immunology to veterinary students at the University of California-Davis for over 30 years, I appreciate the logical order in which the material is presented. . . . a concise, well-written, and well-illustrated book that will be very useful for teaching veterinary immunology."

—Laurel J. Gershwin, DVM, PhD, DACVM, University of California-Davis, in *Vet Med Today: Book Reviews*, JAVMA

AchR	acetylcholine receptor
AD	atopic dermatitis
ADCC	antibody-dependent cell-mediated cytotoxicity
AGD	agar gel diffusion
AIDS	acquired immune deficiency syndrome
AIHA	autoimmune haemolytic anaemia
AINP	autoimmune neutropenia
AITP	autoimmune thrombocytopenia
ALL	acute lymphoblastic leukaemia
ALP	alkaline phosphatase
ANA	antinuclear antibody
APC	antigen presenting cell
APTT	activated partial thromboplastin time
ASIT	allergen-specific immunotherapy
AST	aspartate aminotransferase
BALT	bronchial-associated lymphoid tissue
BCR	B cell receptor
BLAD	bovine leucocyte adhesion deficiency
BLV	bovine leukaemia virus
BNP	bovine neonatal pancytopenia
BoLA	bovine leucocyte antigen
BSA	bovine serum albumin
BVDV	bovine viral diarrhoea virus
CADESI	canine atopic dermatitis extent and severity index
CALT	conjunctiva-associated lymphoid tissue
CAV	canine adenovirus
CD	cluster of differentiation
CDR	complementarity determining region

CDV	canine distemper virus
CFT	complement fixation test
C_H	constant region of the heavy chain
CH_{50}	total haemolytic complement (assay)
C_L	constant region of the light chain
CLA	cutaneous lymphocyte antigen
CLAD	canine leucocyte adhesion deficiency
CLE	cutaneous lupus erythematosus
CLIP	class II-associated invariant chain peptide
CLL	chronic lymphoid leukaemia
CMI	cell-mediated immunity
CNS	central nervous system
ConA	concanavalin A (mitogen)
COX	cyclooxygenase
cpm	counts per minute
CR	complement receptor
CRP	C-reactive protein
CTLA-4	cytotoxic T lymphocyte antigen-4
DAF	decay accelerating factor
DAMP	damage-associated molecular pattern
DAT	direct antiglobulin test
DEA1	dog erythrocyte antigen 1
DLA	dog leucocyte antigen
DNA	deoxyribonucleic acid
DOI	duration of immunity
DTH	delayed-type hypersensitivity
EAE	experimental autoimmune encephalomyelitis
EBP	eosinophilic bronchopneumopathy

ELA	equine leucocyte antigen		**IBR**	infectious bovine rhinotracheitis
ELISA	enzyme-linked immunosorbent assay		**ICAM-1**	intercellular adhesion molecule-1
			IDST	intradermal skin test
ELISPOT	enzyme-linked immunospot		**IEL**	intraepithelial lymphocyte
EPI	exocrine pancreatic insufficiency		**IEP**	immonoelectrophoresis
ER	endoplasmic reticulum		**IFA**	immunofluorescent antibody (test)
ES	excretory–secretory (proteins)		**IFN**	interferon (e.g. IFN-γ)
Fab	antigen-binding fragment (of Ig)		**Ig**	immunoglobulin
FAD	flea allergy dermatitis		*IGHA*	gene encoding the IgA heavy chain
Fc	crystallizable fragment (of Ig)			
FcR	Fc (Ig heavy chain) receptor		**IL**	interleukin
FeLV	feline leukaemia virus		**IMHA**	immune-mediated haemolytic anaemia
FHV	feline herpesvirus			
FIP	feline infectious peritonitis		**IMP**	inosine monophosphate
FISS	feline injection site sarcoma		**IMNP**	immune-mediated neutropenia
FITC	fluorescein isothiocyanate		**IMTP**	immune-mediated thrombocytopenia
FIV	feline immunodeficiency virus			
FLA	feline leucocyte antigen		**iTreg**	induced Treg (cell)
FOCMA	feline oncornavirus-associated cell membrane antigen		**IVIG**	intravenous immunoglobulin (therapy)
GALT	gastrointestinal-associated lymphoid tissue		**JAK**	Janus kinase
			KCS	keratoconjunctivitis sicca
G-CSF	granulocyte colony-stimulating factor		**KIR**	killer cell immunoglobulin-like receptor
GITR	glucocorticoid-induced TNF receptor-regulated (gene)			
			KLH	keyhole limpet haemocyanin
GMCSF	granulocyte-macrophage colony-stimulating factor		**LAD**	leucocyte adhesion deficiency
			LAK	lymphokine-activated killer (cell)
GMP	guanosine monophosphate		**LFA-1**	lymphocyte function-associated antigen-1
GnRH	gonadotrophin-releasing hormone			
			LGL	large granular lymphocyte
GTP	guanosine triphosphate		**LPAM**	lymphocyte Peyer's patch adhesion molecule
GVHD	graft-versus-host disease			
GWAS	genome-wide association study		**LPS**	lipopolysaccharide
HAI	haemagglutination inhibition		**M cell**	microfold cell
HAT	hypoxanthine, aminopterin and thymidine		**M1/M2**	subtypes of macrophages
			MAC	membrane attack complex
HEV	high endothelial venule		**MAdCAM**	mucosal addressin cell adhesion molecule
HGPRT	hypoxanthine–guanine phosphoribosyl transferase			
			MALT	mucosa-associated lymphoid tissue
HIV	human immunodeficiency virus			
HLA	human leucocyte antigen (human MHC)		**MAMP**	microorganism-associated molecular pattern
			MASP	MBL-associated serine protease
IBD	inflammatory bowel disease		**MBL**	mannan-binding lectin
IBH	insect bite hypersensitivity			

MCH	mean cell haemoglobin		**PTH**	parathyroid hormone
MCHC	mean cell haemoglobin concentration		**PVDF**	polyvinylidene fluoride
			PWM	pokeweed mitogen
MCP	membrane co-factor protein		**PWMS**	post-weaning multi-systemic wasting syndrome (of pigs)
MCV	mean cell volume			
MDA	maternally derived antibody		**RAG**	recombination activating gene
MHC	major histocompatibility complex		**RAO**	recurrent airway obstruction
MMP	matrix metalloproteinase		**RBCs**	red blood cells
mRNA	messenger RNA		**RF**	rheumatoid factor
MS	multiple sclerosis		**rHuGCSF**	recombinant human granulocyte colony-stimulating factor
NADPH	nicotinamide adenine dinucleotide phosphate			
			rHuGMCSF	recombinant human granulocyte–monocyte colony-stimulating factor
NALT	nasal-associated lymphoid tissue			
NET	neutrophil extracellular trap			
NF-AT	nuclear factor of activated T cells		**rHuIFN-α**	recombinant human IFN-α
NFκB	nuclear factor κB		**RIG**	retinoic acid inducible gene (receptor)
NI	neonatal isoerythrolysis			
NK	natural killer (cell)		**RNA**	ribonucleic acid
NKT	natural killer T (cell)		**ROS**	reactive oxygen species
NLR	NOD-like receptor		**RT-PCR**	reverse transcriptase polymerase chain reaction
NO	nitric oxide			
NOD1/2	nucleotide-binding oligomerization domain 1/2		**SAA**	serum amyloid A
			SCID	severe combined immunodeficiency/ immunodeficient
NOD	non-obese diabetic mouse			
NOS	nitric oxide synthase			
nTreg	natural Treg (cell)		**SI**	stimulation index
nzb	New Zealand black mouse		**SLA**	swine leucocyte antigen
OLA	ovine leucocyte antigen		**SLE**	systemic lupus erythematosus
OSP	outer surface protein		**SLIT**	sublingual immunotherapy
OVA	ovalbumin		**SNP**	single nucleotide polymorphism
PALS	periarteriolar lymphoid sheath		**SPC**	summary of product characteristics
PAMP	pathogen-associated molecular pattern			
			SRBC	sheep red blood cell
PBMC	peripheral blood mononuclear cell		**SRID**	single radial immunodiffusion
			STAT	signal transducers and activators of transcription
PBS	phosphate buffered saline			
PCR	polymerase chain reaction		**TAM**	tumour-associated macrophage
PCV	packed cell volume		**TAP**	transporter protein (TAP1/ TAP2)
PEG	polyethylene glycol			
PHA	phytohaemagglutinin		**Tc**	T cytotoxic (cell)
PI	persistently infected		**TCR**	T cell receptor
pIgR	polymeric immunoglobulin receptor		**TFh**	T follicular helper (cell)
			Tg	thyroglobulin
PRR	pattern recognition receptor		**TGF**	transforming growth factor
PT	prothrombin time		**Th**	T helper (cell)

TLR	Toll-like receptor	**VEGF**	vascular endothelial growth factor
TMS	trimethoprim–sulphonamide		
TNF	tumour necrosis factor	$\mathbf{V_H}$	variable region of the heavy chain
TNS	trapped neutrophil syndrome	$\mathbf{V_L}$	variable region of the light chain
TPMT	thiopurine methyltransferase	**VN**	virus neutralization
Treg	T regulatory (cell)	**WHO**	World Health Organization
TSH	thyroid-stimulating hormone	**X-SCID**	X-linked severe combined immunodeficiency
UPC	urine protein:creatinine (ratio)		

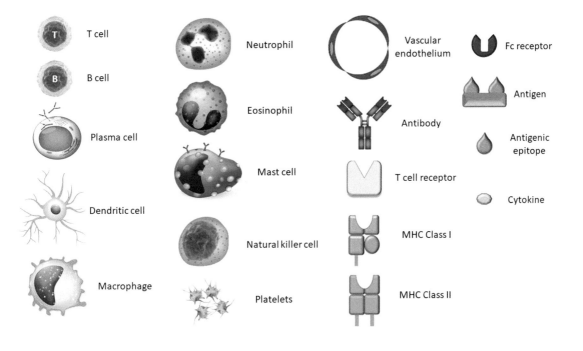

T cell

B cell

Plasma cell

Dendritic cell

Macrophage

Neutrophil

Eosinophil

Mast cell

Natural killer cell

Platelets

Vascular endothelium

Antibody

T cell receptor

MHC Class I

MHC Class II

Fc receptor

Antigen

Antigenic epitope

Cytokine

CHAPTER 1

AN OVERVIEW OF THE IMMUNE SYSTEM

Important Concepts and Principles of Innate and Adaptive Immunity

1

OBJECTIVES

At the end of this chapter, you should be able to:

- Appreciate the relationship between infection and the immune system.
- Describe the strategy used by cells of the innate immune system to detect foreign organisms.
- Explain what is meant by the terms 'antigen' and 'epitope'.
- Describe how lymphocyte development and their selection in primary lymphoid organs play a key role in determining the immune repertoire.
- Describe the adaptive immune response following recognition of antigen.
- Understand why it is necessary to regulate the immune system.
- Define the concept of immunological memory.
- Briefly describe the evolution of the immune system.

1.1 INTRODUCTION

Immunology is a relatively young and rapidly developing science that has established itself as a fundamental cornerstone of human and veterinary clinical medicine. There are many aspects of veterinary clinical activity that require an understanding of the immune system. The husbandry of neonatal domestic livestock and the essential requirements for colostral immunity, the crucial role of vaccination in protecting against infectious disease, the laboratory diagnosis of a wide range of disorders, the consequences of chronic disease for immune function and the myriad of immune-mediated diseases caused by disturbance of immune homeostasis are all examples of how the immune system impacts on the daily practice of veterinary medicine. The aim of this textbook is not only to provide a basic knowledge and understanding of immune function but also to clearly signpost how this relates to veterinary clinical practice.

This first chapter will briefly review the history of immunology and broadly consider the different elements of the immune system and how they interact to create an integrated defence system against infection. The focus here will be on the basic concepts underpinning immunology and will introduce important principles of innate and adaptive immunity, which will be discussed in more detail in subsequent chapters.

1.2 HISTORY OF IMMUNOLOGY

A perusal of the relatively short history of the discipline of immunology reveals just how

DOI: 10.1201/9781003310969-1

integral veterinary immunology has been to the development of this science. This brief summary cannot do full justice to the progressive discoveries that have led to our current state of knowledge. Most texts record the birth of immunology, at least in the Western world, as related to the introduction of the concept of vaccination by Edward Jenner. In 1796 Jenner performed his famous experiment whereby he collected fluid from a cowpox vesicle on the hand of the dairymaid Sarah Nelmes, and inoculated this into the arm of the eight-year-old boy James Phipps. Although James developed cowpox lesions, he was protected when subsequently challenged two months later with virulent smallpox. It has recently been suggested that a similar experiment may have been conducted some 20 years before Jenner by the Dorset farmer Benjamin Jesty. The next major developments in immunology were attributed to

Louis Pasteur, who developed vaccines to fowl cholera (1879), anthrax (1881), swine erysipelas (1892) and rabies (1885). It is noteworthy how many of the early developments in immunology related to animal diseases.

The history of immunological developments can be easily appreciated by studying a list of Nobel Prizes awarded in this discipline (Table 1.1). These chart the progressive recognition of antibody, phagocytic cells and complement through to the current appreciation of the molecular interactions that underpin immune function. Of note is the 1996 award to Peter Doherty, a 1962 graduate of the University of Queensland School of Veterinary Science, who undertook postgraduate studies on ovine 'louping-ill' at the Moredun Research Institute in Scotland, further demonstrating the role that veterinarians have had in the development of immunology. In the current century, veterinary

Table 1.1 **Nobel Prizes in Immunology**		
1901	Von Behring	Discovery of serum antibody
1905	Koch	Immune response in tuberculosis
1908	Metchnikov and Ehrlich	Phagocytosis and antitoxins
1913	Richet	Anaphylaxis
1919	Bordet	Complement
1930	Landsteiner	Blood groups
1951	Theiler	Yellow fever vaccine
1957	Bovet	Antihistamines for allergy treatment
1960	Burnet and Medawar	Immunological tolerance
1972	Edelman and Porter	Antibody structure
1977	Yalow	Development of radioimmunoassay
1980	Benacerraf, Dausset and Snell	Discovery of the major histocompatibility complex
1984	Jerne, Kohler and Milstein	Production of monoclonal antibodies
1987	Tonegawa	Mechanism of antibody diversity
1990	Murray and Thomas	Transplantation
1996	Doherty and Zinkernagel	Major histocompatibility complex restriction
2011	Steinman	Role of the dendritic cell in adaptive immunity
2011	Beutler and Hoffman	Activation of innate immunity
2018	Allison and Honjo	Discovery of cancer therapy by inhibition of negative immune regulation

immunology remains a flourishing field of research with its own journals (e.g. *Veterinary Immunology and Immunopathology*), conferences and societies such as the American Association of Veterinary Immunologists. A major recent achievement in veterinary immunology was the declaration in 2011 that the world was free of rinderpest virus infection following successful vaccination campaigns. This is only the second occasion that a major infectious disease has been eliminated, the first being the eradication of smallpox in 1979.

1.3 THE IMMUNE SYSTEM: AN OVERVIEW

The immune system is responsible for biological defence, providing protection against the diversity of pathogens that cause human and animal infectious diseases. The immune system is capable of reacting, with varying degrees of efficacy, to viruses, bacteria, fungi, protozoa and helminth parasites. In order to achieve this, defensive systems are in place to recognize, react and eliminate foreign organisms that attempt to colonize epithelial surfaces and/or invade and replicate in the host tissues (**Figure 1.1**).

The immune system is intrinsically involved with the inflammatory response and tissue repair processes of the body. In addition to responding to external threats (i.e. foreign organisms), the immune system can also initiate responses to abnormal cells (i.e. altered self) that can arise during neoplastic transformation or against cells/tissues that have been transplanted into the body. Such a potent biological defence system requires careful management and regulatory mechanisms have evolved to ensure immune responses are measured and appropriate, so as not to cause collateral damage to normal body tissues. Unfortunately, this sometimes goes awry and the immune system can initiate inappropriate responses to innocuous environmental antigens, dietary antigens or healthy cells/tissues, leading to immune-mediated allergic and autoimmune diseases that

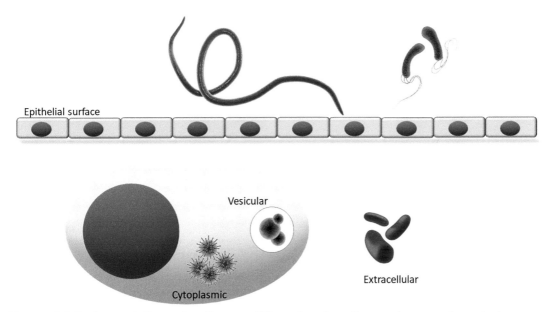

Figure 1.1 Pathogens infect and replicate at different locations. Some pathogens seek to colonize epithelial surfaces of the skin and mucosal epithelium. Others invade and infect tissues, replicating in the extracellular fluid, intracellular fluid or within intracellular vesicles, following endocytosis.

are so important in human and companion animal medicine.

The immune system has evolved from humble beginnings in unicellular organisms to become the complex system of cells, tissues and molecules that we see in mammalian species. The immune system of all animals utilizes a similar 'toolbox' of components that interact with each other in a predictable fashion. Immunology is relatively advanced in human medicine and in the study of rodent species used as experimental models in research, when compared with those species that fall under the remit of veterinary medicine. However, it is possible to extrapolate many of the fundamental immunological principles from the better explored species to domestic animals. That said, there are numer-

ous intriguing species differences that have arisen during evolution and these will be highlighted in the chapters that follow.

The standard building blocks of the immune system are represented in **Figure 1.2**. Traditionally, the immune system is considered to be comprised of two related components: the **innate** and the **adaptive** immune responses. Even relatively primitive life forms have some form of innate immunity and this represents the earliest form of immune defence, which is intrinsically interlinked with the acute and chronic **inflammatory responses**. As evolution progressed on the planet, the adaptive immune system has become more important, to protect against an increasing number and complexity of microorganisms that have adapted to a more

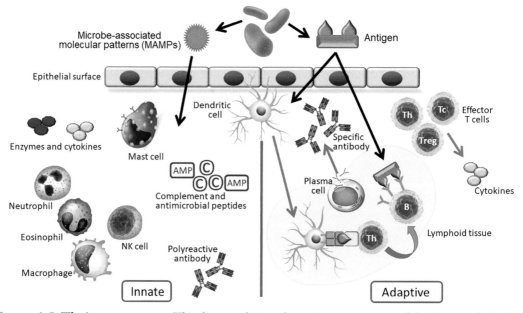
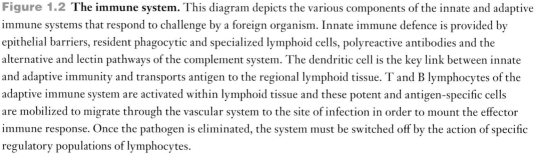

Figure 1.2 The immune system. This diagram depicts the various components of the innate and adaptive immune systems that respond to challenge by a foreign organism. Innate immune defence is provided by epithelial barriers, resident phagocytic and specialized lymphoid cells, polyreactive antibodies and the alternative and lectin pathways of the complement system. The dendritic cell is the key link between innate and adaptive immunity and transports antigen to the regional lymphoid tissue. T and B lymphocytes of the adaptive immune system are activated within lymphoid tissue and these potent and antigen-specific cells are mobilized to migrate through the vascular system to the site of infection in order to mount the effector immune response. Once the pathogen is eliminated, the system must be switched off by the action of specific regulatory populations of lymphocytes.

parasitic lifestyle, gaining a biological advantage to the detriment of other living species. Thus, there has been evolutionary selection pressure for a progressively more complex immune system in order to survive the biological arms race between the host and pathogen. As we shall see in subsequent chapters, the increasing complexity of host immunity was made possible by a process of **gene duplication**, giving rise to related families of immunological molecules, such as pattern-recognition receptors, immunoglobulins and cytokines.

1.4 PRINCIPLES OF INNATE IMMUNITY

Epithelial barriers provide both physical and biochemical defence systems and protect the interface between the host and its environment, in an attempt to prevent infection and invasion by pathogens. The innate immune system is designed as a rapid response team, which is particularly active at those anatomical sites that are most likely to be the first point of contact with potential pathogens, that is, the skin, respiratory tract, gastrointestinal tract, urogenital tract, mammary gland and ocular mucosa. If epithelial barriers are breached, humoral and cellular mechanisms are brought into play in defence of the host. Soluble mediators of innate immunity include the **complement proteins**, which are organized into an amplification system culminating in formation of the **membrane-attack complex**, which damages microbial cell walls, as well as production of various by-products that are inflammatory in nature.

Epithelial and stromal cells can initiate a defensive response through secretion of antimicrobial peptides (AMPs) and alert the immune system by release of **cytokines** (immunological hormones) and **chemokines** (chemoattractants). Subsequently, white blood cells (and their derivatives) are rapidly recruited to the site of infection, further contributing to the inflammatory response. **Neutrophils** and

tissue macrophages are adapted to recognizing bacteria and are engaged in phagocytosis, thereby killing and digesting infecting microorganisms. **Eosinophils** and tissue **mast cells** are adapted to recognizing helminth parasites and react by degranulation, releasing inflammatory, toxic and digestive enzymes onto the pathogen surface.

The innate immune system must have a strategy for detecting the presence of infection as quickly as possible. This is achieved through recognition of certain features of abnormal structure or cellular function (**Figure 1.3**).

The innate immune system utilizes **pattern-recognition receptors** (PRRs) for detection of molecules that are intrinsically foreign. Such structural 'abnormalities' are referred to as **microbe-associated molecular patterns** (MAMPs) and include peptidoglycan (the structural building block of bacterial cell

Figure 1.3 Innate immune strategy for detection of foreign elements. In the example shown, the 'wolf in sheep's clothing' (representing a foreign organism) can be distinguished from the real sheep (representing healthy tissue cells) by virtue of certain noticeable features (representing 'patterns' that are intrinsically foreign).

walls, particularly prominent in Gram-positive bacteria) and lipopolysaccharide (LPS; found in the outer membrane of Gram-negative bacteria). These are molecules that are not synthesized by eukaryotic cells and as such serve as markers to indicate the presence of prokaryotic microorganisms. The PRRs that can detect such MAMPs are broadly reactive and not very specific, that is, recognition of LPS does not discriminate between different Gram-negative bacterial species.

Pattern-recognition receptors can also be utilized for detection of substances released by damaged or dying host cells present at the site of infection. **Damage-associated molecular patterns** (DAMPs) include cytoplasmic contents (such as ATP) that are released into the extracellular fluid as cell membranes break down and other molecules (such as stress proteins) that are synthesized as a result of the cells encountering an adverse tissue environment. Soluble PRRs are present in the plasma and tissue, or they are cell-associated, allowing cells (particularly white blood cells) to sense the presence of MAMPs in the extracellular, intracellular and endosomal compartments, which signal to the cell to deploy their biological defence systems. Triggering of PRRs leads to stimulation of the processes of phagocytosis and inflammation. Phagocytosis is an important defence mechanism against microorganisms, leading to their destruction and digestion. The inflammatory response allows the body to divert blood flow to the site of infection, bringing additional white blood cells and soluble immunological resources (including complement proteins) to the affected tissues.

The innate immune system finds it relatively difficult to detect and counteract viral infection, since viruses have a simple structure, lacking obvious MAMPs. Furthermore, some bacteria have evolved strategies to avoid detection by PRRs, for example by producing an inert polysaccharide capsule, or avoiding downstream effector mechanisms designed for their killing and digestion. Thus, over time, the innate immune system has become less effective in protecting the host from infection and this has generated the evolutionary drive for development of the adaptive immune system.

1.5 PRINCIPLES OF ADAPTIVE IMMUNITY

The **adaptive immune system** has developed more recently than innate immunity in evolutionary terms and is typically the dominant form of immunity in higher species. Although it is more specific and considerably more potent in its effects, compared with the innate immune system, its main drawback is that there is a delay (lag period) between the onset of infection and production of an effector response. Thus, innate immunity has not been made redundant, as it is important for keeping the host alive long enough for an adaptive immune response to take effect and, in fact, many of the outputs of adaptive immunity are designed to enhance rather than replace innate immune mechanisms (such as the complement system).

The adaptive immune system is intrinsically linked to the functionality of **lymphocytes**. Unlike the innate immune detection system for recognition of molecules that are intrinsically foreign (using PRRs), lymphocytes express cell surface receptors designed to detect the presence of antigen, most commonly foreign protein. Pathogens synthesize and express proteins that are different in sequence and (more importantly) shape/structure, compared with those synthesized by host cells. These protein antigens, and the peptides generated from them after digestion, can be recognized as foreign by lymphocyte antigen receptors. Each small element of an antigen that can be recognized by a lymphocyte antigen receptor is termed an 'epitope' (**Figure 1.4**). Thus, even a pathogen as small as a virus will have multiple antigens contributing to its structure and function, and each antigen will have many epitopes that can

Figure 1.4 The elephant analogy of antigenic epitopes. Consider a foreign antigen, in this case represented by the elephant. Each antigen will have a specific shape, with several key features (epitopes), each of which can be recognized by a specific lymphocyte antigen receptor (shown by the Y-shaped molecules).

potentially bind antigen-specific lymphocyte receptors.

There are two main types of lymphocytes, namely B lymphocytes and T lymphocytes that each express antigen receptors designed to detect antigen, as shown in **Figure 1.5**. B cells express antigen receptors designed to detect epitopes of intact antigen on the surface of pathogens in the extracellular space. T cells, on the other hand, recognize digested fragments of antigen that has undergone intracellular processing and which are presented on the surface of other cells in association with a specialist carrier molecule called major histocompatibility complex (MHC).

Lymphocytes must design a uniquely shaped antigen receptor during their development in **primary lymphoid tissues**. All lymphocytes are initially derived from lymphoid progenitors in the **bone marrow** and whereas B cells complete their development here in mammalian species, T cells must leave the bone marrow and complete their development in the thy-

mus, where they also must commit to becoming either CD4+ or CD8+ T cells. During development, antigen receptor diversity is engineered by a process of gene rearrangement. The receptor is constructed from a constant region (which is common to all the receptors) and a variable region (which differs between individual lymphocytes) (**Figure 1.6a**). Through a random process of variable gene segment selection, individualized antigen receptors (with a defined epitope specificity) are generated and a diverse repertoire of antigen receptors is created (**Figure 1.6b**).

Although the adaptive immune repertoire is astonishingly diverse, this is not without its challenges. Firstly, the random generation of antigen-specific receptors inevitably results in receptors that recognize self-antigens. Deletion, inactivation, or suppression of such self-reactive lymphocytes normally avoids reactivity to the host's own cells and tissues (**immunological tolerance**). Secondly, each lymphocyte recognizes a specific antigenic epitope but has

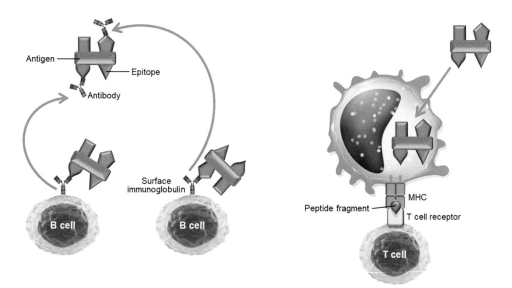

Figure 1.5 **Detection of antigen by B and T lymphocytes.** B cells use their antigen receptor (surface immunoglobulin) to detect surface epitopes of foreign antigen, and react by secreting this molecule as antibody. T cells recognize digested fragments of antigen (i.e. peptide epitopes), displayed on the surface of other cells in association with a specialist carrier molecule called major histocompatibility complex (MHC).

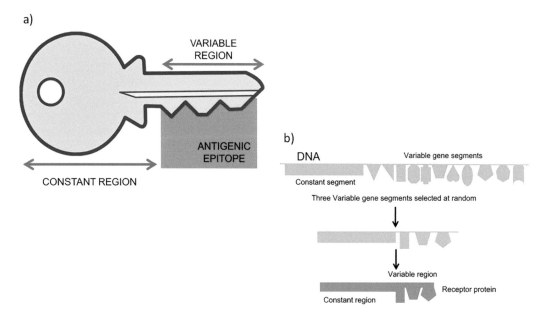

Figure 1.6 **Lymphocyte antigen receptors.** (a) The 'lock and key' model of antigen receptor binding to an antigenic epitope. The receptor protein (the 'key') consists of a constant region that is similar across receptors and a variable region that differs between receptors and which determines the shape of the antigenic epitope (the 'lock') it fits. (b) At the genetic level, the region encoding the constant region is fixed, but there are a number of variable (V) segments available for selection. During lymphocyte development, three V segments are selected at random and combined with the constant segment to encode the receptor protein. Since there are numerous V segments to choose from and the choice is completely random, this generates a huge diversity in receptor specificity.

no way of knowing the port of entry of a pathogen expressing that particular antigen. Thirdly, within the pool of lymphocytes in the body, only a relatively small number will react against a specific pathogen and, on their own, they would be insufficient to deal with the infection. To solve these issues, **secondary lymphoid tissues** are strategically located around the body, which are both sites of accumulation of antigen and preferential sites for migration of **naïve lymphocytes**. Secondary lymphoid tissues include **lymph nodes** that are designed to filter lymph fluid during its return to the circulation, trapping foreign material and acting as a site of deposition of antigen. In addition, there are professional antigen presenting cells, the **dendritic cells**, that acquire antigen in the skin, mucosal epithelial surfaces or within tissues, and subsequently migrate to the nearest lymphoid tissue to present their antigen, thereby alerting T lymphocytes as to their presence. Specialist blood vessels within the lymph node parenchyma allow lymphocytes to leave the blood and inspect the antigens present. In addition to lymph nodes, the **spleen** serves a similar process of antigen filtering of the blood, which is particularly important for blood-borne pathogens, while **mucosal-associated lymphoid tissues**, such as Peyer's patches in the intestine, allow the immune system to 'sample' the antigen coming in through mucosal epithelial surfaces. By using the blood and lymphatic vessels as conduits around the body, lymphocyte recirculation and migration through the various secondary lymphoid tissues, allows effective immune surveillance for the presence of infection.

When an individual lymphocyte encounters its target antigen in a secondary lymphoid organ, it ceases its migratory nature and begins to proliferate, creating many copies of daughter cells with the same antigen specificity as the original reactive clone (**clonal selection and expansion**). Although initially small in number, this process creates an army of lymphocyte

clones that are prepared to engage with the invading pathogen (**Figure 1.7**). We are sometimes aware of this proliferative response during clinical examination of the animal, where we may observe/palpate the presence of enlarged lymph nodes, which can direct the clinician to the likely site of infection.

B cells activated within the secondary lymphoid tissues differentiate to become **plasma cells**. Plasma cells migrate locally within lymph nodes to the medullary cords or may travel a greater distance to the bone marrow, where they reconfigure their antigen receptor to be secreted in the form of **antibody**. These act like molecular 'smart missiles' that target the pathogens bearing their antigen for destruction. Antibody (aka immunoglobulin) is important for neutralization of viruses, enhancing phagocytosis of bacteria and improving mast cell detection of helminth parasites and different types (classes) of antibody can be made to ensure that it is 'fit for purpose' depending on the nature of the pathogen.

After a period of proliferation, the CD4$^+$ T cells differentiate to one of several effector subtypes. CD4$^+$ helper T cells release **cytokines** (immunological hormones) to enhance the activity of other cell types, including B cells and macrophages. The activated, CD8$^+$ killer T cells must be mobilized from the regional lymphoid tissue and sent to the site of infection, where they undertake a search and destroy mission against virus-infected cells. This process involves these cells moving into the lymphatic vessels and blood where they may then interact with the endothelial lining of blood vessels. Once these cells reach the site of infection, they are able to mount a full-scale **'effector' response**, which is considerably stronger than that permitted by innate immunity. As these processes take some time to occur (in the order of four to seven days), there is a delay before adaptive immunity takes over from the innate form of defence.

When the infection has been overcome and the host has recovered, the final components

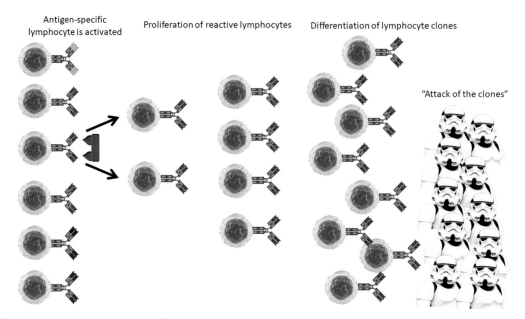

Antigen-specific lymphocyte is activated   Proliferation of reactive lymphocytes   Differentiation of lymphocyte clones

"Attack of the clones"

Figure 1.7 Clonal selection of lymphocytes. Upon exposure to antigen in secondary lymphoid tissues, antigen-specific lymphocytes are activated and undergo a period of proliferation before differentiating to a population of 'clone soldiers' that are ready to engage the pathogen.

of adaptive immunity are the development of **memory** and **regulatory responses**. Not all CD4+ T lymphocytes become helper T cells and towards the end of the response to infection, some differentiate into **regulatory T cells** that function to suppress the response, when it is no longer required (i.e. when the pathogen has been eliminated), so as to avoid collateral damage to healthy tissues. **Immunological memory** is a key feature of adaptive immunity, whereby the adaptive immune system learns from experience and remembers what pathogens look like, in case they seek to re-infect the host in the future. The lymphocyte clones that develop in the lymphoid tissues towards the end of the recovery period, instead of becoming effector cells (front-line soldiers), become quiescent and are known as **memory lymphocytes** (reserve troops). These are long-lived cells and enter the circulation to contribute to immune surveillance, ready to be called into action (**secondary immune response**) if the host encoun-

ters the same pathogen again, even when this is many years later. Immunological memory often provides protection against re-infection without any clinical signs of disease becoming apparent. Thus, once we have recovered, we are said to have developed **long-term immunity**. This phenomenon underpins our use of **vaccination** in clinical medicine, whereby we deliberately educate the immune system as to what it might expect to see in terms of pathogens in the environment and prepare the host to counteract infection more quickly.

A **military analogy** is often used to explain immunological phenomena and it may be helpful to consider the immune system as an army at war with invading pathogens (**Figure 1.8**). In this context, the innate immune system might be considered to be analogous to border security/frontier defences of the body and the dendritic cells to messengers who may be dispatched from the frontier to the military bases (regional lymphoid tissue), where the army of the adaptive

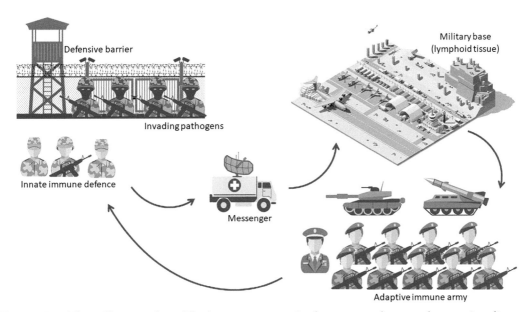

Figure 1.8 The military analogy. The immune response is often compared to a war between invading pathogens and the army of the immune system. In this model, the innate immune defence might be considered as a relatively weak frontier patrol surveying the outer reaches of the body. If an enemy invades this territory, a messenger is dispatched to the closest military base, where a battalion of soldiers that is best equipped to deal with this type of enemy is mobilized and sent back to the place where the frontier has been breached. These reinforcements allow successful elimination of the enemy, following which the army is stood down when instructed by regulatory elements.

immune response is marshalled. The messengers provide information that enables the army to select those troops best equipped to deal with the particular invader, to activate those troops and dispatch them to the front line, where they provide reinforcements to the innate system in the battle against the invading enemy. Once these reinforcements have eliminated the invaders, they are stood down when instructed by the military police (regulatory cells).

1.6 THE EVOLUTION OF THE IMMUNE SYSTEM

The evolution of the mammalian immune system has been widely studied. Even primitive species such as plants and unicellular organisms have **innate mechanisms** that function to protect them from potential pathogens. Organisms such as the amoeba move within their aquatic environment and engulf food particles that they encounter. The discrimination between food and other environmental components may be a receptor-mediated recognition process. These features of amoeba are somewhat similar to those of macrophages and it is suggested that **phagocytic cells** may have had their origin from these unicellular organisms.

The immune system of primitive species relies largely on the activity of a range of non-specific innate **antimicrobial molecules** and the action of phagocytes and cytokine-like proteins. The ability to **distinguish self from non-self** is also an ancestral property of the primitive immune system. Sponges, earthworms and starfish are all capable of 'transplant rejection' when experimentally grafted with tissue from distinct species, but such rejection is mediated by the innate immune system. An excellent example of ancestral immunological molecules that are relatively conserved throughout evolution are the **Toll-like receptors**. These molecules were first discovered in the fruit fly *Drosophila*, and homologues have subsequently been described in most species, from plants to primates.

As host and pathogens co-evolved, a molecular 'arms race' has developed over time, whereby there were continual requirements for a more complex immune system. The emergence of **adaptive immunity** has been described as the immunological 'big bang' and is thought to have occurred in an ancestor of the jawed fish. The key feature of adaptive immunity is the ability to generate diversity in **antigen receptor** molecules. The event that led to development of this ability is thought to relate to integration of a circle of extrachromosomal DNA (a 'transposable element') containing the ancestral **recombination activating genes** (*RAG-1* and *RAG-2*) into a region of the genome containing an ancestor gene encoding a molecule similar to a T or B cell receptor. In parallel, the process of **gene duplication**, as discussed with respect to the immunoglobulin superfamily, allowed creation of a wide diversity in immune system molecules. The broad evolutionary stages of the immune system are summarized in Table 1.2.

Table 1.2 **Evolution of the Immune System**						
	PHAGOCYTIC CELLS AND TLRS	**NK CELLS**	**IG**	**MHC**	**T AND B LYMPHOCYTES**	**LYMPH NODES**
Invertebrates						
Protozoa	+	–	–	–	–	–
Sponges	+	–	–	–	–	–
Annelids	+	+	–	–	–	–
Arthropods	+	–	–	–	–	–
Vertebrates						
Elasmobrachs (sharks, skates, rays)	+	+	+		+	–
Teleost fish	+	+	+	+ (some)	+	–
Amphibians	+	+	+	+ (some)	+	–
Reptiles	+	+	+	+	+	–
Birds	+	+	+	+	+	+ (some)
Mammals	+	+	+	+	+	+

KEY POINTS

- The immune response is divided into innate and adaptive immunity, designed to work together to counteract infection.
- Innate immunity is older in evolutionary terms, is fast-acting and relatively non-specific.
- Innate immunity uses pattern-recognition receptors to detect common microbial components (microbe-associated molecular patterns; MAMPs) or signs of cellular injury (damage-associated molecular patterns; DAMPs) present at the site of infection.
- Phagocytosis and the inflammatory response are important elements of the innate immune response to bacteria, mediated by neutrophils and macrophages.
- Adaptive immunity is slower, but more potent and specific, than the innate immune response.
- The adaptive immune response relies upon B and T lymphocytes and requires dedicated lymphoid tissues to function.
- Lymphocyte antigen receptors are generated through a process of random gene rearrangement to generate receptor diversity.
- Immunological tolerance refers to the mechanisms that avoid reactivity of lymphocytes to the host's own cells and tissues.
- B cells detect antigenic epitopes of surface antigens expressed by pathogens in the extracellular space, whereas T cells recognize peptide epitopes from digested antigen that has undergone intracellular processing and presentation by major histocompatibility complex (MHC) molecules.
- Primary lymphoid organs are important for lymphocyte development, whereas secondary lymphoid tissues are important for immune surveillance, clonal selection and generation of effector cells.
- B lymphocytes differentiate into plasma cells that secrete antibody, which targets extracellular pathogens.
- CD4+ helper T cells produce cytokines that influence the biological activity of other immune cells.
- CD8+ killer T cells seek out and destroy infected cells.
- A regulatory system is required to switch off the immune response when it is no longer required.
- Immunological memory of previous antigenic exposure provides long-term immunity, with the memory (secondary) immune response being more potent that the initial (primary) response.
- Early life forms have a simple innate immune system and evolution of the adaptive immune response occurred through the ability to develop diversity in immune response genes.

OBJECTIVES

At the end of this chapter, you should be able to:

- Describe where and how white blood cells are made, and how this is regulated.
- Describe the function of different white blood cells in the immune response.
- Explain the division of the immune system into primary and secondary lymphoid tissues.
- Describe the basic structure and function of the thymus.
- Describe the microanatomical structure of lymph nodes, spleen and Peyer's patches.
- Explain how lymphocyte migration via lymphatic vessels and blood results in efficient immune surveillance.
- Discuss why it is important that lymphocytes activated at one mucosal surface are able to recirculate to multiple mucosal surfaces.
- Discuss why haematology is important clinically, and how changes in blood cell numbers can be used to indicate disease and pathological processes occurring in the patient.

2.1 INTRODUCTION

This chapter focusses on the key cells of the immune system, which are produced in the bone marrow then circulate in the blood to reach the tissues. White blood cells include neutrophils, eosinophils, basophils, monocytes and lymphocytes. Tissue mast cells are also important immune cells capable of triggering an inflammatory response, particularly in the presence of helminth parasites. Some immune cells undergo further differentiation after they leave the bone marrow. Monocytes for example can differentiate into several cell types, including macrophages and dendritic cells. Lymphocytes proliferate and differentiate in lymphoid tissues in response to antigenic stimulus and various other signals. **Primary lymphoid tissues** are the sites where lymphocyte development occurs and cells undergo initial maturation to a naïve phenotype that are capable of engaging in immune surveillance. **Secondary lymphoid tissues** are where these cells subsequently respond to antigenic stimulation that drives their proliferation and differentiation to effector phenotypes, capable of participating in adaptive immune responses.

DOI: 10.1201/9781003310969-2

2.2 HAEMATOPOIESIS AND PRODUCTION OF WHITE BLOOD CELLS

Before birth, the yolk sac is the first site of hae-matopoiesis (zero to two months in humans), followed by the liver and spleen. In neonatal animals, most bones contain bone marrow, which becomes the primary site of haematopoiesis. In adulthood, bone marrow is found predominantly in vertebrae, ribs, sternum, pelvis and the epiphyses of long bones (as the haematopoietic tissue in the diaphysis is replaced by adipose tissue). The haematopoietic area of bone marrow occupies ~50% of the tissue (the remaining area contains mainly adipocytes) and is highly vascular. The bone marrow forms a suitable environment for stem cell development. It is composed of stromal cells and a microvascular network. The stromal cells include adipocytes, fibroblasts, endothelial cells and macrophages that secrete extracellular matrix and several growth factors necessary for stem cell survival, replication and differentiation. The key cell of the bone marrow is the haematopoietic stem cell, from which multipotent and committed haematopoietic progenitors arise. The latter then differentiate into the blood cells that enter the circulation. The developmental lineages of these cells are summarized in **Figure 2.1**.

Production of the different types of white blood cells (leucocytes) is regulated by various hormones and cytokines in both a paracrine (from the bone marrow stromal cells) and endocrine (from extramedullary tissues) manner. Some mediators, such as stem cell factor and Flt3-ligand act on pluripotent stem cells, whereas others, such as interleukin-3 (IL-3), granulocyte colony-stimulating factor (G-CSF) and granulocyte macrophage colony-stimulating factor (GM-CSF), act on multi-potential progenitor cells and others, such as IL-5, act on committed progenitor cells to drive production of certain cell populations. A com-

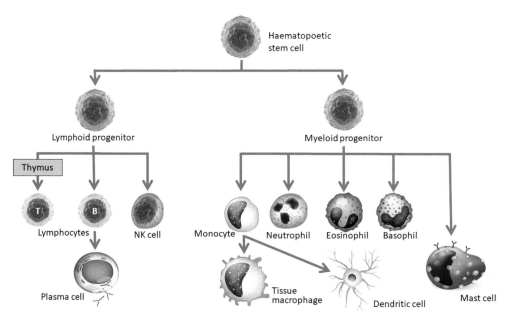

Figure 2.1 Developmental lineages of immune cells. All the cells of the immune system arise from haematopoietic stem cells, which in turn give rise to separate developmental lineages through committed stem cells. Plasma cells are a late maturation stage of B lymphocytes and tissue macrophages and dendritic cells derive from blood monocytes.

mercial product, Imrestor™ (Elanco) contains pegbovigrastim (bovine G-CSF), which is designed to be administered to periparturient dairy cows to stimulate bone marrow production of neutrophils, which help to protect against bacterial infection, for example, mastitis.

There are three main granulocyte populations in the blood, namely neutrophils, eosinophils and basophils. Neutrophils (**Figure 2.2**) (also known as polymorphonuclear neutrophils; PMNs) are one of the main phagocytic cells (alongside macrophages) and are important in the response to bacterial infection. They are one of the first cells recruited to the site of infection during an acute inflammatory response. Neutrophil granules contain various substances, including antimicrobial peptides (AMPs), lysozyme, serine protease enzymes and matrix metalloproteinases. Neutrophilia (raised neutrophil count) is often seen in acute bacterial infections and a 'left shift' may be evident as the bone marrow releases neutrophils before their

nucleus has become fully segmented. However, a mild mature neutrophilia is also a feature of the 'stress leukogram', which is a non-specific finding seen in animals due to the effects of increased corticosteroid (either endogenous or exogenous).

Eosinophils (**Figure 2.3**) are primarily designed to target helminth parasites. Primary (azurophilic) granules contain acid hydrolases and other enzymes, such as cathepsins. The large eosinophilic (specific) granules contain major basic protein, eosinophil peroxidase, eosinophil cationic protein and eosinophil-derived neurotoxin. Upon degranulation, these substances are not only relatively toxic but can also stimulate degranulation of other immune cells, including basophils and mast cells, thus generating a cascade effect at the site of infection. Eosinophilia is often associated with helminth parasite infection, but can also be seen in animals affected with allergic disease. Basophils (**Figure 2.3**) have relatively large, dark-staining

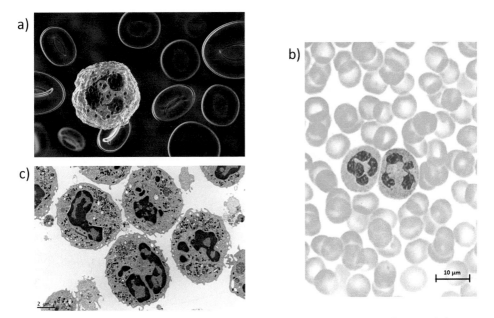

Figure 2.2 Neutrophils. (a) Illustration of a circulating neutrophil. (b) A group of neutrophils as seen on a stained blood smear. Neutrophils have a multi-lobed nucleus and contain primary and secondary granules. (c) Transmission electron micrograph of a group of neutrophils. The characteristic lobulated or segmented nucleus can be observed as well as their intracellular granules.

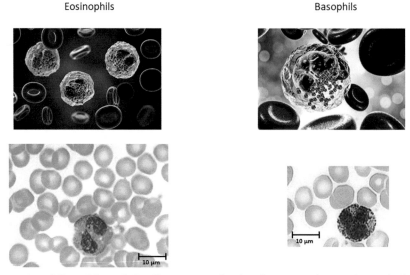

Figure 2.3 Eosinophils and basophils. Illustrations of each cell type are shown, alongside the cells as seen on a stained blood smear. Eosinophils possess a bilobed nucleus and contain primary and specific granules. Basophils are dark-staining cells with a bilobed or tri-lobed nucleus and contain large cytoplasmic granules that occupy most of the cytoplasm.

specific granules, containing histamine, heparin and serine proteases such as chymase and tryptase. The presence of vasoactive mediators aligns with this cell being involved in the acute inflammatory response. The mast cell is somewhat similar to a basophil but is found in the tissues. They arise from committed precursors in the bone marrow that circulate as agranular cells for a relatively short period of time before exiting the bloodstream and taking up residence in the tissues, where they develop similar granules to those seen in basophils. Mast cells express receptors on their surface for immunological molecules (such as immunoglobulin E and complement proteins C3a and C5a), which stimulate their degranulation.

Monocytes (**Figure 2.4**) are agranular mononuclear leucocytes that circulate as immature cells, only fully differentiating once they have undergone extravasation and entered the tissues. Monocytes differentiate into several phenotypes, including macrophages and dendritic cells, although some can become more specialist (non-immunological) cell types in certain tissues, such as the osteoclasts, responsible for bone remodelling (**Figure 2.5**). Even within the tissue macrophage population, there are different subtypes, with M1 macrophages being more pro-inflammatory and having an antimicrobial function, compared with M2 macrophages that are more anti-inflammatory in nature and involved in tissue repair and wound healing.

Lymphocytes (**Figure 2.6**) are found within blood, lymphoid tissue and lymphatic fluid and may be broadly divided morphologically into small and large lymphocytes (lymphoblasts) and plasma cells. **Small lymphocytes** are in the order of 6–9 μm in diameter and have a large, round nucleus with condensed chromatin and a minimal amount of surrounding cytoplasm with few organelles (**Figures 2.6, 2.7**). Small lymphocytes might be either T or B cells and functionally may be either **naïve** (**virgin**) or **memory** cells. Naïve lymphocytes have not been previously exposed to the antigenic epitope they

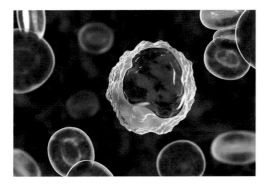

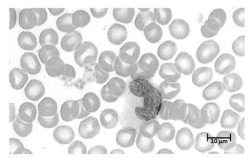

Figure 2.4 Monocytes. An illustration of a monocyte is shown; alongside, the cell is seen on a stained blood smear. These cells are relatively large, agranular and contain a kidney bean-shaped nucleus.

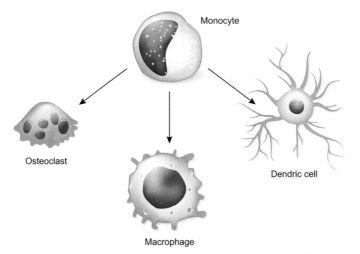

Figure 2.5 Various cell phenotypes derive from circulating monocytes. When monocytes leave the blood, they can become professional antigen presenting cells (dendritic cells), phagocytic immune cells (macrophages) or more specialist mononuclear cells, such as osteoclasts, involved in bone remodelling.

have been selected to recognize via their specific antigen receptor. In contrast, memory lymphocytes have been previously activated during an immune response and re-enter immune surveillance in preparation for any subsequent re-infection by the same pathogen.

Large lymphocytes or lymphoblasts are morphologically larger cells of 12–15 μm diameter (**Figure 2.7**) that have proportionally more cytoplasm with a greater number of organelles and less condensed nuclear chromatin than small lymphocytes. Lymphoblasts are active

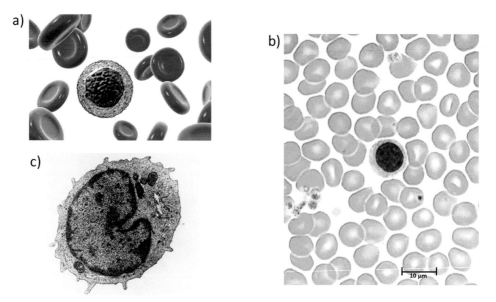

Figure 2.6 Lymphocytes. (a) Illustration of lymphocyte and red blood cells. (b) Blood smear with lymphocyte, red blood cells and a few platelets. (c) Transmission electron micrograph of a single small lymphocyte. The nucleus of this cell is cleaved and there is marginal condensation of chromatin. The cytoplasm contains few organelles. Note the characteristic ruffling of the cell surface.

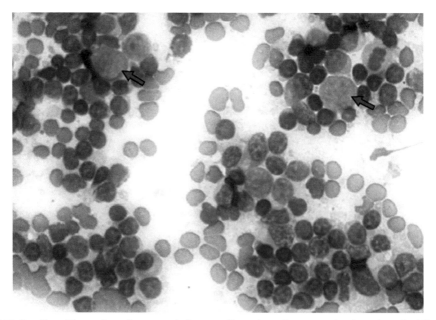

Figure 2.7 Small and large lymphocytes. A fine-needle aspirate from a canine lymph node. The majority of cells in this field are small lymphocytes, which are only marginally larger than the erythrocytes in the background. These cells have a large, round nucleus with dense chromatin and a narrow, pale blue cytoplasmic rim. Several large lymphocytes or lymphoblasts are also present (arrows). These cells are two to four times larger than the small lymphocyte, with a pale nucleus and proportionally more cytoplasm.

cells that have been recently stimulated by antigen and are participating in an active immune response. Within lymphoid tissues, the cells within follicular germinal centres are lymphoblastic, as they are in the process of proliferating and differentiating to their effector forms.

Plasma cells are a late stage of development of B lymphocytes. Although B lymphocytes can synthesize immunoglobulin and display this on their cell membrane, only plasma cells can secrete immunoglobulin in high concentration. Plasma cells have a more oval shape, with a round nucleus placed to one side of the cell (an 'eccentric' nucleus) (**Figures 2.8, 2.9**). The nucleus has characteristic linear condensation of chromatin and this is sometimes referred to as 'clock face' or 'bicycle wheel' chromatin. The cytoplasm of a plasma cell is packed with the machinery of protein production (endoplasmic reticulum and Golgi apparatus) and, even at the light microscopical level, a distinct perinuclear pale region (the Golgi zone) may be observed (**Figure 2.8**). Immunohistochemical labelling can be used to demonstrate the cytoplasmic immunoglobulin within a plasma cell.

2.3 PRIMARY LYMPHOID TISSUES

The development of B and T lymphocytes occurs in primary lymphoid tissues independent of exposure to specific antigens. The production of newly formed lymphocytes is greatest in young animals, before they reach sexual maturity, but continues throughout life. Within the **bone marrow (Figure 2.10)** is a population of **pluripotent stem cells**, which give rise to committed precursors responsible for production of the various haematopoietic lineages, namely erythroid, platelet, myeloid and lymphoid cells.

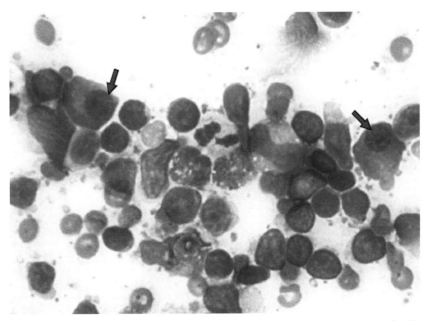

Figure 2.8 Plasma cells. Fine-needle aspirate from the lymph node of a dog. A variety of cells are present including background erythrocytes, small lymphocytes and two central eosinophils. Plasma cells are arrowed. These are oval cells with plentiful cytoplasm and a round nucleus located to one side of the cell. A perinuclear Golgi zone is seen as a pale area within the cytoplasm.

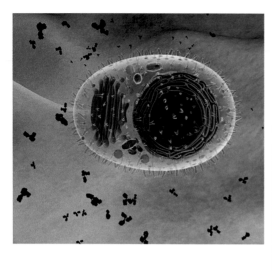

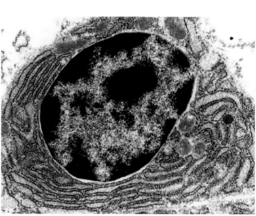

Figure 2.9 **Illustration of plasma cells secreting antibody and cellular ultrastructure.** Transmission electron micrograph of a single plasma cell. The nucleus contains condensed chromatin and sits within abundant cytoplasm containing a large amount of endoplasmic reticulum, an active site of protein synthesis for immunoglobulin secretion.

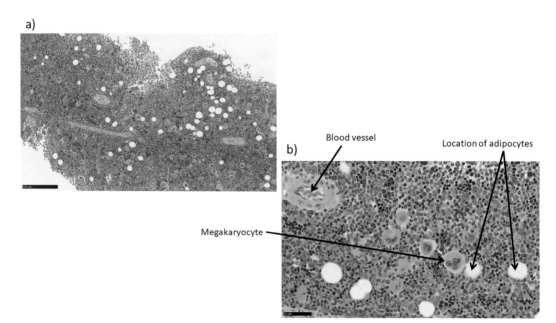

Figure 2.10 **Bone marrow.** Low-power magnification (a) and high-power magnification (b) of normal canine bone marrow comprises a mixture of haematopoietic elements and adipose tissue. The large multinucleate cells are megakaryocytes, the precursors of platelets. Stem cells within the haematopoietic tissue commit to myeloid or lymphoid precursors, which then differentiate to committed granulocytes or lymphocytes, respectively. (Original histology slides prepared by Tanya Hopcroft; digital versions courtesy of Andrew Hibbert, Royal Veterinary College.)

Lymphoid precursor cells commit to innate lymphocyte lineages (including NK cells), B cell or T cell lineages in the bone marrow. The development from stem cells to immature B cells occurs in the bone marrow. The immature B cells leave the bone marrow and undergo further development into mature B cells in peripheral lymphoid tissues under the influence of locally secreted cytokines such as B cell activating factor (BAFF) and the intestinal microbiome. The location and extent of B cell development in the peripheral lymphoid tissues varies by species. In birds, final maturation of B cells takes place in the **bursa of Fabricius**, a lymphoepithelial organ located near the cloaca (**Figure 2.11**). In an early immunological experiment, it was noted that bursectomy (surgical removal of the bursa) in chicks led to failure of B cell development and an inability to produce antibody when they matured into adult birds. In fact, this observation led to the naming of B lymphocytes (from bursa-derived lymphocytes). The bursa reaches a maximum size in young birds and undergoes **involution** (shrinkage) later in life. Lambs and calves have a large amount of lymphoid tissue that occupies the terminal 1–2 meters of the small intestine (**ileal Peyer's patch**). Surgical removal of the ileal Peyer's patch in lambs leads to a transient deficiency in B cells and antibody production, but not as severe as the effect of bursectomy in chickens. Like the bursa in birds, the ileal Peyer's patch decreases in size after sheep reach sexual maturity, being nearly undetectable in adults. Because of the role of these peripheral lymphoid tissues in B cell development, the bursa in birds and ileal Peyer's patch in ruminants are considered primary lymphoid organs. Pigs and dogs also have an ileal Peyer's patch that undergoes involution in adult animals, but it appears to play a less significant role in B cell development. In these species and in rodents and primates, the final maturation phase of B cells occurs in secondary lymphoid tissues including the spleen and lymph nodes.

The **thymus**, an organ found in the cranial mediastinum of the thoracic cavity (**Figure 2.12**), is the primary lymphoid organ for

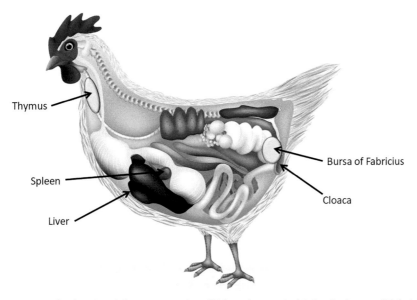

Figure 2.11 Bursa of Fabricius. The primary site of B lymphocyte in birds, the bursa of Fabricius is located close to the cloaca. Birds also have a thymus, in the neck region, which is the primary site of T lymphocyte development.

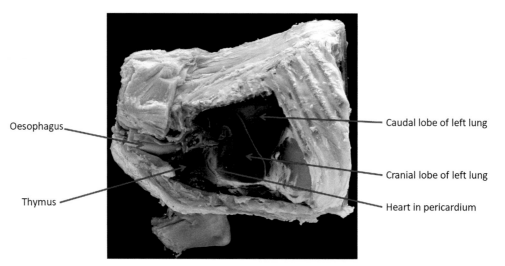

Oesophagus

Thymus

Caudal lobe of left lung

Cranial lobe of left lung

Heart in pericardium

Figure 2.12 Thymus. This anatomical specimen shows the canine thorax, which has been dissected to reveal the internal organs by removal of the cranial ribs and intercostal and neck muscles. The cranial part of the left lung has been removed so the thymus can be seen. (Image courtesy of the Camden Anatomy Museum, Royal Veterinary College.)

T cell development. Precursor cells are exported from the bone marrow to the thymus to undergo development into mature T cells. They were termed 'T lymphocytes' because of the critical role of the thymus in their development. The precursor T cells enter the subcapsular region of the thymus from the blood and initiate their development into functional naïve T cells, during which time they migrate through the outer cortex into the medulla. Upon completion of their development and selection processes, they enter the general circulation as functional naïve T cells. The thymus is of maximal size during the first few months after birth and involutes during puberty to become difficult to identify in adult animals. Although most of the thymus is replaced by adipose tissue in adults, residual thymic tissue does persist, and naïve T cells continue to be produced from the thymus, even into relatively old age. In congenitally athymic animals (i.e. animals born without a thymus) there is a total lack of **cell-mediated immunity (CMI)**, for which T cells are necessary. Interestingly,

this type of genetic mutation is often associated with an inherited inability to form a hair coat, typified by inbred laboratory strains of 'nude' athymic rats and mice. At least one of the hairless canine breeds (the Mexican hairless dog) is also known to have impaired thymic development.

Histologically, the thymus is an **encapsulated** organ consisting of a number of lobules each of which has a distinct outer **cortex** and inner **medulla** (**Figures 2.13, 2.14**). It consists of a network formed by **cortical and medullary epithelial cells** that is closely packed with immature T lymphocytes, with fewer interspersed **macrophages** and **dendritic cells**. **Hassall's corpuscles**, made up of clusters of concentrically arranged epithelial cells in the medulla, are thought to produce growth factors that influence T cell development. There is an incredibly high turnover of thymic T cells, which undergo positive and negative selection processes. Large numbers of cells fail to be selected and die by apoptosis, the cellular debris subsequently being phagocytosed by macrophages (**Figure 2.15**).

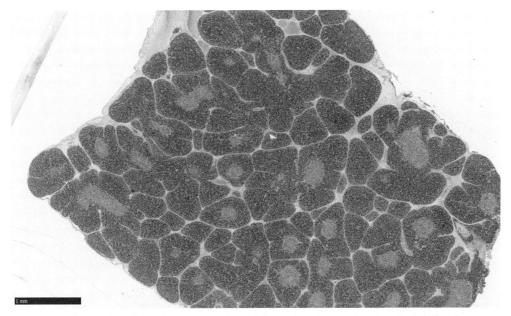

Figure 2.13 Thymic histology. Low-power view of the canine thymus showing the lobulated structure of the organ and the presence of a darker cortical region and paler medullary area within each lobule. (Original histology slides prepared by Tanya Hopcroft; digital versions courtesy of Andrew Hibbert, Royal Veterinary College.)

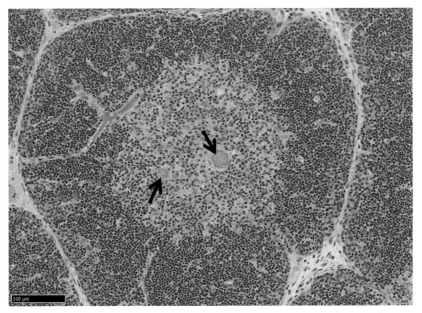

Figure 2.14 Thymic histology. Higher magnification of the canine thymus showing the junction between cortex and medulla. The Hassall's corpuscles of the medulla are arrowed. (Original histology slides prepared by Tanya Hopcroft; digital versions courtesy of Andrew Hibbert, Royal Veterinary College.)

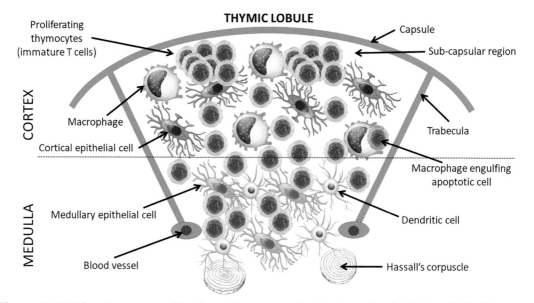

Figure 2.15 Thymic structure. The thymus is an encapsulated primary lymphoid tissue divided into lobules by connective tissue trabeculae. There is an outermost cortex and inner medulla, both densely packed with actively dividing and apoptotic thymocytes (immature T lymphocytes). Interspersed with the lymphocytes are the thymic epithelial cells, dendritic cells and macrophages, the latter phagocytosing debris from apoptotic cells. The squamous epithelial Hassall's corpuscles produce cytokines, which influence T cell development.

2.4 SECONDARY LYMPHOID TISSUES

The secondary lymphoid tissues are broadly divided into those that are **encapsulated** (the lymph nodes and spleen) and those that are **unencapsulated** (the mucosa-associated lymphoid tissues; MALT). A network of **lymph nodes** are strategically located around the body, each receiving tissue fluid/lymph from a particular region (**Figure 2.16**). Lymph fluid, containing antigenic material and antigen presenting cells (e.g. dendritic cells) from the local tissues drains into the lymph node via multiple **afferent lymphatic vessels** (**Figure 2.17**). This fluid first enters the **subcapsular sinus** and then percolates through the lymphoid tissue where it is 'filtered' before reaching the large **medullary sinus**, from where it leaves the node in a single **efferent lymphatic vessel**, located at the hilus. The efferent lymph then travels through a net-work of lymphatic vessels, ultimately returning fluid to the bloodstream via the **tracheal and thoracic ducts** (draining the head and body, respectively). Blood vessels enter and leave the lymph node at the hilus and specialist **high endothelial venules** (HEVs) provide the main point of entry for lymphocytes from the blood into the lymphoid tissue.

The **cortical area** of the lymph node is composed of a closely packed population of lymphocytes arranged as spherical **follicles** within a surrounding **paracortex** (**Figure 2.18**). These two regions contain distinct subpopulations of lymphocytes. The follicles are occupied by B cells and the paracortex consists predominantly of T cells. This **compartmentalization** of the major lymphoid populations can be readily demonstrated by immunohistochemical labelling (**Figure 2.19**). The follicles may take one of two histological forms, depending upon the absence or presence of antigenic stimulation.

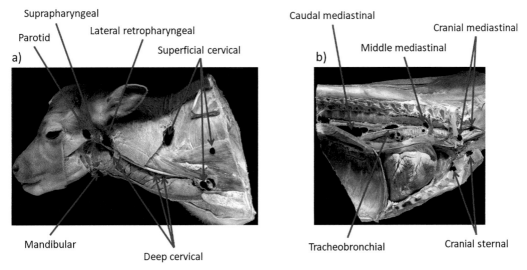

Suprapharyngeal
Parotid
Lateral retropharyngeal
Superficial cervical
a)

Caudal mediastinal
Cranial mediastinal
Middle mediastinal
b)

Mandibular
Deep cervical

Tracheobronchial
Cranial sternal

Figure 2.16 Lymph nodes. (a) This specimen shows the head and neck of a calf, viewed from a left lateral position, with the different lymph nodes identified. (b) This specimen shows the thorax of a calf, viewed from a right lateral position, with the different lymph nodes identified. (Image courtesy of the Camden Anatomy Museum, Royal Veterinary College.)

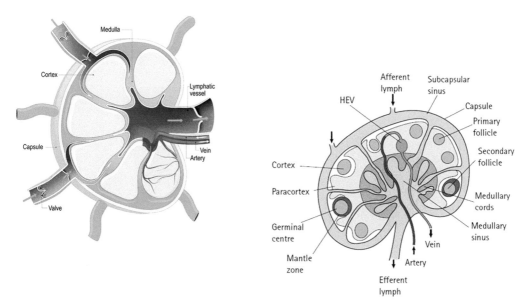

Medulla

Cortex

Capsule

Lymphatic vessel

Vein
Artery

Valve

Afferent lymph
HEV
Subcapsular sinus
Capsule
Primary follicle
Secondary follicle

Cortex
Paracortex

Medullary cords
Medullary sinus

Germinal centre

Mantle zone

Vein
Artery
Efferent lymph

Figure 2.17 Lymph node structure. The lymph node is an encapsulated organ with a cortex and a medulla. Afferent lymph flows into the node from local tissues and carries lymphocytes, antigen presenting cells and antigen. Lymph enters the subcapsular sinus and then percolates through the node to the medullary sinus before leaving in the efferent lymphatic vessel. The lymph node has an arterial and venous blood supply and lymphocytes may enter the node through the high endothelial venules (HEVs). The cortex of the lymph node comprises follicular aggregates of B lymphocytes surrounded by the paracortical T cell zone. Follicles may be primary or secondary, with a mantle zone surrounding a germinal centre. The medullary cords provide a framework for the medullary sinus, containing macrophages and plasma cells.

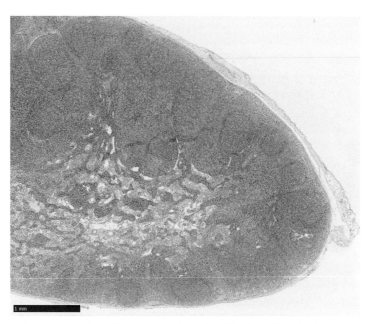

Figure 2.18 Lymph node histology. Low-power view of a feline lymph node showing cortical follicles surrounded by paracortex and a central medullary sinus crossed by interlacing medullary cords. (Original histology slides prepared by Tanya Hopcroft; digital versions courtesy of Andrew Hibbert, Royal Veterinary College.)

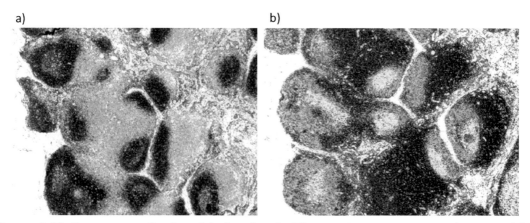

Figure 2.19 Compartmentalization of the lymph node. These two images represent serial sections of the same lymph node immunohistochemically labelled to show the position of B cells in the follicles (a) and T cells in the paracortex (b). Some T helper cells are present within the follicles and are directly involved in B cell activation.

In the absence of antigenic stimulation, lymph nodes are small and contain **primary follicles** which consist of dense aggregates of small naïve lymphocytes. Upon infection or vaccination, the lymph node enlarges and **secondary follicles** develop, which have a distinct **mantle zone** of small lymphocytes surrounding an innermost **germinal centre** of larger, proliferating

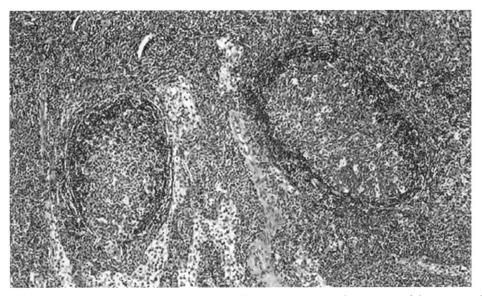

Figure 2.20 Secondary follicles. Two secondary follicles are present at the junction of the cortex and the medulla. The follicles have a pale germinal centre surrounded by a dark mantle zone. The surrounding paracortex is to the top of the image, and the medullary sinus and cords are towards the bottom.

B lymphocytes (**Figure 2.20**). The germinal centre classically includes both a **dark zone** and a **light zone** defined by the staining affinity of the lymphoid cells. The medulla of the lymph node is given structure by a network of **medullary cords** that act as a framework holding the sinus open. As well as macrophages, these cords are also packed with lymphoid cells, particularly **plasma cells**. These plasma cells migrate from secondary follicles to the medullary cords, where they take up residency, secreting their antibody into the efferent lymph.

In the pig, as well as some non-domestic animals including the rhinoceros and hippopotamus, lymph nodes have an inverted/'inside-out' structure, consisting of multiple nodules with an innermost cortex and peripheral medulla. Afferent lymph flows into the centre of each nodule and percolates through closely packed cells (rather than medullary sinuses) to the peripheral medullary areas to reach the efferent vessel. This structure means that in these particular species, lymphocytes rarely leave the node in efferent lymph, but instead migrate directly into the bloodstream via paracortical HEVs.

The **spleen** is a visceral organ that is located in the cranial left quadrant of the abdomen (**Figure 2.21**). It has an arterial and venous blood supply, but no afferent lymph vessels. Whereas the lymph nodes can be considered antigen filters for tissue fluid (lymph), the spleen can be considered to be an antigen filter for the blood. The spleen is encapsulated and, contiguous with the capsular connective tissue, is a network of **trabeculae** that create a framework for the splenic structure. Between the trabeculae, the splenic parenchyma comprises a mixture of red and white pulp (**Figure 2.22**). The **red pulp** is a reservoir of red blood cells and platelets, and the **white pulp** represents the splenic lymphoid tissue (**Figure 2.23**). As in the lymph node, this tissue is compartmentalized into T and B cell zones. The splenic T cells are closely associated with arteriolar blood vessels and form a cylindrical surround to such vessels (akin to the 'lagging' on a water pipe). This T cell region is known as the **periarteriolar lymphoid sheath** (PALS). Splenic B cells are present within primary or secondary **follicles** adjacent to the PALS and both structures are

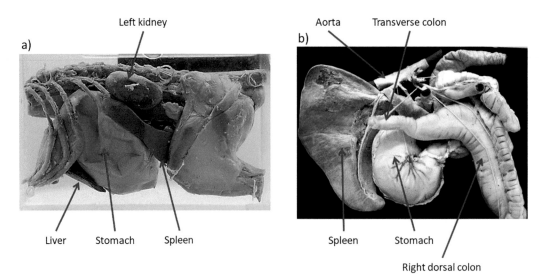

a)
Left kidney

Liver   Stomach   Spleen

b)
Aorta   Transverse colon

Spleen   Stomach

Right dorsal colon

Figure 2.21 Spleen. (a) This specimen shows a left lateral view of feline abdominal organs in situ (with jejunum removed). The abdominal wall has been removed from the ribs caudally to the pelvis to expose the abdominal viscera. (b) This specimen shows the equine juvenile gastrointestinal tract and spleen with its blood supply from the aorta. The gastrointestinal tract and spleen have been isolated from the abdomen of the foal and the small intestines removed to aid visualization of the stomach and colon. Red latex has been injected into the arteries, blue latex into the portal vein, yellow into the transverse colon and dark green latex into the cut ends of the small intestine (removed). (Image courtesy of the Camden Anatomy Museum, Royal Veterinary College.)

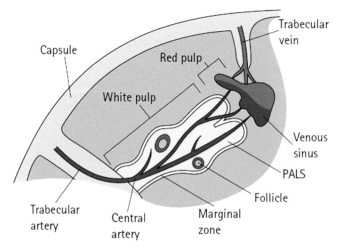

Capsule

Red pulp

White pulp

Trabecular vein

Venous sinus

PALS

Follicle

Trabecular artery

Central artery

Marginal zone

Figure 2.22 Structure of the spleen. The spleen is an encapsulated organ, given structure by a network of connective tissue trabeculae. Lymphoid tissue is localized to the white pulp that lies within the surrounding red pulp. The red pulp consists of blood, which in some animal species (e.g. the dog) is contained by endothelial-lined vascular sinuses. In other species (e.g. the cat), the spleen is non-sinusal. The spleen does not have afferent lymph vessels, but is considered part of the vascular circulation. White pulp lymphoid tissue is compartmentalized. T cells reside in the periarteriolar lymphoid sheath (PALS) and B cells within follicles. A narrow marginal zone delineates the periphery of the white pulp and contains many macrophages.

a)

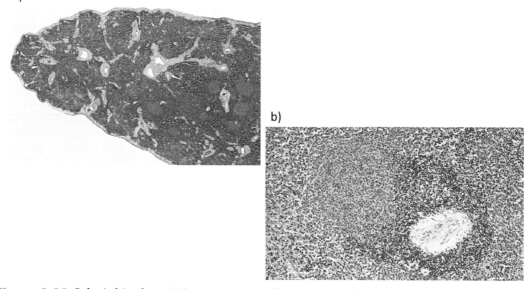

b)

Figure 2.23 Splenic histology. (a) Low-power view of canine spleen showing the outermost capsule, which is continuous with the network of connective tissue trabeculae. The regions of white pulp are seen throughout the surrounding red pulp. (b) Higher magnification of an area of white pulp reveals the periarteriolar lymphoid sheath (PALS) encircling a central arteriole with an adjacent follicular aggregate. The narrow marginal zone is not readily discerned by routine light microscopy.

surrounded by a narrow **marginal zone** that includes a mixture of cell types, including macrophages.

The secondary unencapsulated lymphoid tissues are largely scattered throughout the **mucosal surfaces** of the body and represent a significant proportion of the total lymphoid tissue. These **mucosa-associated lymphoid tissues** (**MALT**) include the **tonsils**, gastrointestinal-associated lymphoid tissue (**GALT**) and bronchial-associated lymphoid tissue (**BALT**). This interlinked network of lymphoid tissues functions to protect the mucosal surfaces, which are particularly susceptible to infection, given their relatively vulnerable simple epithelium. GALT comprises the **inductive** sites of the intestinal immune system (i.e. where alimentary immune responses are initiated) and the **effector** sites (where the outcome of alimentary immune responses take effect; e.g. IgA production). The main inductive sites of GALT are the Peyer's patches of the small intestine

and solitary lymphoid follicles (which are most prominent in the stomach and large intestine). **Palatine tonsils** are aggregates of lymphoid tissue located in the pharynx that serve to initiate immune responses to antigen entering via the respiratory or alimentary tracts (**Figure 2.24**). The **Peyer's patches** consist of unencapsulated lymphoid aggregates comprising B cell-rich follicles (often with germinal centers) surrounded by an intervening zone of predominantly T cells (**Figure 2.25**).

These mucosal lymphoid structures are accompanied by an alteration to the microanatomy of the overlying epithelium and lamina propria, which forms a broad-based **dome** rather than having villous microarchitecture. Within the epithelium covering the Peyer's patch is a specialized population of **microfold** (**M**) **cells**, which sample antigen from the intestinal lumen and transfer this to the underlying lymphoid tissue. The Peyer's patches are linked

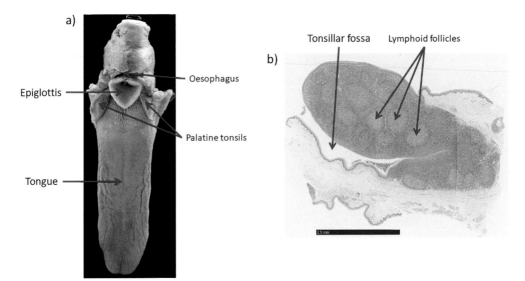

Figure 2.24 Tonsils. (a) This specimen shows a dorsal view of the canine tongue and attached larynx which have been isolated from the head. The palatine tonsils lie on either side of the pharynx. (b) Low-power histology of the feline tonsil showing dense lymphoid tissue containing many lymphoid follicles. (Original histology slides prepared by Tanya Hopcroft; digital versions courtesy of Andrew Hibbert, Royal Veterinary College.)

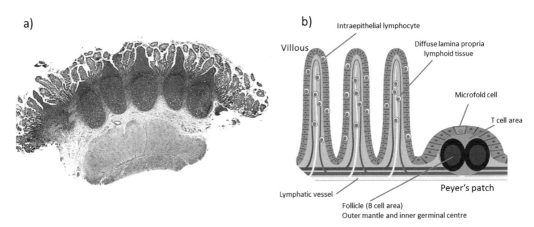

Figure 2.25 Peyer's patch. (a) A full-thickness biopsy of canine intestine showing the mucosa, submucosa and muscularis propria with a Peyer's patch evident in the mucosa-submucosa. (b) Illustration of Peyer's patch showing B cell follicles surrounded by a mixed zone of T cells and antigen presenting cells (APCs). There is modification of the overlying mucosa such that villi are replaced by a broader dome-shaped region. Microfold (M) cells within the epithelium of the dome serve to sample antigen from the intestinal lumen and transfer it to the underlying lymphoid tissue.

to the draining **mesenteric lymph nodes** by afferent lymphatic vessels. Although intestinal immune responses may be amplified in the associated mesenteric lymph nodes, these are not considered part of GALT, as they do not directly sample mucosal antigen.

In addition to these organized lymphoid tissues, there are also diffusely scattered

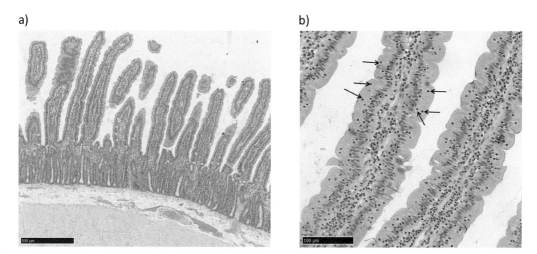

Figure 2.26 **Intraepithelial lymphocytes of the intestinal mucosa.** Low-power magnification (a) and high-power magnification (b) of cat duodenum, showing villous epithelium, with numerous intraepithelial lymphocytes (arrows) scattered within the simple columnar epithelium. (Original histology slides prepared by Tanya Hopcroft; digital versions courtesy of Andrew Hibbert, Royal Veterinary College.)

lymphocytes and plasma cells of the lamina propria and the specialized compartment of intraepithelial lymphocytes (IELs) that reside in the epithelial layer lining the mucosal surfaces (**Figure 2.26**). Although less organized in terms of histological structure, these cells represent a substantial component of mucosal immunity.

2.5 THE LYMPHATIC SYSTEM, LYMPHOCYTE RECIRCULATION AND IMMUNE SURVEILLANCE

Histology images of lymphoid tissues depict a static representation of the immune system. However, this fails to convey the dynamic nature of the cells and molecules that are involved. The immune system utilizes the circulatory system and lymphatic flow around the body, to both detect the presence of foreign invaders and to deploy the necessary resources to the site of infection. The lymphatic system, designed to return excess tissue fluid back to the bloodstream, could be used by pathogens to disseminate around the body from the site of invasion. The immune system turns this potential vulnerability to its advantage by strategically placing lymph nodes within the lymphatic system with responsibility for drainage of specific areas of the body (**Figure 2.27**). Thus, lymph nodes trap and filter out foreign organisms from the afferent lymph and act as a depot of antigen for recognition by lymphocytes during immune surveillance.

In addition to a passive flow of antigen into lymph nodes, the afferent lymph can also be utilized by antigen presenting cells, which sample the tissue microenvironment, then alert the immune system to the presence of foreign antigen. Dendritic cells (DCs) develop from circulating monocytes once they have left the bloodstream and are particularly abundant at epithelial surfaces, such as the epidermis of the skin (where they are known as Langerhans cells) and the mucosal epithelium. At this stage of their development, immature DCs are engaged in phagocytosis and macropinocytosis, taking up antigen from the tissues. They then migrate via the afferent lymphatics to the nearest lymph node, processing the antigen during transit, until they reach the lymph node paracortex. Here, they terminally differentiate, take up residence and present their antigen to T

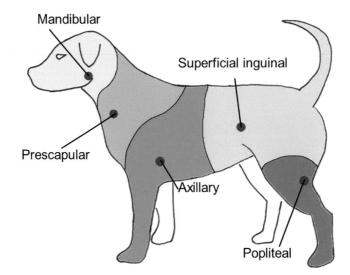

Figure 2.27 Superficial lymphatic drainage of the canine. The diagram shows the main superficial lymph nodes in a dog that are usually palpated during clinical examination. These lymph nodes drain the specific areas indicated by the different colours.

lymphocytes that have specific receptors for the antigenic epitopes (**Figure 2.28**).

Once they have completed their development in primary lymphoid tissues, naïve lymphocytes begin to recirculate, effectively 'visiting' different secondary lymphoid tissues during the course of their travels. Although lymphoid cells may move from tissue to lymph nodes via the afferent lymph, the majority of naïve lymphocytes enter lymphoid tissues via the blood, using the **high endothelial venules** (**HEVs**). The HEVs can be thought of as 'bus stops' in the circulation, that allow the naïve lymphocytes to recognize that they have reached a destination. As we shall see later, targeted extravasation into lymphoid tissues involves interaction between **homing receptors** on the lymphocytes and specific **vascular addressins** expressed on the surface of HEV endothelial cells. Once they have entered the lymph node, B cells migrate to the follicles to inspect any antigen deposited there and T cells spend their time interacting with the DCs in the paracortex, looking for the presence of their specific antigenic epitope. After a period of time, the lymphocytes (having not encountered their antigen in the absence of infection) migrate to the medullary sinus and exit the lymph node via the efferent lymph. As we discussed in Chapter 1, even during an infection, only a very small number of antigen-specific lymphocyte clones will react to a particular pathogen, with the vast majority of those lymphocytes that enter the lymph node being unstimulated and continuing to recirculate. The efferent lymphatics form a network of lymphatic vessels that carry lymph fluid and circulating lymphocytes into the **tracheal ducts** (from lymph nodes in the head) or the **thoracic ducts** (from lymph nodes caudal to the head) (**Figure 2.29**). These then discharge their contents into the venous system, cranial to the heart (e.g. external jugular vein/brachiocephalic vein) (**Figure 2.30**), thus returning both fluid and cells to the systemic circulation and completing **lymphocyte recirculation**.

Circulating lymphocytes may migrate from the blood into any of the secondary lymphoid tissues (lymph nodes, spleen or MALT) (**Figure 2.31**). To give an idea of the scale of this recirculation, in rodents, it is estimated that 4×10^7 lymphocytes enter the blood from the thoracic duct every hour, such that the entire

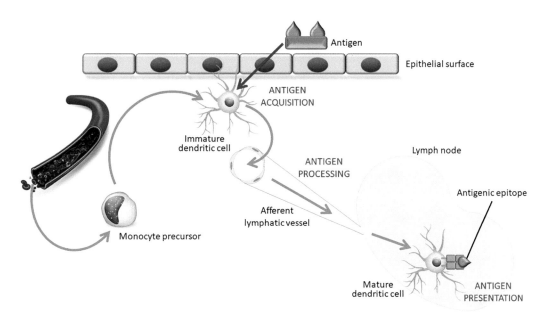

Figure 2.28 Dendritic cell migration. Immature dendritic cells develop from circulating monocytes. They acquire antigen from the tissue microenvironment via phagocytosis and micropinocytosis. Subsequently, they migrate via afferent lymphatics to the regional lymph node where they process and present the antigen.

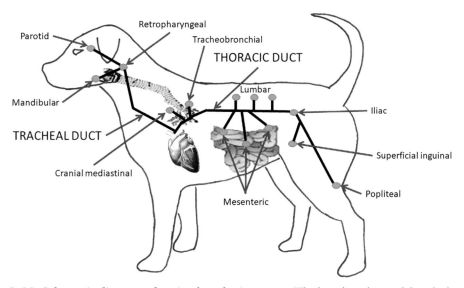

Figure 2.29 Schematic diagram of canine lymphatic system. The lymph nodes caudal to the heart drain into the thoracic duct and those cranial to the heart drain into the left or right tracheal ducts.

blood lymphocyte count can be replaced 10 to 20 times per day. Lymphocyte recirculation is important as each cell recognizes a specific antigenic epitope, but does not have any tacit knowledge as to the likely port of entry of the pathogen bearing that antigen. Therefore, lymphocyte recirculation through secondary lymphoid tissues represents an **immune surveillance** strategy that attempts to maximize the opportunity for lymphocytes to encounter

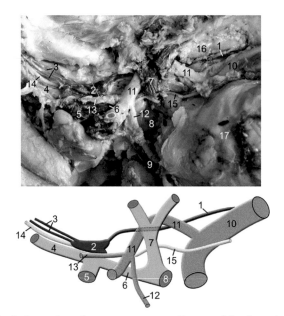

Figure 2.30 Lymphatic drainage into the venous system. Course of the thoracic duct in the pig: 1 = thoracic duct, 2 = ampullary termination of the thoracic duct, 3 = double left tracheal duct, 4 = left external jugular vein, 5 = left subclavian vein, 6 = left brachiocephalic vein, 7 = left costocervical vein, 8 = cranial vena cava, 9 = left internal thoracic vein, 10 = aorta, 11 = left subclavian artery, 12 = left internal thoracic artery, 13 = left superficial cervical artery, 14 = left vagosympathetic trunk, 15 = left vagus nerve, 16 = musculus longus colli sinister, 17 = pericardium. (Reproduced with permission from Gomercic MD et al. *Veterinarni Medicina* 55, 2010 (1): 30–34.)

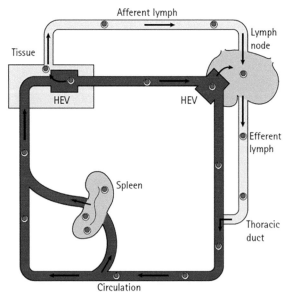

Figure 2.31 Lymphocyte recirculation. This diagram summarizes the routes by which lymphocytes may move throughout the body. While circulating in the blood, the lymphocytes may detour through the spleen or leave the circulation to enter lymphoid tissue (lymph nodes or MALT) through specialized high endothelial venules (HEVs). The cells leave lymph nodes via the efferent lymphatic vessel, which connects to the common thoracic duct which takes them back into the blood.

their cognate antigen, if and when it appears. Those lymphocytes that do react to antigen cease to recirculate and undergo a period of proliferation and differentiation. Many of these effector cells leave the lymphoid tissue and (as a result of changes to their homing receptors) are targeted towards sites of inflammation, likely to contain the infectious organism (**Figure 2.32**).

It is clear that **lymphocyte homing** is important for both the induction phase (allowing naïve lymphocytes to visit sites of antigen deposition) and effector phase (allowing mature lymphocytes to seek out sites of infection) of the adaptive immune response. This process requires interaction between two sets of molecules: **homing receptors**, expressed on the surface of the lymphocytes, and **vascular addressins**, expressed on the surface of endothelial cells. Vascular addressins literally give the 'address' of the particular portion of the vasculature that expresses those molecules. Vascular addressins are generally found on the luminal surface of

a group of modified endothelial cells that protrude into the lumen of the vessel. In lymphoid tissues these are known as **high endothelial venules** (HEVs). The HEVs express addressins such as, for example, glycosylation-dependent cell adhesion molecule-1; (glyCAM-1), which is the ligand for L-selectin, the major homing receptor on the surface of naïve lymphocytes (**Figure 2.33**).

As naïve lymphocytes travel through the bloodstream they can encounter HEVs present in various secondary lymphoid tissues. As the HEVs protrude into the flowing blood, they establish local turbulence in blood flow, and this encourages the circulating lymphocytes to bump up against the endothelial surface of the HEVs, allowing their homing receptors to bind to the vascular addressins present. When such an interaction occurs, the circulating lymphocyte slows down and rolls along the endothelial surface and eventually will be halted. It then has the opportunity to squeeze between

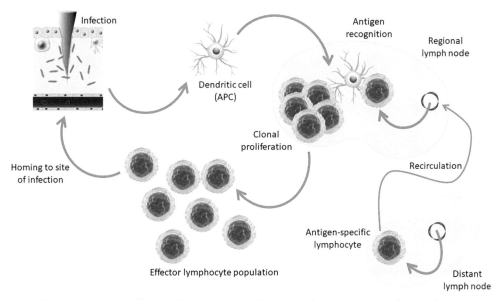

Figure 2.32 Immune surveillance. Foreign antigen from an infected wound is taken to the regional lymph node by a dendritic cell. A lymphocyte able to respond to an epitope on that antigen is currently located within a distant lymph node, but during the course of its subsequent migration it travels through the node of interest and meets the antigen. The adaptive immune response is stimulated and the clonally expanded lymphocytes are sent via lymph and blood back to the tissue site where they are required to fight the infection.

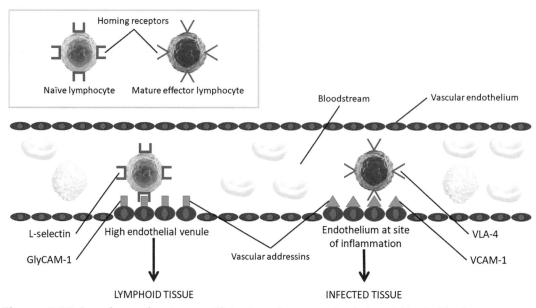

Figure 2.33 Lymphocyte homing specificity. Lymphocytes recirculating within the bloodstream must have the ability to exit the circulation at specific locations, where they may be required. There are focal regions in the circulatory system where specialist blood vessels are located within the lymphoid tissues. These high endothelial venules (HEVs) are also endowed with surface molecules collectively termed vascular addressins (e.g. GlyCAM-1). An inflammatory response in the tissues typically generates cytokines that induce expression of specific vascular addressins (e.g. VCAM-1) that allows recruitment of leucocytes to the area. Circulating lymphocytes bear ligands for vascular addressins, collectively referred to as homing receptors. These differ depending on the status of the lymphocyte, for example naïve lymphocytes express L-selectin, which binds to GlyCAM-1, allowing these cells to undergo recirculation though various lymphoid tissues. In contrast, following clonal expansion and differentiation, effector T lymphocytes express VLA-4 ($\alpha_4\beta_1$ integrin) that binds to VCAM-1 on the surface of endothelial cells within inflamed tissues, which targets these cells to the likely site of infection.

adjacent endothelial cells (**diapedesis**) or, less commonly, through endothelial cells (**transcellular migration**) into the lymphoid tissue beneath.

Lymphocytes change their migratory pathways within the body at different stages of their development. As part of immune surveillance, a naïve (antigen-inexperienced) lymphocyte would want to patrol widely throughout the body in search of its cognate antigen. Once lymphocytes have been activated by foreign antigen, as part of an adaptive immune response, and having undergone a period of proliferation and differentiation, they alter their homing receptors. They are now aware of the fact that there is a

pathogen present somewhere in the body and may need to seek it out. Effector T cells start to express the integrin VLA-4, which is a homing receptor that binds to vascular cell-adhesion molecule-1 (VCAM-1). This particular vascular addressin is induced on endothelial cells that are exposed to pro-inflammatory mediators and therefore is selectively expressed at the sites of inflammation (**Figure 2.33**). It is believed that there is a degree of specificity that is 'imprinted' on the lymphocytes during the process of activation within the lymphoid tissue that determines their subsequent homing pattern. Finally, after the infection has been eliminated, a population of residual memory lymphocytes start to appear in the circulation, which re-enter immune surveillance, alongside their antigen-inexperienced naïve colleagues. However, they do not necessarily revert back to expressing the full range of homing receptors and instead might only be able to access those tissues in which they know they are most likely to re-encounter their cognate antigen (**Figure 2.34**).

One particular example of selective lymphocyte recirculation is that related to the mucosal immune system. Although there is clear specificity in recirculation of lymphocytes activated by mucosal antigen (i.e. cells activated in GALT will preferentially recirculate back to the intestine), it is also possible for these activated cells to migrate to other mucosal sites of the body. Thus, the common mucosal system has a philosophy somewhat similar to Article 5 of the North Atlantic Treaty Organization (NATO),

which states that "any attack on one member is considered an attack on them all". In this way, antigenic stimulation at one mucosal site leads to an 'armed response' in all mucosal tissues. In order for this to be possible, there must be sharing of some common vascular addressins (**Figure 2.35**), thereby permitting lymphocytes activated in one mucosal site to access not only the intestinal tissue where the effector response is required but also the mucosae of the respiratory tract, urogenital tract, eye and mammary gland (**Figure 2.36**). For domestic animals, this system has particular relevance, as it permits lymphocytes with memory of enteric and respiratory pathogens to migrate to the mammary gland during pregnancy and secrete antibodies into the colostrum and milk, which mediate protection of the neonate from infectious disease (see Chapter 13).

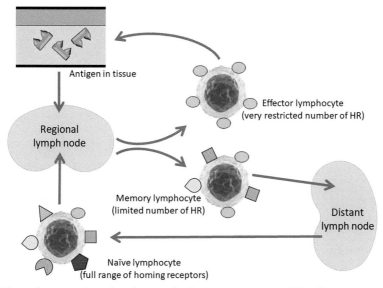

Figure 2.34 **Lymphocyte migration during the immune response.** This diagram proposes a model that might account for differential mobility of lymphocytes at different stages of their life cycle. A naïve lymphocyte that has not previously encountered antigen might have a wide range of homing receptors (HRs), allowing it to access numerous body sites during immune surveillance. Once antigen is encountered, the activated lymphocytes might have a more limited HR expression, restricting it to the tissue in which the antigen was located. This change in HR expression would likely occur during the process of differentiation within the regional lymphoid tissue. The memory lymphocytes that persist once the immune response is completed may have a more limited HR expression compared with the naïve cells, permitting the memory cells to access areas that are most likely to be sites of re-exposure to that antigen.

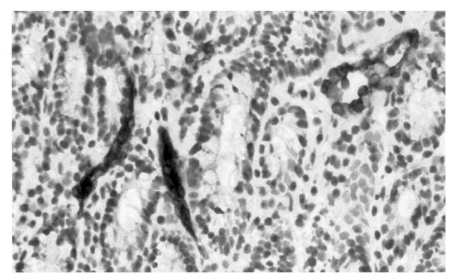

Figure 2.35 Vascular addressins allow specific targeting to mucosal tissues. This section of canine intestinal mucosa is labelled to show the expression of a vascular addressin molecule by capillary endothelial cells. The 'mucosal addressin cell adhesion molecule-1' (MAdCAM-1) is expressed by endothelia at mucosal sites of the body, but not in other tissues, and permits circulating lymphocytes that express the appropriate homing receptor (lymphocyte Peyer's patch adhesion molecule (LPAM), the integrin $\alpha_4\beta_7$) to exit the blood and enter these tissues.

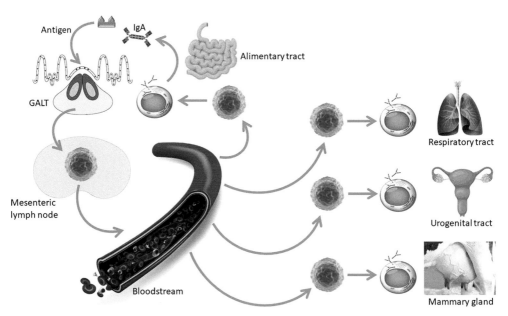

Figure 2.36 Lymphocyte recirculation to mucosal sites. A lymphocyte activated by an enteric pathogen within the inductive site of GALT would be expected to recirculate to the intestine to participate in the effector immune response (e.g. IgA production). However, some of these antigen-specific cells might also enter other mucosal tissues. The greatest relevance of this pathway is to permit antigen-specific cells from the gut and respiratory tract to enter the mammary gland and contribute to the passive transfer of immunity to neonates.

KEY POINTS

- All white blood cells (including lymphocytes) arise from bone marrow stem cells.
- Granulocytes (neutrophils, eosinophils and basophils) circulate in the blood and are important for innate immunity to infection.
- Monocytes are circulating cells that leave the blood and enter tissues where they can differentiate into macrophages (phagocytic cells) or dendritic cells (professional antigen presenting cells).
- Lymphocytes proliferate and differentiate in response to antigenic stimulation, which is critical for adaptive immunity.
- T and B cells may be morphologically small lymphocytes (naïve or memory cells) or lymphoblasts (active).
- Lymphocytes differentiate from lymphoid precursors in the bone marrow (an important primary lymphoid tissue) and commit to either T or B cell lineages, where their maturation pathways diverge.
- B cell maturation occurs in the avian bursa of Fabricius and in other lymphoid tissues in mammals.
- The thymus is the primary lymphoid organ for T cell maturation.
- Secondary lymphoid tissues (lymph node, spleen and MALT) are strategically located around the body for immunological defence of epithelial surfaces and tissues.
- Lymph nodes receive afferent lymph from the tissues and act as 'filters' to capture antigen before allowing lymph fluid to return to the circulation via efferent lymphatic vessels.
- The lymph node structure compartmentalizes lymphocytes, with B cells located in follicles and T cells scattered throughout the paracortex.
- Secondary follicles (with germinal centres) are the sites of activation of B cells, which differentiate to plasma cells that migrate to medullary cords where they secrete immunoglobulin (antibody) into the efferent lymph.
- The spleen has no lymphatic drainage, but has a substantial blood supply, allowing the spleen to act as a 'filter' for blood-borne pathogens.
- A number of lymphoid tissues (known collectively as mucosa-associated lymphoid tissues; MALT) function to protect the mucosal epithelium, which is vulnerable to infection.
- Naïve lymphocytes continually recirculate throughout the body using lymphatic and blood vessels (via high endothelial venules; HEVs) to visit different secondary lymphoid tissues.
- Lymphocyte recirculation allows immune surveillance of the body for antigen.
- Lymphocyte migration is dependent upon expression of homing receptors that interact with specific addressins on the surface of vascular endothelial cells, thus targeting the cells to relevant sites (i.e. naïve lymphocytes into lymphoid tissues, effector lymphocytes into infected tissues).
- Recirculation of lymphocytes that are activated at one mucosal site to other mucosal tissues forms the basis of the common mucosal immune system, and migration of these cells from the gut and respiratory tract to the mammary gland allows passive transfer of immunity to neonatal animals.

INNATE IMMUNITY

OBJECTIVES

At the end of this chapter, you should be able to:

- Recognize that there are physical and biochemical defence systems in place to protect the host epithelial surfaces (skin and mucosal epithelium) from infection.
- Explain how the innate immune system detects the presence of infection via recognition of specific pathogen-associated molecular patterns (PAMPs) or damage-associated molecular patterns (DAMPs) that are sensed by host pattern-recognition receptors (PRRs).
- Describe the different families of PRRs, including Toll-like receptors (TLRs) and NOD-like receptor proteins (NLRs).
- Understand the basis of intracellular signalling to generate a biological response in the host cell, following activation of a PRR.
- Explain the role of the inflammasome in enhancing the inflammatory response to the presence of infection.
- Understand the role of neutrophils and cells of the monocyte lineage in inflammation and phagocytosis of microbial pathogens.
- Explain the stages of phagocytosis and the host molecules involved in microbial killing and digestion.
- Explain the role of natural killer (NK) cells in the host response against intracellular infection.
- Explain the role of $\gamma\delta$T cells in innate immunity.
- Understand how the interferon response is activated and how it contributes to host defence against viruses.
- Describe the main features of acute and chronic inflammation.
- Discuss how the acute phase response enhances innate defence systems including phagocytosis and complement activation.
- Briefly describe the alternative, lectin and classical pathways in terms of the trigger factors that initiate them and the end products produced.
- Describe how the terminal pathway generates the membrane-attack complex.
- Understand how complement activation contributes to antimicrobial defence via cytolysis, opsonization and inflammation.

DOI: 10.1201/9781003310969-3

3.1 INTRODUCTION

The innate immune system is particularly active at those anatomical sites that are most likely to be the first point of contact with potential pathogens: the skin, respiratory tract, gastrointestinal tract, urogenital tract, mammary gland and ocular mucosa. Should any pathogen breach this 'first line' epithelial barrier, then the innate immune system must have backup protective mechanisms. The innate immune system can be thought of as a 'rapid response team' that can detect and react to the presence of infection relatively quickly. These mechanisms must be able to cope with the wide spectrum of potential pathogens that might attempt to enter the body via any of the routes just described, and despite the fact that they are rapidly deployed, they are relatively **non-specific** in their actions. There are a number of cells and molecules that contribute to innate immunity and although these represent reasonably effective countermeasures against many bacterial pathogens, the response to viruses is relatively ineffective at eliminating the infection. Furthermore, some pathogens have evolved to avoid and evade the innate immune system. Thus, the cells and molecules of the innate immune system can provide instant defence from infection, but this may only be able to hold the invading pathogens at bay for a short period of time. In this chapter we shall consider the different elements of the innate immune system and we shall see in later chapters how these systems are upgraded and enhanced (rather than replaced) by the adaptive immune system.

3.2 DEFENSIVE BARRIERS

Animals are exposed to the outside world (containing potential pathogens) at the epithelial surfaces. The skin provides an effective defensive barrier to infection but, unfortunately, the physiological functions of the mucous membranes (gaseous exchange in the respiratory tract and absorption of nutrients in the alimentary tract) means that these epithelial surfaces need to be relatively permeable, making them susceptible to infection. The **epithelial barriers** that cover these surfaces are considered part of the innate immune system. Many of the physical and biochemical barriers have a range of site-specific modifications that further contribute to their ability to exclude pathogens.

The relatively thick stratified squamous keratinized **epithelium of the epidermis** is an inhospitable environment, deterring microbial invasion. Sebaceous and sweat glands also provide secretions onto the surface of the skin that contain substances, such as antimicrobial peptides (AMPs) and fatty acids, the latter maintaining the acidic pH of the skin surface to inhibit microbial growth (**Figure 3.1**). The cutaneous **microflora** (consisting of many species of commensal microorganisms including proteobacteria, actinobacteria, firmicutes and bacteriodetes species) live on the skin surface in healthy animals and compete with any potential cutaneous microbial pathogen for space and nutrients in this environment.

The **respiratory mucosa** is lined by a ciliated columnar epithelium. Interspersed goblet cells produce mucus that traps particulate material in the airway and the cilia beat in unison to transport this material upwards from the lower bronchial tree to the oropharynx, where it can be swallowed (**Figure 3.2**). This transport mechanism is referred to as the **mucociliary escalator**. Glandular secretions released onto the respiratory surface also contain antimicrobial substances. If material reaches the lung, there is a population of **alveolar macrophages** that patrols these spaces, removing any particulate debris by phagocytosis (**Figure 3.3**).

The gastrointestinal tract is protected to some extent by the low pH of the stomach, which kills many microorganisms before allowing food to enter the more vulnerable small intestine. The **small intestinal mucosa** has a simple epithelial barrier, with **mucus-producing goblet**

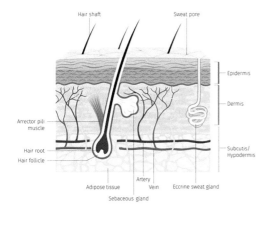

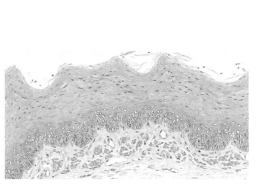

Figure 3.1 Cutaneous innate immunity. Diagram of skin structure and photomicrograph showing stratified squamous keratinized epithelium. The innate immune defences of the skin include the inhospitable nature of stratified squamous keratinized epithelium, the secretion of antimicrobial substances in sebum and sweat, the presence of a cutaneous microflora and a range of different leucocytes including intraepithelial lymphocytes, dendritic cells (Langerhans cells) and phagocytic cells within the dermal microenvironment.

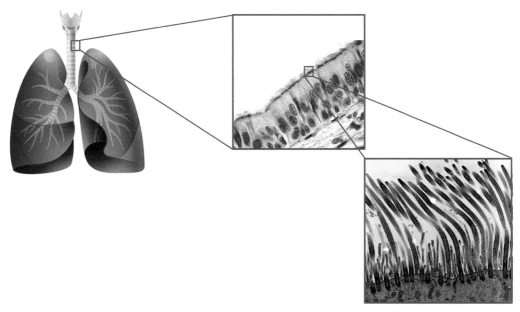

Figure 3.2 Respiratory tract innate immunity. The columnar epithelial cells lining the respiratory mucosa are ciliated. The directional beating of these cilia moves particulate debris and mucus from the lower respiratory tract to the oropharynx, where it is removed by swallowing (the 'mucociliary escalator'). An electron micrograph showing the structure of the cilia. Other innate immune defences include the antimicrobial secretions of mucosal glands and the leucocytes resident in the lamina propria.

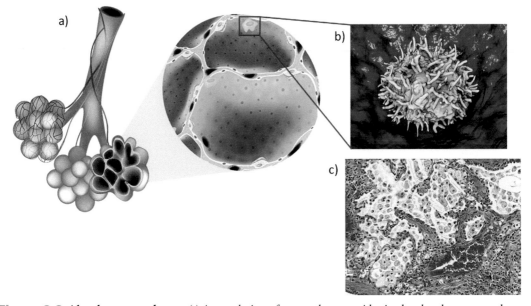

Figure 3.3 Alveolar macrophages. (a) A population of macrophages resides in the alveolar space and phagocytose any particulate debris that might bypass the mucociliary escalator. (b) Image of alveolar macrophage at high magnification. (c) These cells develop a prominent 'foamy' cytoplasm when activated, as can be seen in this histology section.

cells and secretions containing antimicrobial substances, including the various enzymes contained within the bile. The **peristaltic action** of the gut wall provides continued directional movement of luminal content, which discourages local colonization by pathogens. The intestinal tract has a rich **microbial flora** that again competes with pathogens for space and nutrients at this location.

These physical barriers are often complemented by biochemical barriers in the form of secretions enriched in antimicrobial molecules that can cause direct damage to microbes (e.g. digestion of bacterial cell walls). These molecules include enzymes such as **lysozyme** and **phospholipase A** as well as antimicrobial peptides (AMPs) such as **defensins** and **cathelicidins**, produced by Paneth cells in the crypts of the intestine and by epithelial cells of the skin, the oral mucosa and the respiratory and urogenital tracts. Some of the **surfactant proteins** that coat the alveolar surfaces of the lung also have antimicrobial properties. Pathogens coated by surfactant proteins A and D, for example, are more readily removed by the **alveolar macrophages**. Other secreted molecules include the **polyreactive antibodies**, which can potentially bind to a range of foreign invaders, and molecules of the **alternative pathway** of the **complement system**.

3.3 PATTERN RECOGNITION RECEPTORS

The innate immune system must be able to differentiate between self and foreign, if they are to distinguish between healthy cells and pathogenic microorganisms. White blood cells, in particular, are important in this process and these cells express receptors that detect substances that are intrinsically foreign, such as lipopolysaccharide (present in the outer membrane of Gram-negative bacteria), peptidoglycan (abundant in Gram-positive bacterial cell walls) (**Figure 3.4**), repeating mannose sugars (characteristic of prokaryotic carbohydrate

GRAM POSITIVE   **GRAM NEGATIVE**

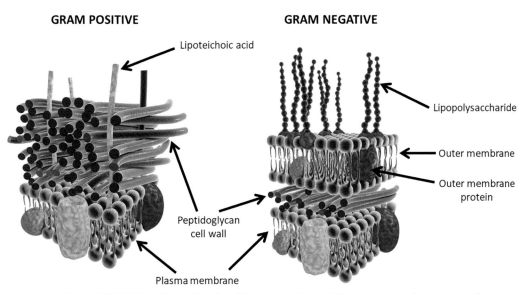

Figure 3.4 Bacterial PAMPs. The cell walls of Gram-positive and Gram-negative bacteria are shown. Lipoteichoic acid, peptidoglycan and lipopolysaccharide are all important PAMPS that can be detected by PRRs of the innate immune system.

molecules) and double-stranded RNA (present during the replication cycle of some viruses). These foreign molecules are collectively known as **pathogen-associated molecular patterns** (PAMPs) or alternatively **microbial-associated molecular patterns** (MAMPs), and the receptors for these are termed **pattern-recognition receptors** (PRRs). PRRs can be found in the cytoplasm, cell membrane, inside endosomal vesicles or as soluble molecules in the plasma or tissue fluid, allowing detection of PAMPs, wherever they may be located. The PRRs are highly conserved throughout evolution, such that similar receptors are found in plants, insects and mammals. A number of such receptors have been identified in domestic animals and their expression has now been studied in several disease states.

An alternative strategy used by the innate immune system is to detect the presence of cellular distress or death, rather than the presence of the foreign organism *per se*. Under adverse conditions, cells induce expression of certain 'markers' on their surface or secrete molecules into the extracellular fluid, including high

mobility group box 1 (HMGB1) and heat shock proteins. In addition, cell lysis or necrosis can lead to release of substances (such as mitochondrial DNA and ATP) that should normally only be present in the cytoplasm, and therefore the presence of these molecules in the extracellular fluid is an indication that something is wrong (so-called **danger signals**).

One of the best characterized families of PRRs are the **Toll-like receptors** (TLRs). The first recognized of these molecules was 'Toll', which is a receptor found in the fruit fly *Drosophila* and which is involved in antifungal host defence. There are at least ten TLRs in mammalian species (e.g. TLR-1 to TLR-10 in humans and most domesticated species and TLR-1 to TLR-13 in mice). The PAMPs with which these receptors interact are highly conserved molecules that are stable in structure and not readily modified (e.g. by genetic mutations). PAMPs include a range of carbohydrate, protein, lipid and nucleic acid molecules. Some examples of TLRs and their ligands are shown in **Figure 3.5**. Other PRRs include **C-type lectin receptors** such as DC-SIGN, Dectin-1 and Dectin-2, which recognize

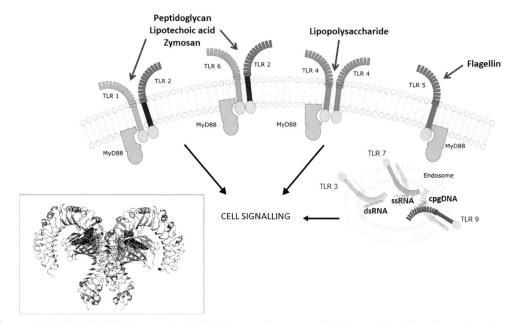

Figure 3.5 PRR–PAMP interaction. This diagram shows several TLRs on the cell surface and within the endosomal compartment. They often form dimers (homodimers and heterodimers), with each one having a different PAMP specificity. MyD88 is an adaptor molecule involved in cell signalling. Since different microorganisms express a different array of PAMPs, they consequently engage different combinations of PRRs on the surface of the cell, leading to differences in signalling and differences in downstream effects. These signals modify the biological response of the innate immune cells, and this can also subsequently impact on their interaction with cells of adaptive immunity. Inset: A molecular model of the TLR4 homodimer interacting with one of its adaptor molecules (MD2; orange) and binding to LPS (grey).

foreign carbohydrates, particularly those with a high mannose sugar content (typical of prokaryote rather than eukaryote carbohydrate molecules) (**Figure 3.6**). Although many PRRs are found on the cell surface, others are found within endosomes and recognize intracellular PAMPs, including pathogen nucleic acid. These include TLRs 3, 7, 8 and 9. Other types of PRRs are located within the cytoplasm and interact with microbial components following intracellular infections or with DNA released from the nucleus in damaged cells. These families of receptors include the nucleotide-binding oligomerization domain receptors (**NOD-like receptors**; **NLRs**) retinoic acid inducible gene (**RIG**)-like receptors (**RLRs**), and the stimulator of interferon genes (**STING**) **pathway**.

The latter is activated when DNA binds and activates the cytoplasmic enzyme **cyclic guanosine monophosphate adenosine monophosphate synthase (cGAS)**, which induces the formation of the dinucleotide cyclic di-GMP di-AMP (**cGAMP**). This dinucleotide binds to STING, a membrane protein of the endoplasmic reticulum.

Engagement of a TLR and its associated adaptor molecules (such as MyD88) triggers activation of cytoplasmic **signal transduction molecules**, which in turn activate transcription factors that migrate into the nucleus of the cell, binding to specific gene promotors and stimulating gene expression (**Figure 3.7**). The best characterized of these, which acts through the MyD88-dependent pathway, is

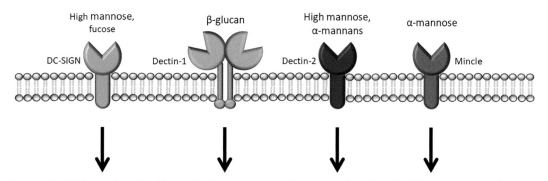

Figure 3.6 C-type lectins. C-type lectins are expressed on the cell surface of cells such as macrophages and dendritic cells and function to detect prokaryotic carbohydrate molecules. Similar to TLRs, they interact with intracellular adaptor molecules to trigger a cell signalling cascade.

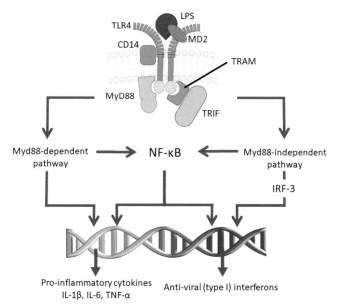

Figure 3.7 PRR signalling. The engagement of PRRs by PAMPs leads to initiation of an intracellular signalling pathway via signal transduction molecules such as MyD88, TRAM and TRIF. The end effect of such pathways is to activate specific genes within the cell, an event which is mediated by transcription factors such as nuclear factor κB (NF-κB) and interferon regulatory factor-3 (IRF-3). In this diagram, TLR-4 interacts with bacterial lipopolysaccharide acting as a PAMP, leading to activation of genes within the cell.

nuclear factor κB. NF-κB binds to a number of immune response genes within the nucleus, including those for pro-inflammatory cytokines (e.g. IL-1β, IL-6 and TNF-α) and costimulatory molecules (e.g. CD80 and CD86). Other transcription factors, such as the **interferon regulatory factor-3** (IRF-3) act through the MyD88-independent pathway, binding to the promotors of type 1 interferons and inducing their expression.

The sensing of danger signals by certain cytoplasmic PRRs (e.g. NLRs) leads to the

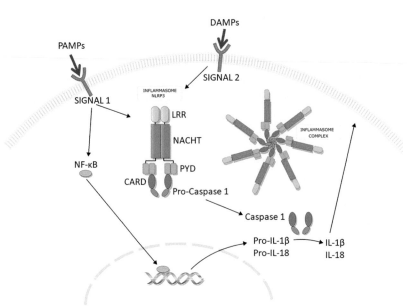

Figure 3.8 The inflammasome. The nucleotide-binding oligomerization domain-like receptors (NLRs) are a family of intracellular sensors of PAMPs and DAMPs. They contain a C-terminal leucine-rich repeat (LRR), which acts as a ligand-binding domain for other receptors (such as TLRs) or microbial ligands, and an N-terminal pyrin domain (PYD), which allows recruitment of pro-caspase 1. The activated caspase 1 cleaves pro-IL-1β and pro-IL-18 into their active forms that are released from the cell.

assembly of a multiprotein complex called the **inflammasome (Figure 3.8)**. Inflammasomes are key components of innate immunity, and structurally different inflammasomes will form depending on the nature of the stimulating signal. The effect of inflammasome formation is to activate the enzyme **caspase 1**, which stimulates the conversion of the inactive forms of IL-1β and IL-18 into their biologically active (pro-inflammatory) state. Thus, cell signalling through PRRs recognizing PAMPs and DAMPs can act in a synergistic manner, with the former stimulating synthesis of pro-cytokines (via the action of NF-κB) and the latter converting these (via release of caspase 1 from the inflammasome) to the active cytokine for release. In addition, caspase 1 induces a form of cell death called **pyroptosis**, which combines features of apoptosis (DNA fragmentation) and necrosis (inflammation and cytokine release).

3.4 PHAGOCYTES AND PHAGOCYTOSIS

The phagocytic cells of the body are the neutrophils and macrophages. Neutrophils are recruited to the sites of infection in a similar manner to that described for lymphocytes in Chapter 2, using dedicated adhesion molecules (**Figure 3.9**). LFA-1 (CD11a/CD18) in particular, is important for neutrophil extravasation and, as we will see in more detail in Chapter 16, genetic mutations impacting on CD18 function in cattle and dogs leads to the syndrome of bovine and canine leukocyte adhesion deficiency. Once they have exited the bloodstream, neutrophils migrate in a directed manner towards the pathogen in infected tissues via a process of **chemotaxis**. There are a number of chemokines (including IL-8 and macrophage inflammatory protein-1α; MIP-1α), typically released

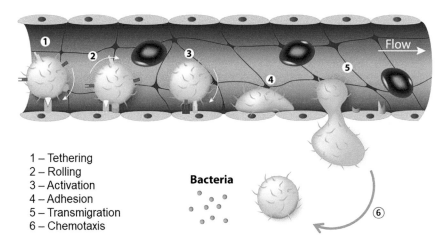

1 – Tethering
2 – Rolling
3 – Activation
4 – Adhesion
5 – Transmigration
6 – Chemotaxis

Bacteria

Figure 3.9 Neutrophil extravasation at the site of infection. The release of proinflammatory cytokines such as TNF and IL-6 induces the expression of sialomucins, and P- and E-selectin on the luminal surface of endothelial cells in inflamed tissues. This allows weak binding of neutrophils to endothelial cells (1) which slows down their movement through the capillaries and venules as they roll along the endothelium (2). Activation of the neutrophils results in a change in the conformation of the integrin LFA-1 (CD11a/CD18) (3), which enables firm binding to ICAM-1 on the surface of vascular endothelial cells (4). This halts the rolling and allows the neutrophils to exit the bloodstream (5). Chemotactic factors guide the neutrophils to the site of infection (6).

by cells in contact with bacteria, designed to ensure the neutrophils reach the area where they are needed. These neutrophils must then recognize and bind the pathogen, a process that involves engagement of their surface receptors (including TLRs, Fc receptors and complement receptors) by molecules expressed (e.g. PAMPs) or deposited (e.g. IgG or complement protein C3b) on the surface of the foreign organism. Following detection, the organism is engulfed by cytoplasmic extensions and drawn inside the neutrophil in a process known as **phagocytosis** (**Figure 3.10**). Once inside the **phagosome**, the microorganism is then killed and degraded via the effects of the **respiratory burst** and subsequent fusion of **lysosomes**, releasing antimicrobial peptides and digestive enzymes into the vacuole.

The respiratory burst represents the oxygen-dependent mechanism of microbial killing. This involves assembly of various enzymes, including the nicotinamide adenosine dinucleotide phosphate (NADPH) complex, in the membrane of the phagosome and utilization of oxygen by the neutrophil to generate large quantities of reactive oxygen species such as hydrogen peroxide (catalysed by superoxide dismutase) and hypohalides such as OCl^-, when myeloperoxidase catalyses the reaction between H_2O_2 and halide ions such as Cl^-.

Lysosomes are cytoplasmic organelles that fuse with the phagosome to form the **phagolysosome**. This process results in the release of antimicrobial peptides (e.g. defensins) and lysosomal enzymes (e.g. lysozyme, proteases and acid hydrolases) into the phagolysosome. Lysosomes also contain lactoferrin, which binds and chelates any free iron, effectively starving the microbe of this essential micronutrient. The antimicrobial peptides act by inserting into and disrupting the structure of bacterial cell walls. Neutrophils are also able to release antimicrobial proteins into the extracellular environment by degranulation, in order to eliminate extracellular organisms.

a)

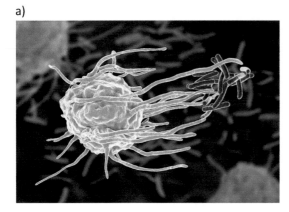

b)

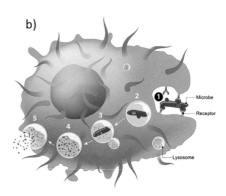

Figure 3.10 Phagocytosis. (a) Model of a phagocytic cell engulfing bacteria. (b) Stages of phagocytosis: 1 = binding and absorption, 2 = formation of the phagosome and microbial killing, 3 = fusion of lysosomes and microbial digestion, 4 = completion of killing and digestion processes, 5 = ejection of microbial digestion products.

As protons are pumped into the phagolysosome, the vesicle becomes increasingly acidic, activating digestive enzymes (such as cathepsins) which digest the microbe, the neutrophil finally ejecting the waste products.

Neutrophils can also destroy extracellular organisms by the release of **neutrophil extracellular traps** (**NETs**). NETs are composed of a core of nuclear DNA to which various proteins (including histone proteins, lactoferrin, cathepsins, myeloperoxidase and neutrophil elastase) are attached. NETs immobilize pathogens, which are subsequently destroyed by antimicrobial substances within the NETs or are phagocytosed by other phagocytic cells such as macrophages.

Circulating monocytes leave the blood and enter the tissues where they differentiate into macrophages. While neutrophils are rapidly mobilized and effective phagocytic cells, they have a short life span and cannot undertake repeated phagocytosis and destruction of

targets. Although macrophages are recruited more slowly into infected tissues, they are more potent and long-lived phagocytes that are capable of multiple phagocytic events. Macrophages also have additional roles in activating adaptive immunity (see Chapter 5), in tissue repair by removing neutrophils that have died by **apoptosis** (preventing inadvertent damage to tissue by viable enzymes within these cells) and necrotic tissue and by releasing enzymes capable of connective tissue remodelling.

Macrophages recognize and phagocytose target organisms in a similar fashion to neutrophils. One of the major enzymatic pathways of these cells involves the generation of **nitric oxide synthase** (NOS), which generates **nitric oxide** (NO) from arginine. NO in turn reacts with superoxide anions to produce substances such as nitrogen dioxide radicals that are highly toxic to phagocytosed organisms. Macrophages able to generate NO are known as **M1** and are important in the inflammatory response.

A second subset of macrophages (**M2**) uses arginase to convert arginine to ornithine. These cells are non-inflammatory, but instead they promote tissue repair and wound healing.

3.5 INNATE LYMPHOID CELLS AND NATURAL KILLER CELLS

The innate immune system utilises a population of cells with a similar morphology to lymphocytes, but which lack expression of antigen receptors. These cells include the **innate lymphoid cells (ILCs)** and the **natural killer (NK) cells**. The ILCs are primarily found in small numbers in non-lymphoid tissues, but little is known about these cells in domestic animals. Studies in mice have shown that ILCs contribute to the regulation of immune responses by the secretion of cytokines. They can be classified analogous to CD4+ T helper cell subsets into ILC1 (secrete IFN-γ similar to Th1 cells), ILC2 (secrete IL-5 and IL-13 similar to Th2 cells) and ILC3 (secrete IL-17 and IL-22 similar to Th17 cells). As these cells lack antigen

receptors, they are activated by other signals including cytokines and neuropeptides. Thus, they seem to recognize 'symptoms' of infection rather than the infectious organism itself.

Natural killer (NK) cells are widely distributed in blood and lymphoid tissues and are able to act rapidly, when required, in a relatively non-specific fashion. NK cells can kill virus-infected cells and tumour cells in a similar manner to that employed by cytotoxic T cells. Upon activation, they release the contents of their granules that induce apoptosis in the target cells. NK cells may recognize a target cell in one of two ways (**Figure 3.11**). The first involves a series of **activating receptors** that are able to bind directly to molecules displayed on the surface of stressed target cells (stress-induced proteins). In addition to activating receptors, NK cells have **inhibitory receptors**, which bind to **MHC class I** molecules expressed on all normal nucleated cells in mammals. This binding inhibits the NK cell and is a mechanism designed to protect healthy cells from cytolytic attack. Any NK cell that binds to a target cell expressing MHC class

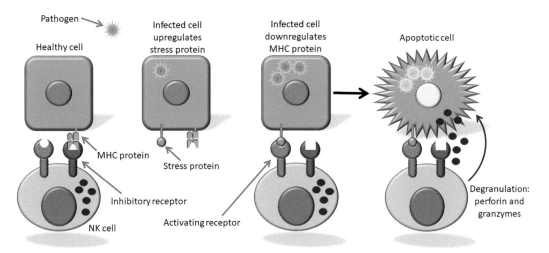

Figure 3.11 NK cell recognition of target cells. The NK receptor families (KIR, Ly49 or NKG2) recognize cell surface molecules (stress proteins) expressed by infected or neoplastic cells, but the NK cell can only be cytotoxic if that target cell also downregulates MHC class I. Where MHC class I is present (i.e. on healthy nucleated cells) this is bound by an inhibitory member of one of the receptor families, which prevents activation of the NK cell.

I will not be able to kill it, even if the activating receptor has recognized a suitable ligand. Therefore, only cells that have downregulated class I expression (e.g. virally infected or tumour cells) can become targets for an NK cell. The NK cell activating and inhibitory receptors are distributed within three families of NK cell receptors. **Killer cell immunoglobulin-like receptors** (KIRs) are the dominant form of human and primate NK receptor. Some KIRs bind MHC class I to inhibit NK function, while others play a role in activating NK cells. In mice and horses, the family of **Ly49 receptors** serve the same roles as KIRs, but these species lack KIRs (while primates and humans lack Ly49 receptors). In contrast, domestic animals other than horses may have both KIR and Ly49 receptors, but KIRs are the predominant and more important form. The third type of NK cell receptors are the **NKG2 receptors**. These molecules recognize MHC class I-like molecules (such as MICA and MICB) that are expressed on stressed target cells.

NK cells also express a receptor for the constant portion (Fc) of IgG. Binding of IgG to viral antigens or tumour antigens on cells allows binding of the Fc-receptor of NK cells, which results in NK cell activation. This mechanism of cell killing is called antibody-dependent cell-mediated cytotoxicity (ADCC) and is one of the effector mechanisms of the humoral immune response. Finally, NK cells can secrete IFN-γ, which activates macrophages to control intracellular pathogens during the early phase of infections.

3.6　LYMPHOCYTES WITH RESTRICTED DIVERSITY OF ANTIGEN RECEPTORS

The majority of B and T cells express highly diverse antigen receptors which allow recognition of a wide variety of pathogenic microorganisms. However, subpopulations of B and T cells have receptors with much more limited diversity. These include the B1 cells and marginal zone B

cells, T cells that express an alternative form of TCR composed of γ and δ chains (**γδ TCR-bearing cells**; γδT cells) and NKT cells. Here, we will discuss the γδT cells and NKT cells.

The γδT cells undergo **intra-thymic development** similar to the conventional T cells that express the TCRαβ, and it is suggested that the γδ TCR may be an evolutionarily older form of immunological receptor. The δ chain genes are clustered together with those that encode the TCR α chain, whereas there is a distinct cluster of genes encoding elements of the γ chain. There is a **single δ chain constant region gene**, but there are **species differences in the number of γ chain constant region genes** ranging from two in humans to five or six in ruminants. In most species there are **few variable region genes** in both gene clusters, meaning that γδ TCRs have **limited diversity**. However, in ruminants and pigs these receptors have greater diversity, as there are more variable, diversity and joining region genes.

γδ T cells are highly enriched at the **mucocutaneous surfaces** of the body and often reside within epithelial barriers (**Figure 3.12**) where they are well-located to provide one of the first points of contact with the immune system. γδ T cells are in fact readily activated by exposure to bacteria that colonize such surfaces (e.g. *Listeria, Escherichia coli, Salmonella, Mycobacterium*). These cells generally do not express CD4 or CD8 and they are not MHC restricted. There are many subsets of γδ T cells, recognizing various ligands, but some are activated by stress-induced non-classical MHC class 1b molecules, including **MICA**. Such proteins are upregulated when cells are distressed (e.g. during infection) and therefore the γδ T cells can react and eradicate those cells considered to be a risk to the host (**Figure 3.13**).

There are particular species differences in the distribution of γδ T cells. These cells constitute a small percentage of T cells in the peripheral blood and lymphoid tissues of primates, rodents, and dogs and cats, but tend to be enriched in the skin and mucosal surfaces.

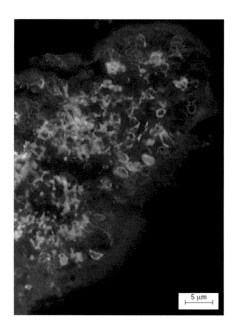

Figure 3.12 γδ T lymphocytes. This section of a canine intestinal villus is labelled with two fluorescent antibody markers that show the location of T cells expressing the αβ TCR (green) within the lamina propria, and T cells expressing the γδ TCR (red) within the epithelial barrier. The latter population is well situated to be able to provide first-line defence of the intestine from pathogens that attempt to colonize the surface or invade. (From German AJ, Hall EJ, Moore PF et al. (1999) Analysis of the distribution of lymphocytes expressing the αβ and γδ T cell receptors and expression of mucosal addressin cell adhesion molecule-1 in the canine intestine. *Journal of Comparative Pathology* **201**: 249–263, with permission.)

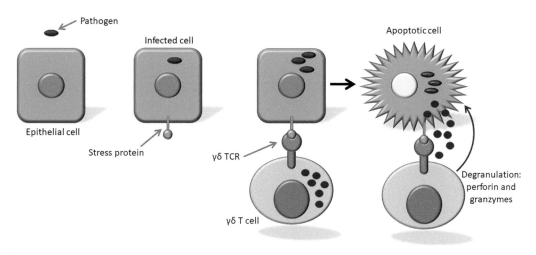

Figure 3.13 Recognition of stress proteins/non-classical MHC molecules by γδ T cells. The γδ TCR does not interact with classical MHC molecules (class I or class II) and instead recognizes stress proteins that are upregulated on the cell surface when it experiences an adverse event (e.g. infection by a pathogen). Thus, the γδ T cells can eradicate any infected cells that are a potential threat to the host.

However, γδ T cells contribute a significant proportion of circulating T cells in blood of cattle, sheep, pigs and chickens, especially in young animals. The percentage decreases as the animals age. Porcine γδ T cells have broad antigen specificity at birth but develop more restricted specificity with age. Bovine γδ T cells comprise two separate subsets. The WC1⁺ subset is con-

sidered part of innate immunity, although it shows a Th1-like function by producing IL-12 and IFN-γ, whereas cells that do not express WC1 have a more regulatory function. In dogs, γδ T cells may comprise up to one-third of T cells in the spleen.

Natural killer T (NKT) cells, so named because they share molecular markers in common with both NK and T cells, are a subset of lymphocytes that express TCRαβ, but are distinct from CD4+ and CD8+ T cells. Two NKT subsets are reported. **Type I NKT cells** express a TCR that comprises an **invariant α chain** combined with a limited number of β chains. **Type II NKT** cells have **greater variability in α chain usage**. Both cell types recognize lipid **antigens presented by the CD1d** molecule, such as glycolipids derived from bacterial cell walls. Type I NKT cells have been described in the dog.

3.7   THE INTERFERON RESPONSE

Viruses are relatively difficult to detect as (unlike bacteria) they do not express structural PAMPs. Viral nucleic acid, often in the form of double-stranded RNA, can be used as a marker of intracellular infection, and there are Toll-like receptors (TLRs 3, 7, 8 and 9) and cytoplasmic helicases (retinoic acid-inducible gene 1; RIG-1, melanoma differentiation-associated protein 5; MDA-5) capable of signalling to the cell that virus infection and replication is taking place. Signalling via these receptors activates interferon regulator factors (IRF3, IRF7) as well as other transcription factors such as NF-κB that migrate into the nucleus, switching on type I interferon genes, especially IFN-α and IFN-β, through their binding to specific promoter elements (**Figure 3.14**).

Type I interferons are secreted somewhat altruistically by the infected cell, as it has predominantly a paracrine effect, alerting its cellular neighbours to the presence of the pathogen, preparing them for exposure to new virus particles when they are released and attempt to infect other cells within the target tissue. Activation of type I interferon receptors leads to a

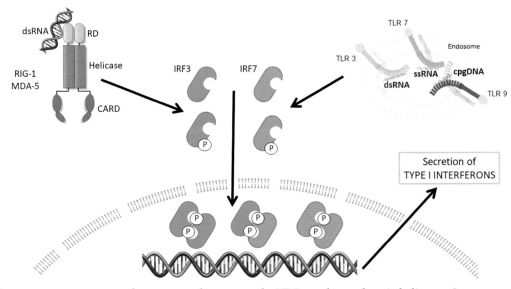

Figure 3.14 Activation of type 1 interferon genes by TLRs and cytoplasmic helicases. Some TLRs recognize the presence of foreign nucleic acid molecules (e.g. double-stranded RNA). There are also cytoplasmic helicases such as RIG-1 and MDA-5 that can detect double-stranded RNA. These signal to induce phosphorylation of specific interferon regulatory factors (IRF3 and IRF7) that subsequently form homodimers and heterodimers. The IRF dimers migrate into the nucleus and bind to promotor elements of the type I interferon genes (typically IFN-α and IFN-β), stimulating gene expression. Following their synthesis, these proteins are secreted into the extracellular fluid.

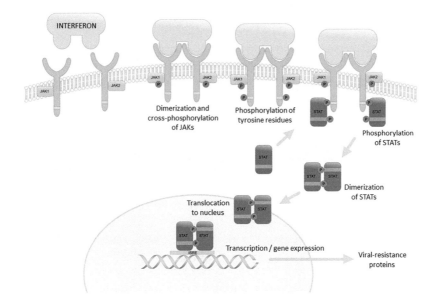

Figure 3.15 Cell signalling via the interferon receptor. Binding of interferon to its receptor leads to dimerization and phosphorylation of the associated intracellular Janus kinases (JAKs). These, in turn, phosphorylate signal transducer and activator of transcription proteins (STATs), which dimerise and move into the nucleus, acting as transcription factors for inducing expression of viral resistance genes.

classical cytokine receptor JAK-STAT response with the transcription factors binding to interferon stimulated response elements (ISRE) within the promoters of key virus resistance genes (**Figure 3.15**). Interferon-stimulated cells are enabled to resist viral replication, with increased capability for degradation of viral mRNA (via production of RNAse L), reduced synthesis of viral proteins (via production of Mx proteins) and enhanced antigen presentation. In particular, production of Lmp2 and Lmp7 subunits modifies the structure and function of the proteasome (see Chapter 5), so that the cell becomes more effective at generating foreign peptides for antigen presentation.

Interferon omega (Virbagen® Omega, Virbac), a type I interferon, is available commercially for antiviral therapy of companion animals. It may slow down viral replication in the animal and reduce the viral load, and it is sometimes used clinically to treat cats with chronic/persistent virus infections such as

FeLV, FIV and FIP. Some viruses produce non-structural proteins that inhibit the interferon response, either in the secreting cell by interfering with signalling from the PRRs or in the target cell by interfering with the IFN receptor signalling pathway.

3.8   THE INFLAMMATORY RESPONSE

The collective effect of the molecules and cells of the innate immune system in responding to insult by pathogens, trauma or a local immune response is termed **inflammation**. The **inflammatory response** is initiated by tissue resident (or sentinel) cells of the innate immune system, including tissue macrophages, dendritic cells and mast cells. These cells bear PRRs that allow them to respond to structural or secreted components of pathogens or to molecules released by damaged or dying cells (collectively referred to as **alarmins**). Engagement of the sentinel cell receptors leads to those cells releasing an array of soluble mediators, which are responsible for the ensuing inflammatory response. These mediators include **derivatives of arachidonic acid**, **cytokines** and **chemokines** (see Chapter 7) and vasoactive molecules such as histamine. An inflammatory response involves the accumulation of fluid, plasma proteins and leucocytes at the site of the insult (**Figure 3.16**). The aims of the inflammatory response are to neutralize the inciting cause (e.g. a pathogen), to contain pathogens and prevent their systemic spread and to promote tissue repair and return to homeostasis, once the pathogen has been eliminated. The inflammatory response has two distinct types known as acute and chronic inflammation.

    Acute inflammation begins within minutes of tissue insult and is characterized by the presence of the five **cardinal signs** of inflammation, namely redness, heat, swelling, pain and loss of function of the affected tissue. These changes relate to alterations in the capillary vasculature of the affected tissue. Blood vessel dilation (**vasodilation**) leads to increased local blood flow and reduced velocity of the leukocytes, while increased permeability leads to the loss of fluid from blood into the tissues (**oedema**) and associated leakage of plasma proteins. The **endothelial cells** lining the blood vessels express surface **adhesion molecules** (see Chapter 2), which permit the adhesion of leucocytes and the migration of these cells between the stretched endothelial cells into the tissue (**Figure 3.17**). The first leucocytes to undergo such **extravasation** are the **neutrophils** and this migration occurs within the first few hours of the acute inflammatory response (**Figure 3.9**). In parallel with these changes, the stretching of the vessel wall exposes the collagen beneath the endothelium to the flowing blood and initiates **coagulation** involving the platelets and coagulation factors. This provides

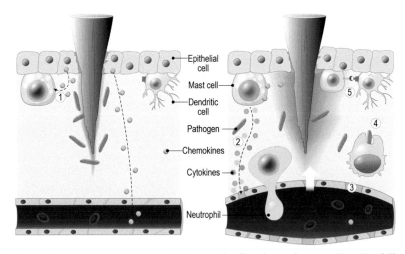

Figure 3.16 The inflammatory response. (1) Damaged cells release alarmins/DAMPs following tissue damage/injury. (2) These (in addition to PAMPs) can be detected by mast cells and they degranulate, releasing the vasoactive mediator, histamine. Cytokines may also be released by tissue macrophages. (3) The inflammatory mediators stimulate vasodilation (redness and heat) and may act on local nerve endings (pain). Plasma leaks into the tissues (swelling) bringing soluble factors to the site of injury/infection. Upregulation of vascular addressins recruits leukocytes to the site of inflammation. (4) Extravasation of neutrophils and monocytes (the latter differentiating to macrophages) leads to phagocytosis of microbes within the infected tissues. (5) Migration of lymphocytes into the region may supplement the innate immune response.

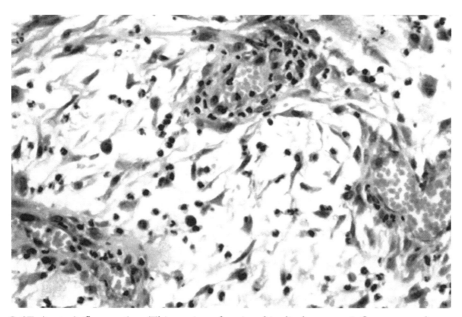

Figure 3.17 Acute inflammation. This section of canine skin displays acute inflammatory changes. There is vasodilation and tissue oedema with margination and egress of leucocytes (chiefly neutrophils) from vessels into the tissue. Those neutrophils distant from the vessel may be migrating along a chemotactic gradient towards the trigger of the inflammatory response (e.g. a local bacterial infection).

a means of preventing spread of pathogens into the bloodstream.

The acute inflammatory response takes place against the background of a complex network of soluble inflammatory mediators that are released at the site of inflammation. These include the initiating alarmins, **histamine** (derived from mast cells and a potent vasodilator), the **kinins** (such as bradykinin), chemokines (involved in the recruitment of leucocytes from the bloodstream) and cytokines (involved in recruitment and activation of leucocytes), molecules of the alternative pathway of **complement**, **coagulation factors** and the vasoactive lipids (collectively **eicosanoids**). Eicosanoids are derived from arachidonic acid released from phospholipids of damaged cell membranes by the action of phospholipases (**Figure 3.18a**). Arachidonic acid may be converted to **leukotrienes** (e.g. leukotriene B_4) by the action of lipoxygenase, or to a range of **prostaglandins** (e.g. prostaglandin E_2), **thromboxanes** (e.g. thromboxane A_2) and

prostacyclins (e.g. prostaglandin I_2) by the **cyclooxygenases** COX-1 and COX-2. Some of these molecules interact with nervous system sensors (pain receptors; nociceptors), and neuropeptides may be released from nerve endings within inflamed tissue.

The second stage of the inflammatory response begins 24–48 hours after the initiating insult and is characterized by the recruitment of **monocytes** from the blood into the tissue. These cells differentiate into **tissue macrophages** and provide the second line of innate immune defence. The local inflammatory response may be accompanied by systemic signs of illness in addition to local tissue inflammation. This effect largely relates to the release of a series of **pro-inflammatory cytokines** from activated macrophages that have autocrine, paracrine and endocrine effects (**Figure 3.18b**). These include interleukin (IL)-1, IL-6 and tumour necrosis factor alpha (TNF-α) (see Chapter 7). They interact directly with nerve

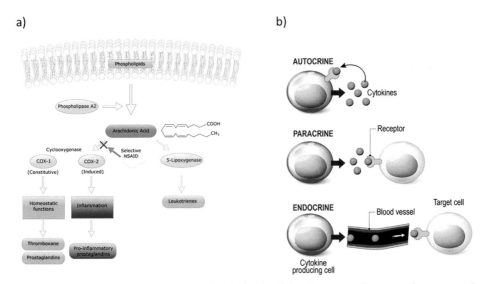

Figure 3.18 Inflammatory mediators. (a) Phospholipids of the plasma membrane can be converted to arachidonic acid by the action of phospholipase A2. Cyclooxygenases subsequently act on arachidonic acid to form various lipid mediators of inflammation. COX-2 selective non-steroidal inflammatory drugs (NSAIDs) can block this arm of the inflammatory response. (b) Pro-inflammatory cytokines are synthesized by leukocytes that have local (autocrine and paracrine) as well as distant (endocrine) effects.

cells in damaged tissue and can enter the circulation where they have effects on the brain, liver and bone marrow. These '**endogenous pyrogens**' bind to receptor molecules in the hypothalamus that mediate **pyrexia** (altering the body temperature set mechanism), **lethargy** (promoting sleep-inducing molecules) and **anorexia** (acting on the satiety centre). In the liver, these cytokines stimulate production of acute phase proteins (see later in this chapter). Other cytokines (colony-stimulating factors) are released from macrophages at the site of infection that circulate to the bone marrow and increase the production and release of white blood cells.

The overall effect of the acute inflammatory responses, often in concert with adaptive immunity, is to destroy and remove pathogens and promote tissue repair and return to homeostatic function. There are, however, some initiators of inflammation that are very difficult to eliminate. These include intracellular pathogens able to subvert the protective mechanisms

described earlier (e.g. *Mycobacterium bovis*), large structures (e.g. tissue migrating parasites) or inert irritants such as metal particles, suture materials or certain vaccine adjuvants. These substances tend to induce persistent and chronic inflammatory foci that become 'walled off' from normal tissue by formation of a **granuloma** (**Figure 3.19**). A granuloma may have a necrotic core, containing the foreign material with some neutrophils or eosinophils, that is surrounded by macrophages that may fuse together to form **multinucleate giant cells** (**Figure 3.20**). These cells are in turn surrounded by a zone of fibrous connective tissue. A 'sterile' granuloma (e.g. induced by suture material) has this typical composition, whereas an 'infectious' granuloma containing antigenic material may also include many lymphocytes. Chronic inflammation also occurs when the immune system is repeatedly or continuously exposed to the inciting agent such as in allergic and autoimmune diseases.

The final stage of the inflammatory response is tissue repair, which may largely be regulated

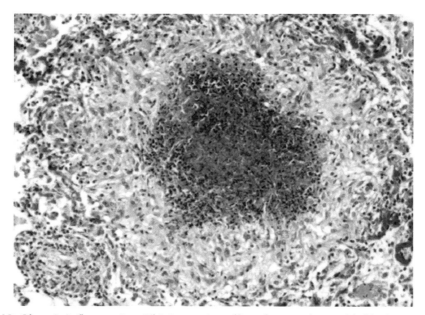

Figure 3.19 Chronic inflammation. This is a section of lung from an alpaca with *Mycobacterium* infection. The lesion is a granuloma with a necrotic core, containing degenerate neutrophils and bacteria (not visible), that is walled-off by a layer of macrophages with scattered lymphocytes.

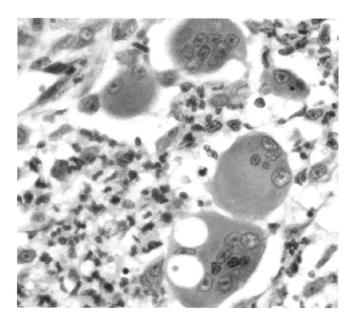

Figure 3.20 Multinucleate giant cells. Section of skin from a cat with localized pyogranulomatous dermatitis. Within the inflammatory infiltrate there are multinucleate giant cells representing fused macrophages. The cell at the base of the image contains two large cytoplasmic vacuoles. No specific infectious cause for this reaction was identified and it may have been triggered by a penetrating foreign body.

by M2 macrophages. These cells produce fibroblast and blood vessel growth factors, including platelet-derived growth factor, and cytokines such as **transforming growth factor beta** (TGF-β), which promote the depo- sition of collagen and stimulate angiogenesis to replace areas of necrotic tissue. The fibrous repair (scar tissue) may be remodelled by **matrix metalloproteinases** (MMPs) derived from macrophages. Scar tissue may be devoid of normal tissue elements in tissues that are incapable of regeneration. Other subtypes of M2 macrophages may have a more immunosuppressive role, via production of the cytokine IL-10.

3.9   THE ACUTE PHASE RESPONSE

Pro-inflammatory cytokines (such as IL-1β, IL-6 and TNF-α) produced at the site of infection have an endocrine effect on the liver, stimulating production of a number of **acute phase proteins** (**Figure 3.21**). These include molecules such as **mannose-binding lectin** (MBL), **C-reactive protein** (CRP), **serum amyloid A** (SAA), **fibrinogen** and **haptoglobin** (which binds iron and thereby sequesters it from bacteria that have an obligate metabolic requirement for this element). The acute phase proteins may be detected in the blood and these biomarkers are useful indicators of the presence of inflammation. The acute phase proteins act as opso-

nins as they can bind to the surface of bacteria and enhance their uptake by phagocytic cells. In addition, some components (e.g. MBL and CRP) can trigger the complement system (see next section).

3.10   SYSTEMIC INFLAMMATORY RESPONSE SYNDROME

Excessive production of cytokines, including IL-1, IL-6 and TNF but also others, can lead to severe systemic disease. The umbrella term for this condition is **systemic inflammatory response syndrome** (**SIRS**). A typical example is a systemic bacterial infection in which bacteria enter the blood. This is referred to as bacterial sepsis or sometimes simply as sepsis. However, sepsis can also occur with viruses, fungi and protozoal pathogens. In addition, there are non-infectious causes of SIRS, such as extensive tissue damage caused by burns or trauma. SIRS can occur in all species, and is a common

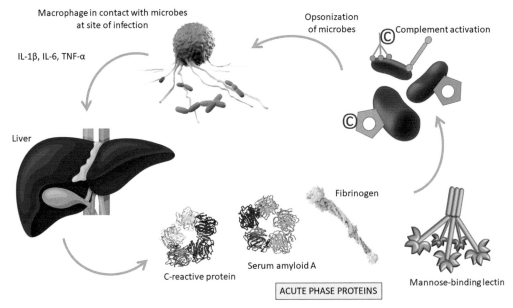

Figure 3.21 **The acute phase response.** Cytokines produced by macrophages at the site of infection have an endocrine effect on the liver and stimulate production of acute phase proteins. These can bind to the surface of microbes and enhance their phagocytosis. In addition, some components (e.g. MBL and CRP) can trigger the complement system.

complication of colic in horses as a result of extensive damage of intestinal tissues and systemic release of intestinal bacteria. The clinical signs of SIRS are increased body temperature, heart and respiratory rate, and an increased number of immature neutrophils (band cells) in the blood with an overall increase or decrease in number of leucocytes. SIRS results from potent and systemic stimulation of PRRs by the presence of large amounts of PAMPs and DAMPs, resulting in massive release of cytokines, which is often called a **cytokine storm**. This initiates the acute phase response, but also vascular changes with hypoxia in tissues resulting in endothelial cell injury. Damaged endothelial cells activate platelets and the coagulation cascade, resulting in formation of blood clots which further impair the delivery of oxygen to tissues. Disseminated intravascular coagulation (DIC) is one of the hallmarks of advanced stages of SIRS. It results in organ failure and is often fatal. At the same time, the release of anti-inflammatory cytokines causes immunosuppression and makes the animal more susceptible to opportunistic bacterial infections which further aggravate the condition. In these cases, antibiotics, corticosteroids to suppress cytokine production and supportive therapy such as intravenous fluids are used to attempt to prevent the patient's death.

3.11   THE COMPLEMENT SYSTEM

The **complement system** consists of a family of approximately 30 plasma proteins. When activated, they interact sequentially, forming a self-assembling enzymatic cascade, generating biologically active molecules that mediate a range of antimicrobial and inflammatory processes. The basic principle of this cascade system is, in general terms, somewhat similar to the coagulation system involved in secondary haemostasis. Intrinsic in such pathways is the presence of a regulatory system that can switch off the cascade when no longer required, in

order to avoid inappropriate damage to normal tissue.

There are four complement pathways, known as the **alternative**, **lectin**, **classical** and **terminal** pathways. The first three share a common end-point (conversion of inactive C3 to its active components C3a and C3b), which in turn is the start of the shared terminal pathway, leading to formation of the end product, the **membrane-attack complex** (MAC). Complement components generated by the activation of these pathways mediate the key biological effects of the system, namely cytolysis, opsonization and inflammation (**Figure 3.22**).

The complement system is present in all animal species and the constituent components are relatively conserved. These components are mostly described using the abbreviation 'C' (for complement), with a number to indicate the specific component (e.g. C4) and a lower case letter to indicate a subunit of that component (e.g. C4a and C4b). The numbering of the components does not always follow a logical sequence, as each component was numbered in the order in which it was discovered, rather than by the position it holds in the hierarchy of the system. Some components and regulatory proteins do not conform to the 'C' nomenclature. To add further confusion, there are some minor differences in nomenclature used in North America and Europe (for example, the classical pathway C3 convertase is C4bC2b in Europe compared with C4bC2a in North America) and readers of other texts should be aware of this, as the European system is presented here.

3.11.1   The Alternative Pathway

The **alternative pathway** of the complement system is the oldest in evolutionary terms and is considered to be integral to innate immunity, independent of the adaptive immune system. The alternative pathway has two distinct phases. The first of these continually cycles at a low level in clinically healthy animals and is often referred to as the **'tick over' phase**.

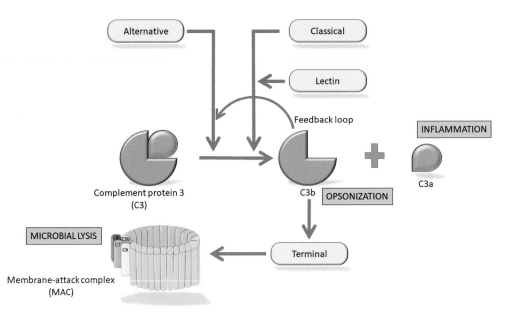

Figure 3.22 The complement pathways. The four complement pathways interact as depicted in this diagram. The end-point of the alternative and classical pathways is the generation of the key molecule, complement factor C3b. The lectin pathway shares elements of the classical system. A variant unique to the alternative pathway is the feedback amplification loop. The common terminal pathway begins with factor C3b and ends with polymerization of C9 monomers into the C9 polymer, termed the 'membrane-attack complex' (MAC). Various elements of the complement system are involved in microbial lysis (MAC), opsonization (C3b) and inflammation (C3a).

The key feature of this phase is that it occurs within the extracellular fluid and is not associated with the presence of infection. The second phase reflects full activation of the system and requires an appropriate substrate, which in the case of the alternative pathway is the presence of a '**trigger surface**' that permits deposition of molecules of the enzymatic cascade. Such trigger surfaces are provided by microbes (particularly bacteria or yeasts), by abnormal tissue cells (e.g. virus-infected or neoplastic cells), by aggregates of immunoglobulin or by foreign material (e.g. asbestos) (**Figure 3.23**). The molecules on trigger surfaces responsible for activation of the alternative pathway (and the lectin pathway—see later) are either pathogen-associated molecular patterns (PAMPs) or damage-associated molecular patterns (DAMPs); thus those elements of the complement system that initiate

the cascade (e.g. C3 and mannose-binding lectin) can be considered to be soluble pattern recognition receptors (PRRs).

The tick over phase is initiated by **C3** in the extracellular fluid, which undergoes spontaneous hydrolysis to form C3a and C3b. In the tick over pathway, the majority of the C3b that is generated undergoes spontaneous hydrolysis and inactivation. Should any C3b become deposited onto the surface of healthy cells, it will be displaced/inactivated by one of several cell surface proteins, including CD46 (membrane cofactor protein; MCP), CD55 (decay-accelerating factor; DAF) and CD59 (**Figure 3.24**). Thus, host cells are protected from complement-mediated damage. Xenotransplantation (e.g. pig organs into humans) leads to acute tissue rejection, as a result of the porcine cell surface molecules failing to completely protect the transplanted

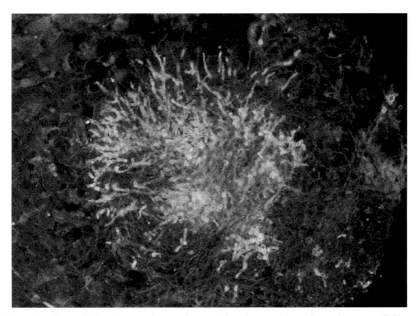

Figure 3.23 Activation of the alternative pathway. The alternative pathway becomes fully activated in the presence of an appropriate trigger surface that allows sequential deposition of the constituent complement molecules. Microbial surfaces provide an ideal trigger for this pathway. This is a colony of *Aspergillus* fungi growing within the pancreas of a German shepherd dog with disseminated *Aspergillus* infection. The fungi appear green as they have been specifically labelled with an antiserum conjugated to fluorescein. The antiserum detects complement C3 deposition on the surface of the hyphae. (From Day MJ, Penhale WJ (1991) An immunohistochemical study of canine disseminated aspergillosis. *Australian Veterinary Journal* **68**:383–386, with permission.)

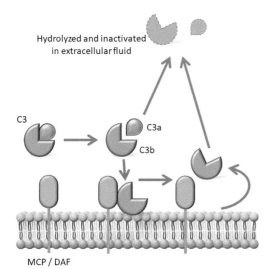

Figure 3.24 Alternative pathway 'tick over' phase (simplified). This complement pathway continually cycles at a low level in the extracellular tissue fluid of clinically normal individuals. Any C3b that is deposited is displaced by a range of membrane protective molecules found on the surface of normal tissue cells (e.g. membrane cofactor protein; MCP and decay accelerating factor; DAF) where it is degraded and inactivated in the fluid phase.

tissue cells from complement-mediated damage by the recipient. Transgenic pigs expressing the three major human complement-regulating factors (CD46, CD55 and CD59) have now been produced in an attempt to mitigate against this mismatch.

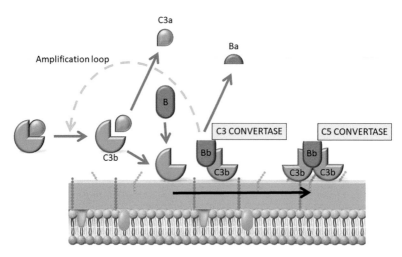

Figure 3.25 The alternative pathway of complement activation (simplified). The presence of an appropriate trigger surface permits deposition of the C3b generated from the tick over phase. C3b complexes with Factor B, which is degraded to become C3Bb, a C3 convertase, thus generating more C3b. The end-product of the pathway is the formation of the C3bBbC3b complex, which is a C5 convertase.

The same initiating sequence occurs during microbial infection, but in this instance a trigger surface (the microbial cell wall) permits the deposition of the C3b molecule. **Factor B** can then bind to the C3b and is acted on by Factor D, which fragments the molecule to the two components, Bb (which remains) and Ba (which is cleaved off). Once the alternative pathway C3 convertase (C3bBb) has been generated, it becomes associated with the molecule **properdin** (P), which stabilizes the complex. This stable C3 convertase initiates the **feedback amplification loop**, which generates large quantities of C3b locally, that becomes deposited on the trigger surface. The final step in the activation phase of the alternative pathway is the addition of further C3b to the C3 convertase (**C3bBbC3b**) to form the alternative pathway **C5 convertase (Figure 3.25)**.

3.11.2 The Lectin and Classical Pathways

The **lectin pathway** of complement activation is the most recently discovered, but is part of the innate immune system. Mannose-binding lectin (**MBL**) is an acute phase protein that associates with the enzymes MBL-associated serine protease (MASP)-1 and MASP-2. MBL initially binds to bacterial surface carbohydrates (particularly where there are repeating mannose sugars) and is then able to activate the MASPs. Activated MASP-2 acts on **C4**, and splits this into two subunits, C4a and C4b. The C4b fraction attaches to the surface of the microbe and binds **C2**, which is also cleaved by MASP-1 or MASP-2, to generate C2a and C2b. The C2b fragment remains associated with the C4b fraction and this complex has now become the classical pathway **C3 convertase** (C4bC2b). As this name suggests, the next stage of the sequence is that the C3 convertase acts on **C3** to split this molecule into C3a and C3b. Following release of C3a, C3b is deposited adjacent to the C4bC2b complex to form a new complex of C4bC2bC3b, which is a **C5 convertase (Figure 3.26)**. This is the end-point of the lectin pathway.

The classical pathway is so called as it was the first one discovered, although in evolutionary

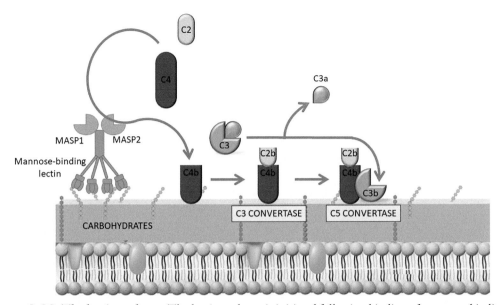

Figure 3.26 **The lectin pathway.** The lectin pathway is initiated following binding of mannose-binding lectin to microbial carbohydrates expressing repeating mannose sugars. A C3 convertase is generated through activation of MASP-2 on C4 and C2 proteins. Generation of C3b allows formation of the C4bC2bC3b complex, which is a C5 convertase. This sequential deposition of complement components occurs on the surface of the microbe.

terms it probably represents the most recent adaptation of the complement system. This pathway allows the adaptive immune response to utilize the power of the complement system, by enabling antibody binding to trigger the reaction. This is facilitated by the molecule **C1**, which can be thought of as a 'bolt-on', allowing adaptive and innate systems to interact. In the case of some bacteria, C1 might bind directly to a cell wall structural component (e.g. lipoteichoic acid of the wall of Gram-positive bacteria), or it may bind to C-reactive protein, which is in turn attached to a bacterial polysaccharide. However, most often, C1 attaches to an immune complex of **antigen and antibody** (IgG or IgM), by binding to the Fc region of the immunoglobulin molecule. C1 consists of three subunits, **C1q, C1r** and **C1s**. C1q attaches to the immunoglobulin molecule, causing a conformational change in the associated C1r–C1s complex. This in turn stimulates the enzymatic activity of C1r, which subsequently activates the C1s enzyme.

C1s (which has a similar role to MASP-2 in the lectin pathway) acts on C4 and C2 to form the C3 convertase complex (C4bC2b), which then associates with the C3b that is generated to form a **C5 convertase** (C4bC2bC3b) (**Figure 3.27**). The generation of this C5 convertase is the final stage in the classical pathway.

Complement molecules are highly susceptible to heat and, in vitro, may be inactivated by heating a serum sample to 56°C for a short period of time (a process known as 'heat-inactivation'). In the body, a system of **regulatory control** is built into the complement system to inactivate it when it is no longer required. Complement components have a relatively short half-life when generated, therefore avoiding prolonged effects, once the system has ceased to be operational. Additional means of controlling the classical pathway relate to the presence of a number of specific inhibitory factors that act at different points in the pathway. The **C1 inhibitor** cleaves C1r from C1s, thereby disrupting

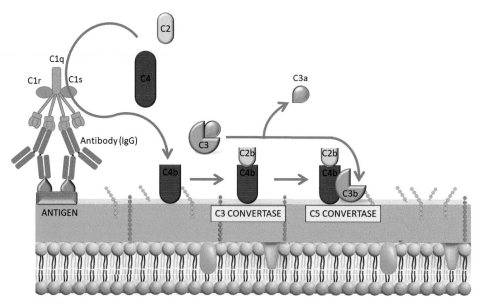

Figure 3.27 The classical pathway. The classical pathway is usually triggered by the binding of antibody to antigen and attachment of C1, with activation of C1s. The reaction subsequently proceeds through similar stages as the lectin pathway to generate the C5 convertase.

the activity of this particular complex. The **C4-binding protein** displaces C2b from C4b and works in combination with **Factor I**, which subsequently cleaves C4b into two inactive subfractions, C4c and C4d. Factor I is also able to cleave C3b into the inactive subfractions C3c and C3d.

3.11.3 The Terminal Pathway

The **terminal pathway** is initiated by C5 convertases (either C3bBbC3b from the alternative pathway or C4bC2bC3b from the lectin or classical pathways) that function to split **C5** to its subfractions, C5a and C5b. C5a has potent biological activity similar to that of C3a, but these effects occur distant to the surface of the microbe (see later). Activated C5b recruits C6 and C7 to form a complex of **C5bC6C7**, which associates with the membrane of the target cell. This complex in turn binds to **C8**, which penetrates the cell membrane and recruits a number of **C9** molecules that insert into the membrane and polymerise to form a 'doughnut-like' transmembrane pore known as the **membrane attack complex** (**MAC**) (**Figure 3.28**). The formation of the MAC represents the end-point of the terminal pathway.

3.11.4 Biological Consequences of Complement Activation

Once the complement system has been fully discharged, the biological consequences are enacted, which engage both local and somewhat distant antimicrobial defences (**Figure 3.29**). Activation of the terminal complement pathway generates thousands of MACs that lodge within the membrane of the target microbe (**Figure 3.30**). Thus, the surface of this target cell becomes riddled with holes and an **osmotic imbalance** between the cell cytoplasm and extracellular fluid is established such that there is a net influx of water into the cell. The cell swells and subsequently bursts in a phenomenon known as **osmotic lysis**. If that target cell is a bacterium or an abnormal tissue cell, then clearly this mechanism of cellular lysis (cytolysis) is beneficial to the host and is a valuable part of the protective immune response.

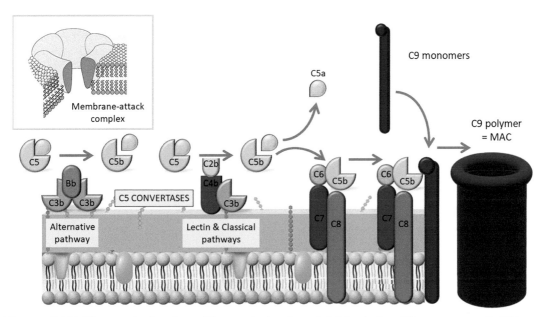

Figure 3.28 The terminal pathway. The terminal pathway is initiated when C5 convertases split C5 into C5a and C5b. C5b recruits C6 and C7 and this complex associates with the membrane of the cell that is the target of the complement system. Subsequent recruitment of C8 and polymerization of C9 monomers leads to the formation of a transmembrane pore known as the membrane attack complex (MAC).

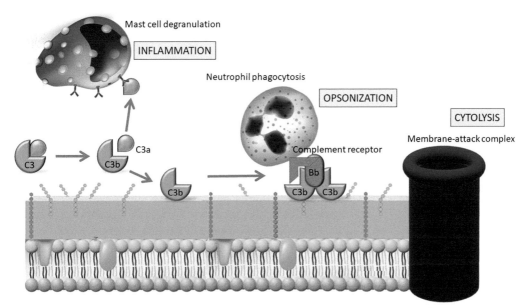

Figure 3.29 Overview of biological consequences of complement activation. Complement activation leads to downstream effector mechanisms such as cytolysis (via MAC), opsonization (via C3b) and inflammation (via C3a and C5a).

a)   b)

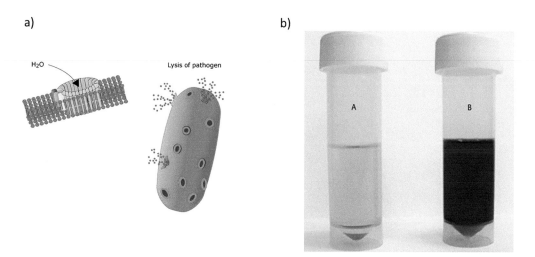

Figure 3.30 **The role of complement in cytolysis.** (a) A diagrammatic representation of the lipid bilayer of the membrane of a cell that has been targeted by the complement system. Numerous MACs insert through the membrane, establishing an osmotic imbalance with a net influx of water into the cell cytoplasm. The target cell will swell and burst. (b) This phenomenon is readily seen when erythrocytes become the target of complement-mediated cytolysis, as rupture of RBCs leads to release of free haemoglobin. These two test tubes contain erythrocytes that were suspended in saline. In tube A the normal red cells have settled to the base of the tube, leaving clear saline above. In tube B the action of antibody and complement (mixed with the saline) on the cells has led to release of free haemoglobin that colours the saline red; there is also an absence of a significant pellet of erythrocytes at the base of the tube.

The second biological consequence of complement activation is the removal and destruction of complement-coated cells by phagocytic cells such as neutrophils and macrophages (**Figure 3.10**). The interaction between target and phagocyte can be enhanced by the complement system by the processes of opsonization and immune adherence. 'Opsonization' comes from the Greek 'to make tasty for the table' and the analogy is often used that coating a microbe with C3b is akin to covering dinner with gravy! The phenomenon of **opsonization** arises because phagocytic cells express a series of receptors for the complement molecule C3b (**complement receptors [CR]1–CR4**). CR1, which binds C3b, is the best characterized of these receptors (**Figure 3.31**).

A related event is that of **immune adherence**, a process that is of great importance in clearing particulate antigen from the blood-stream. Immune adherence arises from the fact that erythrocytes express CR1, which permits circulating antigen that is coated with C3b to attach to the surface membrane of red blood cells. As these antigen-laden red cells pass through the hepatic sinusoids and spleen, they encounter phagocytic cells resident in those tissues that bind via their own CRs and remove the particle from the red cells for internalization and destruction (**Figure 3.32**).

The third major consequence of complement activation is the generation of those fragments of complement that trigger an inflammatory response. The two most important of these are **C3a** and **C5a**. C5a is more potent than C3a, but much larger quantities of C3a are generated. These molecules are sometimes known as anaphylatoxins, named after the fundamental effects that they mediate. C3a and C5a have key roles in the tissue inflammatory response via a

TUBE A   TUBE B

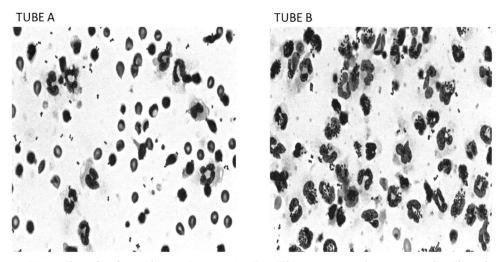

Figure 3.31 The role of complement in opsonization. This experiment demonstrates the effect of opsonization. Neutrophils were isolated from the blood of a dog and placed into two tubes. A suspension of staphylococcal bacteria was placed into tube A (left) and a suspension of bacteria that had been previously incubated with dog serum was placed into tube B (right). Most of the neutrophils in tube B have numerous bacteria within their cytoplasm, but only low numbers of organisms are present in the cytoplasm of some neutrophils in tube A. The bacteria in tube B had been opsonized by coating with antibody and complement present in the dog's serum, and this has led to much more efficient phagocytosis of these targets.

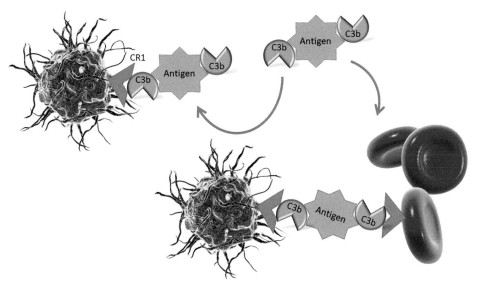

Figure 3.32 Removal of particulate material. The interaction between antigen and phagocyte can be greatly enhanced if the antigen is pre-coated with C3b, as the phagocyte bears membrane receptors for these molecules. Circulating erythrocytes bear CR1 receptors that permit binding of particulate antigen within the bloodstream that is coated by C3b. This antigen may be transferred to macrophages in the liver or spleen via the CR1 molecules expressed by these cells and subsequently destroyed by the phagocyte. This process of immune adherence is important in the removal of such material from the circulation.

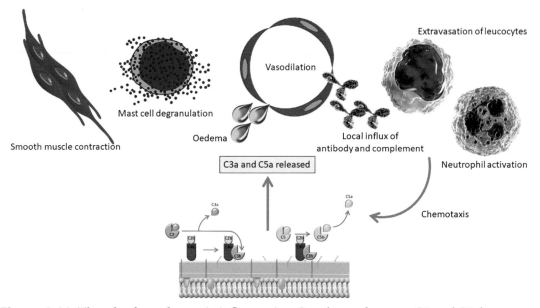

Figure 3.33 **The role of complement in inflammation.** Complement fragments C3a and C5a have profound effects on the local tissue inflammatory response. These molecules mediate vasodilation with associated tissue oedema and extravasation of leucocytes and plasma proteins into tissue. These events may be amplified through mast cell degranulation, which may also be directly mediated by C3a and C5a. These complement fragments may further activate the recruited leucocytes and form chemotactic gradients that direct these cells to the location of infectious agents or damaged cells within the tissue.

number of mechanisms (**Figure 3.33**). These molecules induce mast cell degranulation which results in **vasodilation** of blood vessels within the tissue in which they are generated. Dilated vessels with leaky endothelial cell junctions permit the egress of fluid, protein and cells from the circulation into the tissue. If the generation of C3a and C5a has resulted from infection or damage within that tissue, this process has a clear benefit to the animal. Vascular fluid loss leads to **tissue oedema**, which may be important in diluting locally produced toxins. Leakage of blood proteins (immunoglobulins and complement molecules) may be beneficial if these molecules participate in a local immune response. Migration of leucocytes, particularly phagocytic cells such as neutrophils and macrophages, is likely to be of benefit in the removal of infectious agents or tissue debris. C3a and C5a may additionally have direct activating

effects on neutrophils in order to enhance their phagocytic function and release of further inflammatory mediators. The **chemotactic** role of C3a and C5a may further enhance the value of local tissue recruitment of leucocytes. After migrating from the circulation into tissue, phagocytic cells may be directed towards the location of pathogens or damaged cells by moving up a '**chemotactic gradient**'. This gradient is formed by complement molecules that are at highest concentration at the point at which they were generated and at progressively lower concentration the further away from that source they are within that tissue.

In some circumstances the actions of C3a and C5a may become exaggerated and induce a pathological inflammatory response by virtue of their effect on mast cells. For example, degranulation of mast cells in the respiratory tract releases histamine, leading to contraction of bronchiolar smooth muscle (**bronchoconstriction**), which may be detrimental to the host. There are several mechanisms designed to prevent inadvertent complement activation, including inhibitors of the enzymatic activity of the convertases, others that accelerate removal of activated complement elements in addition to CD55 (decay-accelerating factor; DAF), which is expressed on host cells to prevent complement mediated damage. In the case of the latter, human CD55-transgenic pigs have been created in an attempt to try to circumvent complement-mediated destruction of tissues/organs in a human recipient of xenotransplantation.

KEY POINTS

- The structural and functional properties of epithelial surfaces helps to prevent colonization and invasion by pathogens.
- The skin, respiratory tract and alimentary tract have specific elements that are designed to impede infection.
- Biochemical factors are released by host cells that have antimicrobial properties.
- Host epithelial defences are supplemented by the presence of commensal microorganisms that can themselves outcompete invading pathogens and secrete antimicrobial substances.
- The innate immune system recognizes the presence of infection either via the presence of molecules that are intrinsically foreign (PAMPs) or via evidence of cellular damage (DAMPs).
- Pattern recognition receptors are broadly reactive at the molecular level and are not particularly pathogen specific.
- PRR signalling typically leads to activation of intracellular transcription factors that induce gene expression of inflammatory mediators (via NF-κB) or antiviral defence molecules (via IRF-3).
- Activation of the inflammasome enhances the inflammatory response via conversion of pro-cytokines to their biologically active form.
- Neutrophils and macrophages (derived from circulating monocytes) are the main phagocytic cells that leave the circulation at sites of inflammation and travel to the site of infection.
- Microbes are ingested by phagocytic cells by phagocytosis or receptor-mediated endocytosis, where they are subjected to the oxygen-dependent mechanism of microbial killing.
- Intracellular organelles called lysosomes are important for enhancing intracellular killing and digestion of microbes.
- NK cells recognize infected cells when they upregulate expression of stress proteins and downregulate their expression of MHC, stimulating degranulation and induction of apoptosis of the target cell.
- γδ T cells recognize infected cells when they upregulate expression of stress proteins and react by degranulation, inducing apoptosis of the target cell.

- The presence of viral nucleic acid can be detected by intra-cellular PRRs, leading to synthesis and secretion of type 1 interferons that act on neighbouring cells to upregulate their antiviral defence mechanisms.
- Acute inflammation involves a series of vasoactive events allowing recruitment of neutrophils to the site of infection.
- Chronic inflammation involves the subsequent recruitment of mononuclear cells (monocytes and lymphocytes) to the site of infection.
- The systemic signs of inflammation occur through an endocrine effect of cytokines acting on the bone marrow, liver and hypothalamus.
- Production of acute phase proteins by the liver and their attachment to microbial surfaces enhances phagocytosis and the complement system.
- The complement system is an important component of innate immunity.
- The alternative pathway of complement activation requires conversion of the complement protein C3 to its subunits and deposition of C3b onto the microbial surface where it forms a C5 convertase with other complement components.
- The lectin pathway (initiated by mannose-binding lectin) and the classical pathway (initiated by antibody binding to antigen and attachment of C1) share a common route to generating their C5 convertase.
- The terminal pathway is initiated by conversion of C5 to its active components (via a C5 convertase) and ends with formation of the membrane-attack complex.
- The products of complement activation are beneficial to the host through the mechanisms of cytolysis (via the MAC), opsonization (via C3b) and inflammation (via C3a and C5a).

LYMPHOCYTE DEVELOPMENT

OBJECTIVES

At the end of this chapter, you should be able to:

- Describe the difference in how B and T lymphocytes recognize antigen.
- Describe the basic structure of lymphocyte antigen receptors.
- Summarize the development of B cells in mammals and birds.
- Understand how B cell receptor (BCR) diversity develops in mammals and birds.
- Understand how diversity in the T cell receptor (TCR) is achieved.
- Describe how B cells develop in the primary lymphoid tissues.
- Describe the intra-thymic development of T cells and the processes of positive and negative selection.

4.1 INTRODUCTION

Lymphocytes are the primary cells of the adaptive immune system. Unlike the innate immune system, which recognizes common microbial components (i.e. PAMPs) or the presence of cellular distress and destruction (i.e. DAMPs), lymphocytes express cell surface receptors that react to **antigen** (typically, although not exclusively, foreign proteins). More specifically, lymphocyte antigen receptors react to small regions, known as **antigenic epitopes** (either on the surface or generated by antigen processing). This strategy creates a challenge, since there is a huge variety of pathogens that might infect the host and each one may have multiple antigens, themselves with multiple epitopes. Therefore, the lymphocyte antigen receptors, unlike say TLR-4 that reacts to LPS from a variety of Gram-negative bacteria, must be highly specific

and there must be a mechanism for generating a diverse repertoire of different shaped antigen receptors within the lymphocyte population. In this chapter, we will focus on how antigen receptors are made during lymphocyte development in primary lymphoid tissues and explore how lymphocytes are selected to ensure that only those that are useful are allowed to enter the lymphocyte circulatory pool for immune surveillance.

4.2 ANTIGEN RECOGNITION BY LYMPHOCYTES

Lymphocytes are able to respond to antigenic stimulation, but this happens in an entirely distinct manner comparing B cells and T cells. The B cell receptor (**BCR**, also known as surface immunoglobulin; sIg) interacts with a relatively large area of an **intact antigen** on the

DOI: 10.1201/9781003310969-4

surface of the pathogen, which generally has **conformational or planar (linear) structure**. Consequently, the B cell has no requirement for antigen processing and may directly recognize antigen in the extracellular fluid. Although antigen may sometimes be bound to the surface of another cell (e.g. a follicular dendritic cell within lymphoid follicles), the epitopes remain intact and are detected by the BCR without any requirement for prior processing (**Figure 4.1a**). In contrast, the T cell receptor (**TCR**) recognizes fragments of digested (processed) antigen, typically a small peptide that is presented to them in association with a specialized antigen-presenting molecule called major histocompatibility complex (MHC) (**Figure 4.1b**).

In Chapter 1, we discussed the 'lock and key' model of antigen receptor binding to antigen. This analogy allows us to appreciate how diversity can be generated, using the same basic template, in other words, although we all have keys

in our pocket that look somewhat similar, each one is highly specific, in terms of which lock it opens (**Figure 4.2**). Diversity in antigen receptors of B and T cells is generated by randomly selecting gene segments that will encode the amino acid sequence at the N-terminal region of the protein (the variable region, responsible for antigen binding) and combining this unique combination of segments with a constant gene segment, which encodes the rest of the molecule.

Before we consider the mechanism for generating diversity in antigen receptors, we need to compare and contrast the structures of the BCR and TCR (**Figure 4.3**). The BCR is a tetramer made up of two heavy chains and two light chains. Each chain contains a C-terminal constant region and an N-terminal variable region, the latter determining the antigenic epitope that it binds. The TCR is a heterodimer consisting of an alpha and a beta subunit,

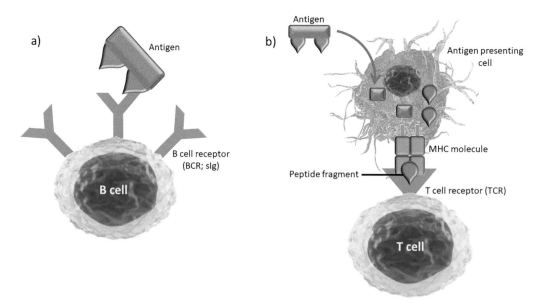

Figure 4.1 Recognition of antigen by B and T cells. (a) B cells use their B cell receptor (BCR, surface immunoglobulin; sIg) to detect intact antigen, typically on the surface of a pathogen in the extracellular fluid. (b) In contrast, the T cell receptor (TCR) recognizes digested antigen, processed by other cells (antigen presenting cells) into peptides that are presented by specialist carrier molecules, the major histocompatibility complex (MHC).

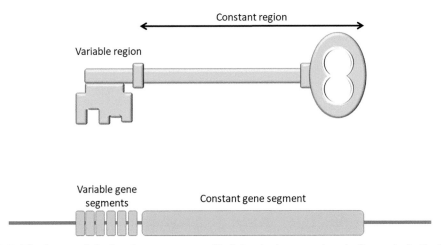

Figure 4.2 The key model of antigen receptors. Each key looks somewhat similar as the bulk of its structure is made from the template (constant region). However, a huge diversity can be generated by altering the shape at the end (the variable region), which determines the key's specificity for its lock. In biological terms, this diversity is encoded within the genome by a number of variable gene segments that are randomly selected and recombined to create the variable region of the antigen receptor.

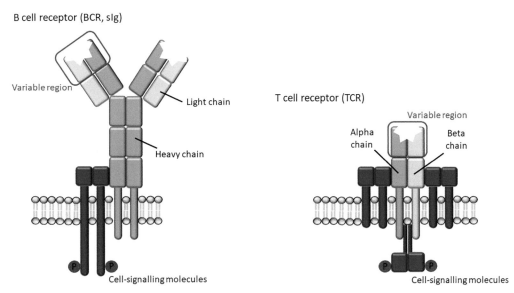

Figure 4.3 B cell and T cell receptors are somewhat similar, but are adapted to their mechanism of antigen recognition. The BCR is a tetramer made up of two heavy and two light chains. The TCR is a heterodimer consisting of an alpha and a beta chain. Each of these elements has a constant region and a variable region. The antigen receptors are associated with other molecules that are responsible for transmitting an intracellular signal upon antigen binding.

with an N-terminal variable region for binding to the MHC-peptide complex. Thus, there are four elements that make up lymphocyte antigen receptors, immunoglobulin heavy chain, immunoglobulin light chain, TCR alpha chain and TCR beta chain, and each requires genetic recombination in order to generate a variable region capable of antigen binding.

4.3 DEVELOPMENT OF B LYMPHOCYTES

The adaptive immune system must carry a large **repertoire of antigen-specific B cells** to account for all the possible interactions with antigen that might occur throughout life. Developing B cells create diversity by selecting from a series of genes encoding different regions of the immunoglobulin heavy and light chains, and diversity in the variable region of these chains is achieved in a number of ways. The **heavy chain variable region** is encoded by three segments: **variable** (V), **diversity** (D)

and **joining** (J) regions. There are in the order of ~40 V region gene segments, 23 D region gene segments and 6 J region gene segments for the human immunoglobulin heavy chain (**Figure 4.4**). One each of these gene segments is selected and the intervening sequence is looped out and deleted. This process involves interaction between different nuclear enzymes that cleave and subsequently ligate the DNA. The complex is collectively known as the **V(D)J recombinase** and includes two important enzymes encoded by the **recombination activating genes *RAG-1*** and ***RAG-2*** (**Figure 4.5**). There are two gene clusters encoding the human immunoglobulin **light chains (IGK and IGL**, encoding the kappa light chain and the lambda light chain, respectively), with each containing **V and J segments** (but no D region) and either C_κ (single segment) or C_λ (one of four to five different segments) (**Figure 4.4**). In many animal species the C_λ gene is preferentially used, so that most immunoglobulins contain a lambda light chain.

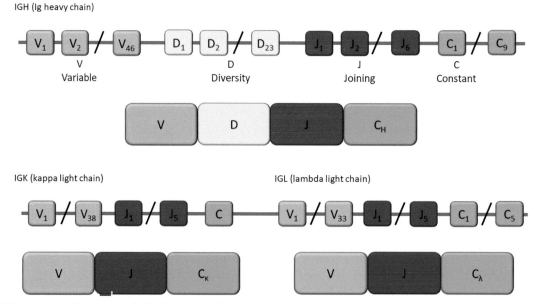

Figure 4.4 Immunoglobulin heavy and light chain genes. The BCR consists of heavy chains and light chains, each with a variable region encoded by a combination of V (variable), D (diversity—heavy chain only) and J (joining segments). The numbers shown are representative of the human loci.

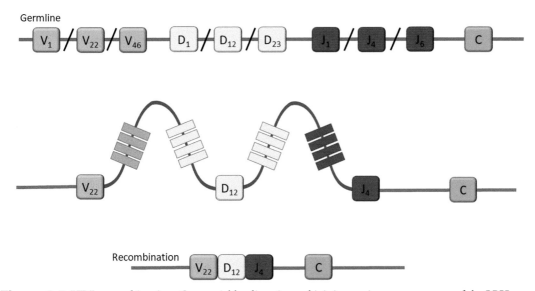

Figure 4.5 VDJ recombination. One variable, diversity and joining region gene segment of the IGH locus is selected and the intervening intronic sequence is looped out and deleted, with the remaining DNA recombined to form a distinct VDJ combination. Following transcription, the selected constant region is added by RNA splicing. VJ recombination occurs in a similar way for the IGK and IGL (light chain) loci.

The theoretical calculation of the possible **germline repertoire** for BCRs produces a figure of approximately 2×10^6 permutations and combinations of these gene segments. However, B cells can further increase diversity in their receptors by the processes of **alternative junctional recombination** and **somatic mutation** (in secondary lymphoid tissues—see Chapter 9). When gene segments are ligated back together again, during VJ or VDJ recombination (for light and heavy chains, respectively), the enzymes responsible introduce some degree of variability at the boundary between V-J and V-D-J segments, known as junctional diversity (**Figure 4.6**). This creates additional variability, as these somewhat random bases impact on the amino acid sequence of the immunoglobulin variable region. Somatic mutation refers to **single nucleotide substitutions** (point mutations) in a gene sequence that again would result in an immunoglobulin polypeptide of different sequence. These two additional processes expand the theoretical germline repertoire to a factor of approximately 10^{11}, but in reality it has been calculated that the **recirculating B cell pool** probably includes around 10^9 BCR specificities at any one time.

In domestic animals, there are a limited number of V, D and J heavy chain and V and J light chain gene segments, and these are rearranged to achieve a restricted range of VDJ combinations in immature bone marrow B cells. These precursor B cells, either within bone marrow or after migration to other lymphoid tissues such as the bursa of Fabricius, ileal Peyer's patch or the spleen, express surface immunoglobulin and start to proliferate. During proliferation, a process of '**gene conversion**' allows the introduction of multiple short areas of DNA sequence from a series of upstream **V region pseudogenes** into the expressed V region gene segment (**Figure 4.7**). This process generates diversity in the V region sequence of the heavy chain. BCR diversity can also be further expanded by somatic mutation. If gene conversion and somatic mutation

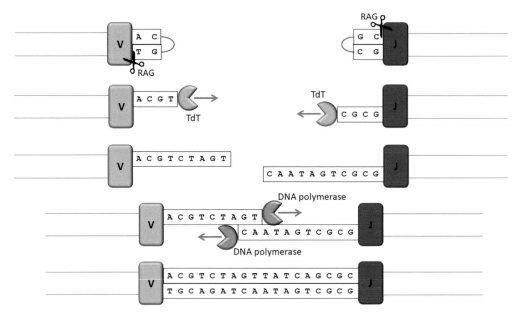

Figure 4.6 Junctional diversity. Further variability is introduced into the variable region of immunoglobulin genes by a process of imprecise joining of gene segments, creating base pair differences at the V-J and V-D-J boundaries. The RAG enzyme complex opens up the hairpin at the ends. A variable number of random bases is inserted by the terminal deoxynucleotidyl transferase (TdT) enzyme. Pairing of base pairs occurs and a DNA polymerase complex fills in the gap to generate double-stranded DNA encoding a variable amino acid sequence between the V and J boundary.

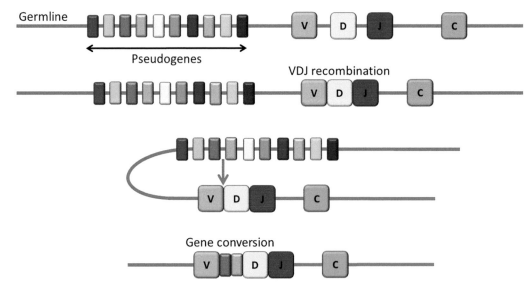

Figure 4.7 Generation of diversity in the BCR of animals and birds by gene conversion. Formation of the BCR differs in domestic animals and birds, in which there are restricted V, D and J heavy chain gene segments. Within the bone marrow or spleen, these segments are combined within immature B cells to form limited primordial variable regions. These immature B cells express the BCR formed of those limited variable regions and start to proliferate. During proliferation there is incorporation of small segments from upstream V region pseudogenes into this V region by the process of 'gene conversion'.

produce an immunoglobulin protein that can be expressed on the surface of the B cell, that cell will be maintained within the B cell pool; if not, the cell will undergo apoptosis. The relative contribution of somatic recombination, gene conversion and somatic hypermutation to the diversity of the pool of naïve B cells varies between animal species.

As for all haematopoietic cells, B lymphocytes develop from bone marrow stem cells that commit to becoming lymphoid precursors. The initial stages of B cell maturation occur within the bone marrow, while the final stages of maturation are thought to occur in extramedullary locations such as the ileal Peyer's patch in ruminants, the bursa of Fabricius in birds or the spleen in humans (see Chapter 2). The earliest form of B cell recognized (the **pre-B cell**) is a precursor that produces the μ heavy chain within the cytoplasm. In the next stage of development, complete IgM monomers are synthesized and the **immature B cell** displays these on its plasma membrane. Any B cell that

expresses an IgM receptor that binds strongly to self-antigen in the bone marrow is induced to undergo apoptosis, although this self-tolerance mechanism is relatively inefficient at eradicating autoreactive B cells (see Chapter 14). Alternatively, the B cell may undergo a process known as **receptor editing**, whereby further gene rearrangements are made within the light chain, so that a modified BCR can be expressed. If this new receptor is still self-reactive, the cell will undergo apoptosis, but, if not, it will leave the bone marrow for final maturation in the periphery. B cells that react with self-antigens during this final maturation step may become anergic or undergo apoptosis. The final step in development is the co-expression of membrane-bound **IgM and IgD** molecules on the surface of **naïve B lymphocytes** by a process of alternate RNA splicing (**Figure 4.8**). At this stage, the B cells become part of the recirculating lymphocyte pool, surveilling the secondary lymphoid tissues for their cognate antigens (**Figure 4.9**).

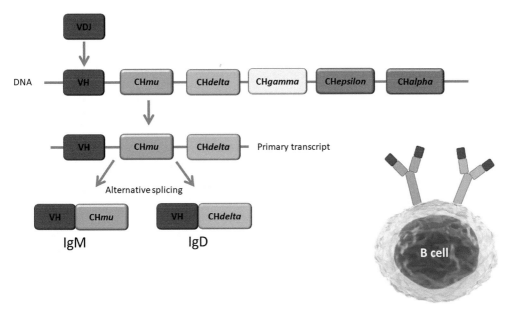

Figure 4.8 **Naïve B cells express surface Ig in the form of monomeric IgM and IgD.** Using a process of alternative splicing of RNA, naïve B cells are able to express IgM and IgD as antigen receptors, with the same antigen specificity.

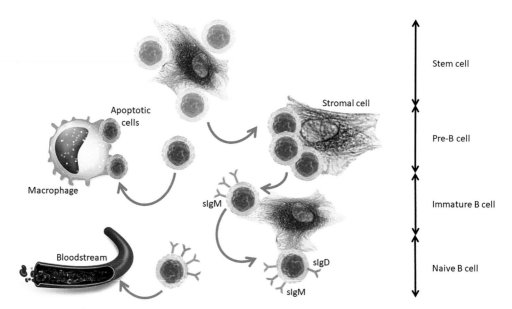

Figure 4.9 Development of B lymphocytes. B cells originate from the bone marrow stem cell and undergo a series of maturation events involving interactions with stromal cells that supply stimulatory cytokines such as IL-7. The pre-B cell expresses cytoplasmic μ heavy chain, while immature B cells express surface membrane IgM. The final maturation step results in a mature naïve B cell that expresses both IgM and IgD. The immature B cell development occurs in the bone marrow, while the final maturation occurs in secondary lymphoid tissues or in the ileal Peyer's patch in ruminants or, in avian species, the bursa of Fabricius. The naïve B cells, bearing surface IgM and IgD, are allowed to enter the blood and commence immune surveillance.

4.4 T CELL DEVELOPMENT AND MATURATION

Once a lymphoid progenitor has committed to the T cell lineage, it is obliged to leave the bone marrow and travel via the blood to the thymus to complete its development. Within the thymus, a number of important decisions and assessments have to take place before any T cell is permitted to leave and enter the general circulation. Firstly, the T cell must generate the cell surface molecules required for recognition of antigen. Once an antigen receptor has been generated, the T cell must decide whether it is to become a CD4+ T cell or a CD8+ T cell, through a process of **positive selection**. Having negotiated that stage, T cells subsequently migrate to the thymic medulla and undergo negative selec-

tion, where they are screened for reactivity to self-antigen. The majority of T cells that have generated a TCR that is autoreactive undergo apoptosis (**clonal deletion**) and fail to make it to the circulating lymphocyte pool. Those T cells that are successful in the thymic selection processes are allowed to enter bloodstream and start to recirculate through secondary lymphoid organs as part of their role in immune surveillance. For the remainder of this chapter, will consider these different stages of T cell development and selection in more detail.

Similar to the situation with B lymphocytes, the vast number of conventional T cells (i.e. those that express the alpha/beta TCR, rather than those that express the gamma/delta TCR) within the immune system must demonstrate almost unique antigen specificity, in order to provide protection from the multitude of pathogens that may potentially enter the body. Each individual animal must therefore have a vast 'repertoire' of different TCRs to make this possible. TCR diversity is generated in a similar manner to that described for the BCR, in terms of rearranging a series of gene segments that encode different parts of the TCR α and β chains. The TCRαβ heterodimer requires additional molecules in order to function. The TCR is directed towards one of the two classes of MHC molecule by either CD4 (for MHC class II) or CD8 (for MHC class I). In addition, there is a group of signalling molecules, known as the CD3 complex, that are required to generate an intracellular signal after the TCR has bound to its target antigenic epitope presented to it by MHC molecules (**Figure 4.10**).

The **TCR α chain** consists of a variable domain (V_α) and a constant domain (C_α) linked together by a joining region (J_α). At the **TRA gene locus**, there are numerous possible Vα genes (around 34 in dogs) and numerous possible J region genes (around 61 in dogs), but only a single constant region gene (Cα) (**Figure 4.11**). Genes encoding elements of the TCR δ chain (utilized by γδ T cells: see Chapter 3) are integrated with the α chain genes. As T cells mature and commit to the αβ lineage, the TRD genes are excised and following re-ligation are permitted to progress to VJ recombination to generate their TCR α chain. The residual element (containing the TRD genes) circularizes to form the signal-joint T cell receptor excision circle (sjTREC). Since this does not replicate upon cell division, it can be used as a biomarker of recent thymic emigrants. In humans, sjTREC assays have been used to study immunosenescence (ageing of the immune system), and in dogs, it has been shown that the sjTREC assay can be used to study thymic output, which declines with age and seems to vary between dog breeds (**Figure 4.12**).

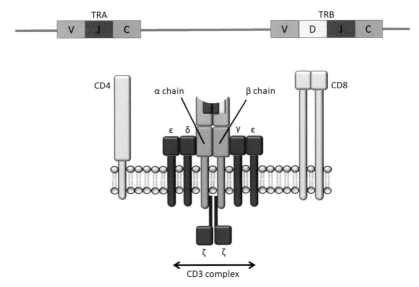

Figure 4.10 The TCR complex. This diagram depicts the structure of the αβ TCR–CD3 complex and the T cell molecules CD4 and CD8 on the cell surface. Note that co-expression of CD4 and CD8 occurs only on immature T cells within the thymus, and T cells are required to become single positive (CD4 or CD8) during thymic selection. Shown above the membrane proteins is the TRA and TRB locus, illustrating the gene segments that contribute to the alpha and beta chains, with the VJ (alpha) and VDJ (beta) segments encoding the variable region, which will contact the MHC-peptide complex.

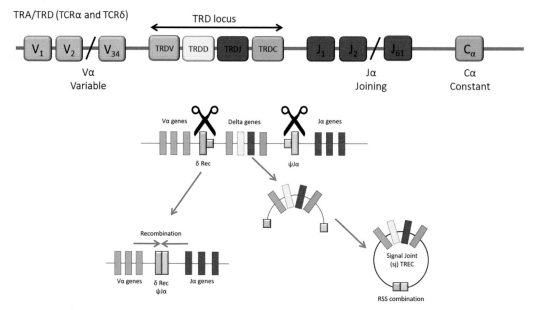

Figure 4.11 The canine TRA/TRD (TCRα and TCRδ) locus. The canine TRD locus is embedded between the TRA variable and joining gene segments. In order for VJ recombination to take place, the TCR delta gene must be excised. This is achieved by enzymatic cleavage at specific recognition sites (δRec and φJα). The recombination signal sequences (RSS) at either end of the excised portion allows circularization and ligation to form the signal-joint T cell receptor excision circle (sjTREC). The genomic DNA is re-ligated to bring together the Vα and Jα segments and VJ recombination can then proceed to generate the TCR alpha chain.

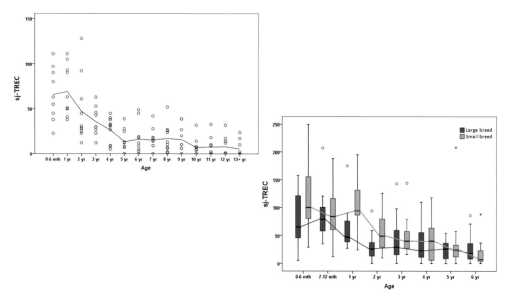

Figure 4.12 The sjTREC can be measured to study thymic output. Since the sjTREC does not replicate during cell division, it can be used as a marker for recent thymic emigrants (i.e. naïve T cells) in the blood. In this experiment, blood samples from Labrador dogs of various ages (top panel) were analyzed showing a reduction in thymic output with increasing age. When small breed dogs were compared with large breed dogs (bottom panel), there was evidence of a premature decline in thymic output in the latter group. (Data from Holder A, Mella S, Palmer DB, Aspinall R, Catchpole B. (2016) An age-associated decline in thymic output differs in dog breeds according to their longevity. *PLoS One* 11:e0165968.)

The series of genes (TRB locus) encoding the TCR β chain is more complex because at some point during evolution there was duplication of a portion of this area of the genome. Although there is only one set of Vβ genes (around 21 in dogs, but can be over 130 in cattle), there are two sets of genes encoding the diversity, joining and constant regions (D–J–C cluster 1 and 2) in human, mice, dogs and horses (**Figure 4.13**). In other species, such as pigs, cattle and sheep, there are three such replicates of the D–J–C cluster. It is possible to interrogate the T cell repertoire by polymerase chain reaction with sense primers designed to be specific for individual Vβ gene segments, paired with a common antisense primer which anneals to the sequence of the C gene segment. Lymphocytes in the blood can then be assessed for their level of diversity, which may deteriorate with increasing age (**Figure 4.14**).

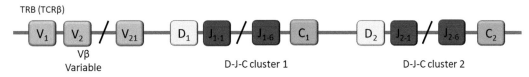

Figure 4.13 The canine TRB (TCRβ) locus. This locus consists of numerous variable (Vβ) genes with a duplication of the diversity, joining and constant gene segments (D-J-C cluster 1 and 2). Each of these two clusters has a single diversity and constant region gene segment, with a limited number of joining gene segments.

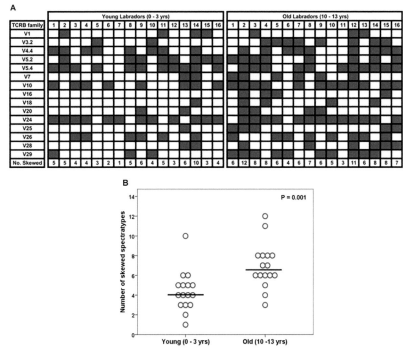

Figure 4.14 T cell receptor diversity. In this experiment, blood samples from young (n=16) and old (n=16) Labrador retrievers were assessed for TCR diversity by 'CDR3 spectratyping', a method whereby individual Vβ genes are amplified and evaluated for their expression profile. The data indicates that there are significantly more abnormal/skewed spectratypes (shaded boxes) in the older dogs than in the young dogs. (Data from Holder A, Mirczuk SM, Fowkes RC, Palmer DB, Aspinall R, Catchpole B. (2018) Perturbation of the T cell receptor repertoire occurs with increasing age in dogs. *Developmental and Comparative Immunology* **79**:150–157).

Once T cells have generated their TCR via V(D)J recombination, they undergo a series of screening processes within the thymus to ensure that the antigen receptor created by each individual cell will be functional within the adaptive immune system. Some TCRs fail to interact with MHC molecules and since this is critical for detection of foreign antigen, these T cells are effectively non-functional and are deleted. In contrast, some other TCRs not only engage with MHC but also react to the presence of self peptides. Such autoreactive T cells are potentially dangerous if released into the circulation, so they are also deleted. Thymic maturation is a very wasteful process, as some 99% of immature T cells that enter the thymus fail to progress to completion.

An early event in T cell maturation that occurs shortly after expression of the TCR is cell surface expression of both CD4 and CD8 molecules. These immature T cells there-fore convert from being 'double negative' (i.e. expressing neither CD4 nor CD8) to 'double positive' (i.e. expressing both CD4 and CD8). They then undergo a screening process, called **positive selection**, whereby the immature T cell must prove that it has a TCR capable of interacting with MHC molecules. Within the thymic cortex are numerous **cortical epithelial cells** that display MHC class I and II molecules on their surface. Each immature T cell must 'test out' its TCR to ensure that it can interact with one or the other of the MHC molecules (**Figure 4.15**). If their TCR favours MHC class I, the CD8 molecule co-ligates and the cell loses its CD4 molecule. If their TCR favours MHC class II, the CD4 co-ligates and the cell loses its CD8 molecule. T cells that have a functional TCR 'pass the test' of positive selection, become 'single positive' (i.e. expressing either CD4 or CD8) and receive survival signals (including cytokines such as

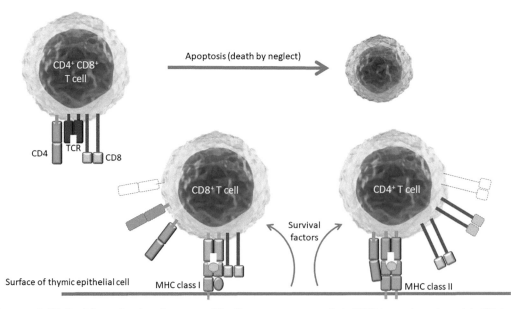

Figure 4.15 Positive selection. Immature T cells start to express their TCR in conjunction with CD4 and CD8 (double positive). These cell surface receptors are tested for their interaction with MHC on the surface of thymic epithelial cells. Those with TCRs that favour MHC class I become single positive CD8+ T cells, whereas those with TCRs that favour MHC class II become single positive CD4+ T cells. T cells whose TCR fails to engage any MHC molecules undergo apoptosis (death by neglect).

IL-7). Cells that fail to mount such an interaction with any MHC molecules undergo 'death by neglect'.

Those T cells that are positively selected move into the medulla of the thymic lobule where they are subjected to the second assessment, called **negative selection (Figure 4.16)**. It is at this stage that the immature T cells meet **thymic dendritic cells and medullary epithelial cells** displaying MHC class I or II molecules containing peptides derived from a wide array of **self-proteins**. The transcription factor 'autoimmune regulator' (**AIRE**) is promiscuous and induces low level expression of many proteins in epithelial cells, including those that are considered to be relatively tissue specific. Indeed, within the thymus it is known that there is expression of peptides derived from tissues such as the brain, endocrine pancreas and joint. In this assessment, the T cell must prove that its TCR is incapable of responding to these self-antigens. Cells that fail as a result

of expressing TCRs with high affinity for self-peptide (i.e. are potentially autoreactive) will be induced to undergo **apoptosis**. The threshold level for determining how reactive is too reactive is a somewhat difficult decision for the thymus to make. If the bar is set too low, then the individual will be unlikely to develop autoimmune disease in later life, but the immune repertoire might be compromised, impacting on the ability to respond to foreign antigen. In contrast, if the bar is set too high, this maximizes the immune repertoire against infection but increases the risk of immune-mediated disease occurring. In particular, it is tricky to know what to do with T cells that are 'borderline' reactive cells, as the outcome is somewhat black and white (survive or killed). In fact, there is a 'middle ground' whereby these borderline cells are rescued on condition that they differentiate to a specific phenotype, known as **natural regulatory T cells**. These cells exit the thymus with evidence that they have encountered their

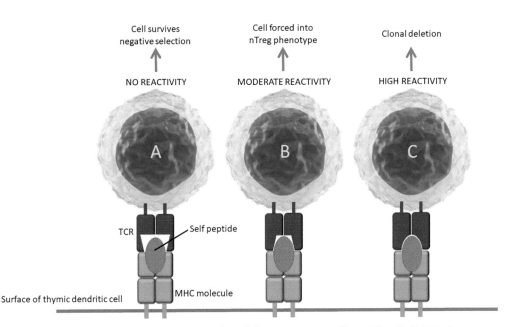

Figure 4.16 Negative selection. TCRs are selected for reactivity to self-peptides. Cell A binds to MHC but does not react to self-peptide and is selected to enter the circulating lymphocyte pool. Cell C is autoreactive and is deleted by induced cell death (apoptosis). Cell B is somewhat self-reactive and is rescued from apoptosis, but forced into become a natural T regulatory cell (nTreg).

cognate antigen already (as they express the CD25 activation marker), but they also express markers associated with a downregulation in their activity (such as CTLA-4) and carry a transcription factor, known as FoxP3, which prevents synthesis of pro-inflammatory mediators and instead directs the cell to produce more anti-inflammatory/suppressive cytokines should they subsequently meet their target antigen in the periphery. As such, these potentially self-reactive T cells are 'made safe' rather than deleted by the thymus. We will revisit these cells in Chapter 14, when we discuss immunological tolerance in more detail.

Cells that successfully negotiate both positive and negative selection processes are allowed to progress to the final stage of maturation and 'graduate' from the thymus, entering the blood to join the recirculating lymphocyte pool. The high failure rate for these two examinations accounts for the marked loss of T cells during intra-thymic development. Although only 'single-positive' T cells are exported from the thymus, there are some unique species differences in the phenotype of circulating T cells. For example, the majority (approximately 60%) of circulating T cells in the pig are double-positive, with most of the remainder being double-negative (CD4⁻CD8⁻) and a significant proportion of circulating T cells in young cattle are also double-negative and expresses the γδ TCR. Thus, although the principles of lymphocyte development are broadly applicable to most mammalian species, one needs to be mindful of inter-species differences and careful not to extrapolate too much information from human to veterinary species.

KEY POINTS

- Lymphocytes express antigen receptors that are made up of a constant region and a variable region.
- The BCR recognizes large conformational determinant of antigen, typically on the surface of a pathogen in the extracellular fluid.
- The TCR recognizes digested fragments of antigen (peptides) following antigen processing and presentation by MHC molecules on other cells.
- BCR diversity is achieved by different means in man, domestic animals and birds.
- In man, VJ (immunoglobulin light chain) and VDJ (immunoglobulin heavy chain) recombination generates a diverse repertoire of BCRs.
- The repertoire of BCRs is further increased by alternative junctional recombination and somatic mutation.
- In some animal species VJ and VDJ recombination does not generate huge diversity, but this is achieved by subsequent gene conversion and somatic hypermutation.
- B cells develop in the bone marrow from lymphoid precursors, but may undergo final maturation in extramedullary tissues in some species.
- Naïve B cells co-express surface membrane IgD and IgM, which act as their antigen receptors for immune surveillance.
- The TCR requires additional molecules (CD3 complex, CD4 and/or CD8) in order to function.
- Generation of diversity in the T cell repertoire is achieved by rearrangement of genes encoding segments of the α and β chains of the TCR in a similar manner to that of the BCR.

- T cells must travel to the thymus to complete their development, but only a relative few successfully negotiate thymic selection and enter the circulation as naïve T cells.
- Double-positive (CD4$^+$CD8$^+$) T cells with a functional TCR, capable of interacting with MHC molecules are positively selected during thymic development.
- Those cells with a TCR that favours MHC class I become CD8$^+$ single positive T cells, whereas those that favour MHC class II become CD4$^+$ single positive T cells.
- Most T cells with a high-affinity autoreactive TCR are negatively selected during thymic development and undergo clonal deletion.
- Some relatively self-reactive T cells are rescued but forced to develop into natural T regulatory cells before they are allowed to leave the thymus.

OBJECTIVES

At the end of this chapter, you should be able to:

- Understand how major histocompatibility complex (MHC) molecules are required for antigen detection by T cells.
- Describe the structure of MHC class I and MHC class II molecules.
- Explain how MHC molecules interact with peptides.
- Describe the MHC class I processing pathway for antigen presentation to CD8+ T cells.
- Describe the MHC class II processing pathway for antigen presentation to CD4+ T cells.
- Understand that CD1 molecules are capable of presenting lipid antigens.
- Discuss the role of dendritic cells as the major professional antigen presenting cell (APC) for stimulation of naïve T cells.
- Describe the role of macrophages and B lymphocytes as APCs.
- Appreciate how the MHC is organized at the genetic level.
- Explain why polygeny, polymorphism and co-dominant expression is necessary for optimal MHC class I and class II function.
- Discuss the concept of MHC–disease association and how it applies to veterinary medicine.
- Describe how an incompatible tissue graft is rejected by the recipient immune system.

5.1   INTRODUCTION

This chapter considers the structure and function of a key set of immunological molecules encoded by a set of genes known as the **major histocompatibility complex** (**MHC**). The primary importance of these molecules is in **antigen presentation** to T cells, which is essential for initiation of **cell-mediated immunity**. This chapter addresses an important interaction between the innate and adaptive immune responses and considers how the use of two different MHC pathways effectively communicates the location (intracellular versus extracellular) of infectious organisms to the adaptive immune system. Previously, we considered how foreign antigen entering the body is translocated from the site of infection to regional lymphoid tissue for initiation of the adaptive immune response, and how this process is mediated by antigen presenting cells (APCs), such as dendritic cells. Here we consider how antigen is processed and

DOI: 10.1201/9781003310969-5

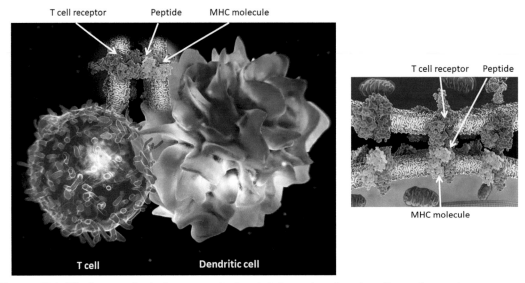

Figure 5.1 **The immunological synapse.** Antigenic information about invading pathogens is communicated between antigen presenting cells, such as dendritic cells and T cells, via presentation of foreign peptides bound to MHC molecules, which is detected by the T cell receptor.

presented to the immune system by MHC molecules so that (antigenic) information about invading pathogens can be relayed to the adaptive immune system, the so-called **immunological synapse** (**Figure 5.1**). We will also consider how the MHC genes that are inherited by individual animals can have a profound effect on their susceptibility to infection, the response to vaccination and their susceptibility to immune-mediated disease.

5.2 THE MAJOR HISTOCOMPATIBILITY COMPLEX

Major histocompatibility complex (MHC) molecules fall into one of two classes: MHC class I and MHC class II. They are expressed on the cell surface, anchored into the cell membrane with a transmembrane domain and bearing a cytoplasmic tail (**Figure 5.2**). **MHC class I molecules** are expressed on **all nucleated cells** of the body and consist of a large α chain paired with a smaller protein known as β_2 **microglobulin**. The α chain has three extracellular domains (α_1, α_2 and α_3). **MHC class II** molecules consist of two transmembrane chains, an α **chain** with two extracellular domains (α_1 and α_2) and a β **chain** with two similar domains (β_1 and β_2). MHC class II has a more limited cellular distribution, being normally restricted to **antigen presenting cells (APCs)**, including dendritic cells, macrophages and lymphocytes (particularly B cells), although they are inducible on the surface of other cells, given the correct cytokine signals.

The β_2 microglobulin and α_3 domain of MHC class I molecules and the α_2 and β_2 domains of MHC class II molecules have a relatively conserved amino acid sequence, whereas the α_1 and α_2 domains of MHC class I, and the α_1 and β_1 domains of MHC class II molecules have much greater variation in their amino acid sequence. This basic structural arrangement is very similar to that previously described for the heavy and light chains of immunoglobulin. In fact, the constant domains of immunoglobulins, MHC proteins and a number of other key immune molecules are all related to each other, and are considered to be part of the **immunoglobulin superfamily**.

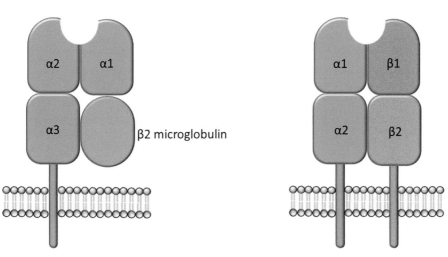

Figure 5.2 **The structure of the major histocompatibility complex.** MHC class I molecules consist of a large alpha chain with three globular domains (α_1, α_2 and α_3) and an associated β_2 microglobulin protein. MHC class II molecules consist of an alpha and beta chain, each having two globular domains (α_1, α_2 and β_1, β_2).

Figure 5.3 **MHC structure: peptide-binding groove.** (a) Molecular model and diagrammatic representation of an MHC class II molecule (side view) showing the peptide (yellow) sitting within the groove generated by the alpha helices of the α_1 and β_1 domains (alpha chain in red, beta chain in blue). (b) Molecular model and diagrammatic representation of an MHC class II molecule (surface view) showing the peptide (yellow) sitting above the beta-pleated sheet (green) and flanked by the alpha helices (red). (Molecular models generated in iCn3D, from PDB ID: 1AQD.)

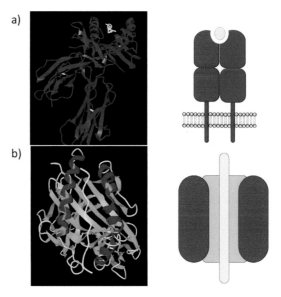

The α_1 and α_2 domains of MHC class I, and the α_1 and β_1 domains of MHC class II molecules come together to create a **peptide-binding groove**. This region of the MHC molecule is folded, such that a beta-pleated sheet forms the 'floor' of the groove, with alpha helices forming the 'sides'. This allows a peptide, generated from processing of antigen, to bind to the MHC molecule (**Figure 5.3**). Variability in the amino acid sequences of the α_1 and α_2 domains of MHC class I, and the α_1 and β_1 domains of MHC class II molecules has functional consequences, as it impacts the peptide-binding specificity of the MHC molecules.

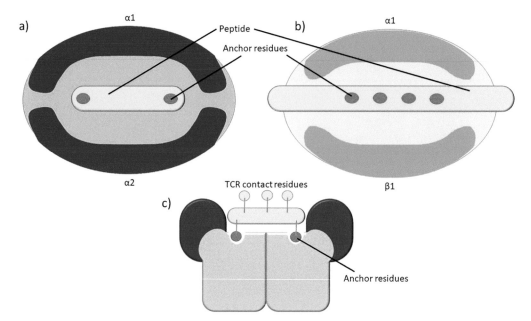

Figure 5.4 Peptides of different sizes bind to MHC class I versus class II. (a) MHC class I binds to peptides of eight to ten amino acids in length, restricted owing to the somewhat 'closed' ends of the peptide-binding groove. (b) In contrast, the peptide-binding groove of MHC class II is more open and can accommodate peptides of varying length, typically between 13 and 25 amino acids. (c) Peptide binding requires interaction between the MHC molecule and specific 'anchor residues' of the peptide, which for MHC class I are towards the end of the peptide (being more evenly spread out for MHC class II).

The peptide-binding groove of MHC class I is somewhat restricted at its ends, which means that it typically will only accept peptides of eight to ten amino acids in length. In contrast, the MHC class II peptide-binding groove is more open ended and peptides of varying length (typically between 13–25 amino acids) are allowed to bind (**Figure 5.4**). In the next sections we will consider how peptides are generated during cellular processing of antigen and how the two types of MHC molecules allow compartmentalization of antigen presentation from intracellular versus extracellular sources.

5.3   THE MHC CLASS I PROCESSING PATHWAY

MHC class I molecules specifically present peptides from '**endogenous**' antigen, that is, those that are found within the cytoplasm

of the cell, and do not require phagocytosis. Endogenous antigens include self-antigens (autoantigens), tumour antigens, alloantigens and viral antigens, derived from the expression of nascent viral proteins synthesized during virus replication in the cytoplasm of an infected cell. These proteins are first tagged by a cytoplasmic molecule known as **ubiquitin**, which directs them to the **proteasome**, a large cytoplasmic enzyme complex. Degradation of the antigen into small peptide fragments occurs within the proteasome and these peptides are then released (**Figure 5.5**). The next stage involves translocation of the peptides from the cytoplasm into the **endoplasmic reticulum** (ER), which is facilitated by specific '**transporter proteins**' (TAP1 and TAP2). Within the ER, the peptides encounter newly synthesized MHC class I molecules and, if the peptide contains appropriate anchor residues,

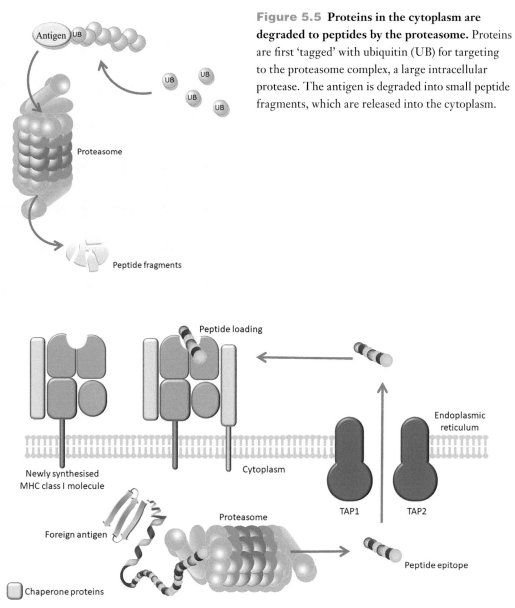

Figure 5.5 Proteins in the cytoplasm are degraded to peptides by the proteasome. Proteins are first 'tagged' with ubiquitin (UB) for targeting to the proteasome complex, a large intracellular protease. The antigen is degraded into small peptide fragments, which are released into the cytoplasm.

Figure 5.6 Peptide loading onto MHC class I molecules. Peptides generated from the proteasome are imported into the endoplasmic reticulum by TAP transporters, where they may bind to newly synthesized MHC class I molecules. This process is facilitated by chaperone proteins including calnexin, calreticulin and tapasin.

it will bind (**Figure 5.6**). The next stages in the sequence involve the peptide-loaded MHC class I molecule transiting through the **Golgi apparatus**, into secretory vesicles, then passing to the cell surface. Cells expressing MHC class I peptide can be detected by CD8⁺ T cells. The CD8 molecule binds to the α3 subunit, allowing the TCR to screen the bound peptide for complementarity of contact residues (**Figure 5.7**).

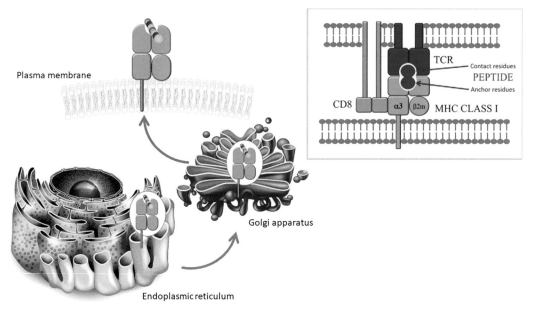

Figure 5.7 The MHC-peptide complex is translocated to the cell surface. Once the MHC-peptide complex has formed in the ER, it enters the secretory pathway (via the Golgi apparatus) and is expressed on the cell surface. Inset: the MHC class I-peptide complex can be recognized by a CD8⁺ T cell.

5.4 THE MHC CLASS II PROCESSING PATHWAY

The majority of foreign antigens are extracellular and must be actively taken up into endosomal vesicles, before they can be processed and presented via the MHC class II pathway. This is a somewhat specialist process and only a limited number of cell types can perform this task. These are known as antigen presenting cells (APCs), which include dendritic cells, macrophages and B lymphocytes. These APCs are capable of synthesizing MHC class II molecules in the endoplasmic reticulum in preparation for antigen presentation. Since both MHC class I and class II molecules are produced in the ER and there is a need to keep endogenous and exogenous pathways separate, newly synthesized MHC class II molecules associate with a chaperone protein called the **invariant chain**. This latter protein effectively 'blocks' the peptide-binding groove, thus preventing endogenous

peptides from gaining access (**Figure 5.8**). This chaperone protein also enables the MHC class II molecule to traffic out of the ER, through the Golgi apparatus, and into a specialist vesicular structure, known as the **MIIC compartment**, where they await antigen.

Exogenous antigens come from a wide array of infectious agents (e.g. bacteria, viruses, fungi, protozoa, helminths) or environmental substances (e.g. pollens, dust mites, foodstuff) to which the body is exposed. It used to be thought that uptake of exogenous antigens by the APC by **phagocytosis**, **pinocytosis** or **macropinocytosis** was a relatively non-specific event that might be enhanced by the process of opsonization from antibody or complement deposition. It is now known that the interaction between cell-surface expressed PRRs (see Chapter 3) and the PAMPs associated with the microorganisms, has a major effect on how the antigenic component is presented (**Figure 5.9**). In particular, the '**danger signals**' generated via the PRRs allows

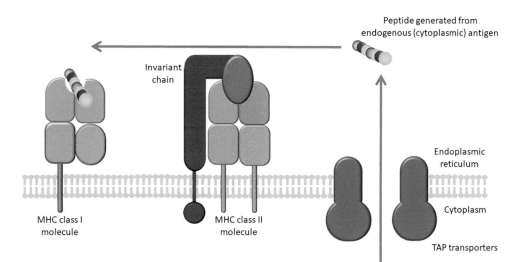

Figure 5.8 **MHC class II synthesis.** MHC class II α and β subunits are synthesized in the endoplasmic reticulum, where they come together to form a heterodimer that associates with a chaperone protein, called the invariant chain. The latter has a globular head, which binds to the peptide-binding groove, blocking this against endogenous peptides, destined for binding to MHC class I molecules.

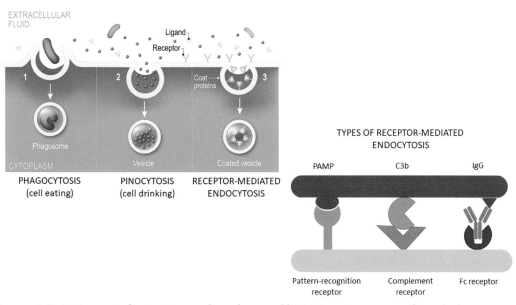

Figure 5.9 **Antigen uptake.** Antigen can be endocytosed by APCs using non-specific methods (phagocytosis and pinocytosis) or more efficiently by receptor-mediated endocytosis, utilizing innate and adaptive 'markers' of foreign material including PAMP-PRR, C3b-complement receptor and antigen/antibody-Fc receptor interactions.

additional information to be transmitted via the APC to the T cells via the 'immunological synapse' in addition to a recognition signal via the TCR (see Chapter 7).

Antigen in the extracellular fluid is engulfed into a membrane-bound cytoplasmic compartment known as the **phagosome** or **endosome**, where it may encounter further PRRs and then enter the second stage of antigen processing. This is the means by which a large, complex antigen is broken down into small constituent elements that are of a suitable size to be presented. Lysosomes (containing acid hydrolases such as cathepsins) fuse with the early endosomes and the resultant phagolysosome becomes increasingly acidic, allowing the enzymes to degrade the protein antigen into **small peptides** in the order of **10–30 amino acids** in length. At this stage, the two pathways merge in the MIIC compartment, bringing together MHC class II molecules and foreign peptides. Within the MIIC, the invariant chain is degraded, leaving the globular head in position over the peptide-binding groove. This remnant of the invariant

chain is termed the class II-associated invariant chain peptide (CLIP), which can be subsequently displaced by a foreign peptide (provided it has suitable anchor residues) generated from antigen processing (**Figure 5.10**). Once a stable MHC-peptide complex has been achieved, it is transported to the plasma membrane of the APC, where it is made available for recognition by CD4⁺ T cells (**Figure 5.11**).

The MHC processing pathways are often considered distinct and compartmentalized for exogenous and endogenous antigens, but this is not always the case, and sometimes exogenous antigen can enter the endogenous processing pathway (this is known as **cross-presentation**). In this manner, peptides from exogenous antigens may occasionally be presented by MHC class I (when peptides generated within the endosome escape into the cytoplasm and enter the proteasome). This is particularly important when viruses do not directly infect professional APCs. Indeed, since only professional APCs such as dendritic cells can activate naïve T cells, if a virus does not infect and replicate

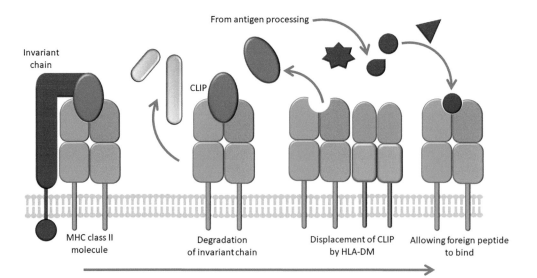

Figure 5.10 MIIC compartment. Within the MIIC compartment, the invariant chain is degraded, leaving behind the class II-associated invariant chain peptide (CLIP). Merging of the endocytic pathway delivers processed foreign antigen in the form of peptides that may displace the CLIP (facilitated by HLA-DM) for binding to the MHC class II molecules, provided the peptide has suitable anchor residues.

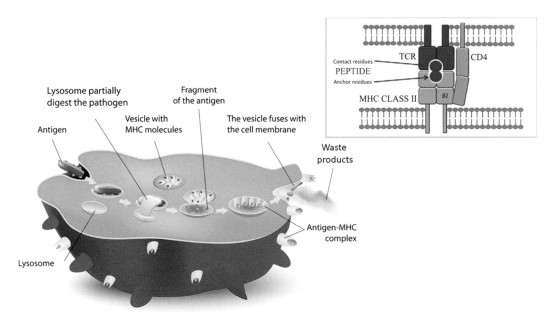

Figure 5.11 Following antigen processing, MHC-peptide complexes traffic to the cell surface where they can interact with CD4+ T cells.

within it, there would be no way to activate the naïve CD8+ T cells, as foreign peptides would not be processed via the endogenous pathway. The solution is to allow exogenous antigen to cross over from the exogenous pathway to the MHC class I pathway in these professional APCs. That way, the immune system can present foreign peptides associated with MHC class I, even if viral antigen is only available to professional APCs from the extracellular fluid. Furthermore, endogenous peptides may sometimes be presented by MHC class II molecules (when cytoplasmic material enters the endosome following the process of **autophagy**). Autophagy is a means by which a cell can remove unwanted or misformed proteins or organelles from its cytoplasm (**Figure 5.12**). The structure to be removed is first enclosed within a cytoplasmic double membrane (the autophagosome), which in turn fuses with a lysosome containing enzymes that digest the structure. The building blocks of the structure (e.g. amino acids, fatty acids and nucleotides) are released to the cytoplasm of the cell to be reused. The autophagy

pathway may also be used to remove invading infectious agents (e.g. intracellular bacteria), and their antigenic peptides may be expressed on MHC class II in addition to MHC class I.

5.5 ANTIGEN PRESENTION BY CD1 MOLECULES TO NKT CELLS

Conventional antigen presentation requires protein antigen to be processed to generate peptides that can be detected by the T lymphocytes when they associate with MHC molecules on the surface of cells. A separate and distinct antigen-presenting pathway is used for the presentation of **lipid antigen**, which is a major constituent of some pathogens such as *Mycobacterium spp.* and *Rhodococcus equi* and therefore of relevance to veterinary medicine (**Figure 5.13**). Such antigens are presented by a CD1 molecule that has an evolutionary relationship to MHC. CD1 is also a transmembrane glycoprotein with three extracellular domains that associates with β2 microglobulin. The CD1 molecules are encoded by a series of genes that lack the

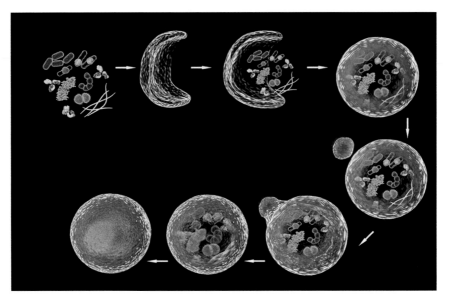

Figure 5.12 Autophagy can 'feed' endogenous antigen into the MHC class II pathway. Autophagy is a process whereby vesicles form around cytoplasmic organelles (or microorganisms) that then can enter the MHC class II processing pathway.

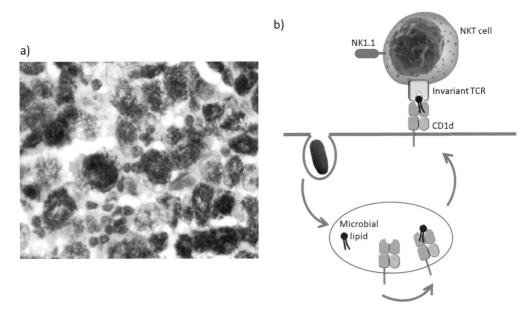

Figure 5.13 CD1 antigen presentation. (a) This section of spleen comes from a dog with disseminated mycobacterial infection and has been stained by the Ziehl–Neelsen method. The macrophages are distended by the large number of acid-fast bacilli within their cytoplasm. (b) Lipid mycobacterial antigen is expressed on the surface of these cells by members of the CD1 family of antigen-presenting molecules and can be detected by NKT cells, which express NK markers as well as an invariant TCR.

polymorphism that characterizes the MHC. The loading of CD1 with lipid antigen probably occurs within an endosomal compartment, with subsequent translocation of the lipid-laden CD1 molecule to the surface of the APC. These CD1/lipid complexes can be detected by T cells with diverse TCRαβ receptors, but also by a population of natural killer T (NKT) cells with an invariant TCR that are part of innate immunity and respond rapidly *en mass*, when CD1d molecules expressing foreign lipids are detected.

5.6 ANTIGEN PRESENTING CELLS

There are three types of APCs: **professional**, **non-professional** and **induced**. Professional APCs are purpose-designed for this role and include dendritic cells, macrophages and B lymphocytes. The most important of this group are the dendritic cells that can activate naïve T cells in secondary lymphoid tissues during a primary adaptive immune response in a host, who has not previously been exposed to the triggering antigen. Dendritic cells are considered much more potent than macrophages in this respect. Dendritic cells are named due to their morphology and are characterized by having multiple elongate cytoplasmic processes (dendrites), which may extend between cells in the surrounding matrix, providing a large surface area for contact with potential antigen and other cells (**Figure 5.14a**).

Several subtypes of dendritic cell have now been defined. **Conventional dendritic cells** arise from a bone marrow dendritic cell progenitor (or pre-dendritic cell) and include two subcategories: **migratory** and **lymphoid** dendritic cells. Migratory dendritic cells are widely distributed in body tissues where they capture antigen and migrate to regional lymphoid tissue in order to induce adaptive immune responses. In contrast, lymphoid dendritic cells remain within lymph nodes but are also important in the induction of T cell responses. **Non-conventional dendritic cells** include three further subcategories: **plasmacytoid**, **monocyte-derived** dendritic cells and Langerhans cells. Plasmacytoid dendritic cells have a pre-dendritic cell origin and are found in both lymphoid and non-lymphoid organs (e.g. liver and lung). Plasmacytoid dendritic cells express PRRs for foreign nucleic acid (e.g. dsRNA) and are an important source of type 1 interferon production during viral infection. Monocyte-derived dendritic cells are primarily induced in inflamed tissues from where they take up antigen and migrate to regional lymphoid tissue. **Langerhans cells** are dendritic cells in the epidermis that initiate immune responses to antigens that penetrate the skin barrier.

Dendritic cells are prominent at mucosal sites such as the respiratory and alimentary tracts (**Figure 5.14b**) and these cells are known to be able to extend their dendrites between enterocytes to 'sample' antigen passing through the lumen. The epidermal **Langerhans cell** is perfectly located to be able to interact with antigens that penetrate the epidermal barrier (**Figure 5.14c**). Once dendritic cells have captured antigen, they carry this to the regional lymphoid tissue via the afferent lymphatics. While in these lymphatic vessels, the cells undergo a morphological transformation whereby the cytoplasmic dendrites are replaced by broader and shorter cytoplasmic projections and they are known as **veiled** dendritic cells. Once these cells have arrived in the lymph node they localize to the paracortex where they demonstrate a typical dendritic morphology in order to optimize their surface area for contact with T lymphocytes (**Figure 5.15**).

Follicular dendritic cells are located within the B cell follicles of the lymph node. These cells are not bone marrow derived and, unlike the dendritic cells described previously, they are non-phagocytic and do not express MHC class II. Instead, these cells allow adherence of immune complexes of antigen and antibody on their surface and shed them in the form of bead-like structures (iccosomes), which can be

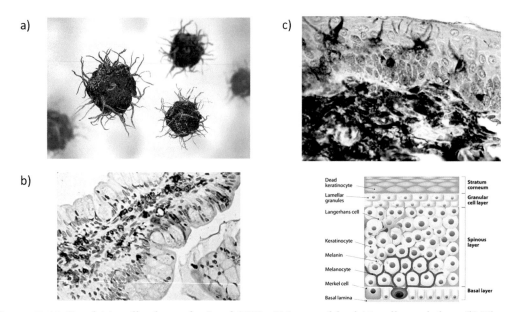

Figure 5.14 Dendritic cells, the professional APCs. (a) Image of dendritic cell morphology. (b) The dendritic cells within the lamina propria of this intestinal villus are labelled to show expression of MHC class II. Some linear labelling is also present within the enterocyte layer and this likely represents the extension of dendritic processes through this barrier to sample luminal antigen. (From Waly N, Gruffydd Jones TJ, Stokes CR et al. (2001) The distribution of leucocyte subsets in the small intestine of normal cats. *Journal of Comparative Pathology* **124**:172–82, with permission.) (c) Langerhans cells are the dendritic cells that reside in the epidermis of skin and are a first point of contact with antigen that penetrates this cutaneous barrier. Note the long cytoplasmic processes that arise from these cells and insinuate between keratinocytes. These cells are labelled to show that they express MHC class II antigens. (From Carter J, Crispin SM, Gould DJ et al. (2005) An immunohistochemical study of uveodermatologic syndrome in two Japanese Akita dogs. *Veterinary Ophthalmology* **8**:17–24, with permission.)

taken up by B cells that process and present the antigen to T cells. Follicular dendritic cells may act as a 'reservoir' of antigen that can persist for many months on the cell surface in this form.

When monocytes leave the circulation and enter the tissues, they can differentiate into a number of different cell types (see **Figure 2.5**). One of these, the macrophage is not only an important phagocytic cell but also acts as an APC. Tissue macrophages are widespread throughout the body and at mucocutaneous surfaces. Macrophages in the marginal zone of the splenic peri-arteriolar lymphoid sheaths (PALS) and Kupffer cells within the liver sinusoids have a similar roles in engulfing antigen from the bloodstream. Macrophages are large cells (15–30 µm) with an oval to kidney-shaped nucleus and plentiful cytoplasm that expands and becomes vacuolated when the cells are 'activated' (**Figure 5.16**). Activated macrophages may fuse together to form very large **multinucleated giant cells** that are characteristic in chronic inflammatory lesions, such as granulomas. Macrophages typically process and present antigen via MHC class II to recruit T cell help and in return receive cytokine signals (e.g. interferon-gamma) that allow them to function more effectively.

B lymphocytes are also capable of presenting antigen via the MHC class II pathway. Antigen is initially bound by the BCR and internalized into a cytoplasmic endosome in preparation for

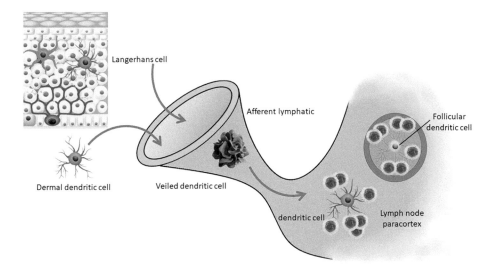

Figure 5.15 Dendritic cell migration. Dendritic cells carry antigen from the site of entry into the body to the regional lymphoid tissue in order to activate the adaptive immune response. In the skin, both epidermal Langerhans cells and dermal dendritic cells may migrate in afferent lymph to the local lymph node. During migration the cell undergoes a morphological change to become a 'veiled' dendritic cell. In the lymph node, the cells localize within the paracortex. An unrelated population of follicular dendritic cells resides within the follicle.

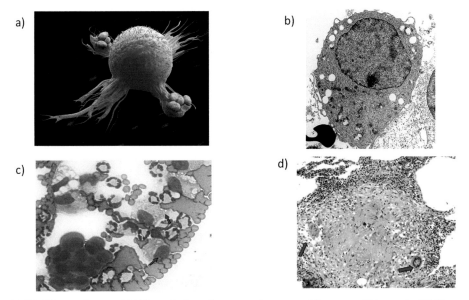

Figure 5.16 Macrophages. (a) 3D rendering of a macrophage engulfing coccoid bacteria. (b) Transmission electron micrograph of a canine macrophage showing the abundance of cytoplasm and the presence of a number of cytoplasmic compartments. Foreign antigen is incorporated into these compartments for processing. (c) The dominant cell type within this sample of abdominal fluid from a cat with feline infectious peritonitis (FIP) is an activated macrophage with abundant and finely vacuolated cytoplasm (examples arrowed). Some of these cells also have phagocytosed erythrocytes. Neutrophils and a cluster of mesothelial lining cells are also present. (d) Granuloma with the presence of multi-nucleate giant cells (examples arrowed).

processing and presentation (**Figure 5.17**). This allows antigen-specific B cells to interact with antigen-specific CD4+ T cells in order to receive T cell help for antibody production. Although the antigen recognized by the B and T cells is often the same, the epitopes they detect are often different, as the B cell recognizes surface conformational epitopes in the context of intact antigen, whereas the T cell recognizes linear peptide epitopes following digestion and presentation of the antigen (by the B cell).

Under certain circumstances, a range of other cells that do not normally present antigen can be induced to do so. In some species (but not all), T cells themselves can upregulate MHC class II expression and can present antigen to other T helper cells. **Epithelial**, **endothelial** and other **stromal** cells such as fibroblasts can all be stimulated to express MHC class II and can present antigen to T cells. The key signal required to induce such non-professional APCs to express MHC class II is provided by the cytokine **interferon-γ** (IFN-γ). For example, the enterocytes lining the intestine may express MHC class II and may act as functional APCs. Canine enterocytes normally express MHC class II on their basolateral surfaces, while such expression is induced on feline enterocytes in the presence of intestinal inflammation or neoplasia (**Figure 5.18**). As these non-professional APCs are located in non-lymphoid tissues and do not express costimulatory molecules, they are unlikely to activate naïve T cells, and instead may only interact with effector T cells, although their role in such immune responses is uncertain.

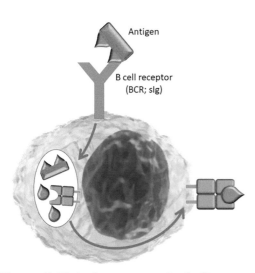

Figure 5.17 Antigen presentation by B cells. Antigen is bound by the BCR (surface immunoglobulin) and internalized within a cytoplasmic endosome for enzymatic digestion. The MHC class II pathway is employed to allow presentation of peptide fragments of this antigen on the surface of the cell, to allow engagement with a CD4+ T cell.

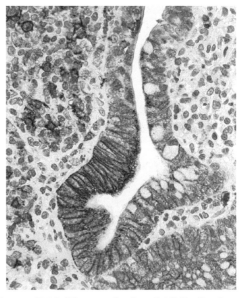

Figure 5.18 Non-professional APCs. Section of dog small intestine showing the basal region of a crypt. The enterocytes lining this structure express MHC class II molecules. It is suggested that these cells are able to present antigen to underlying T cells. Professional APCs expressing membrane MHC class II are also present within the adjacent lamina propria.

5.7 GENETICS AND INHERITANCE OF THE MHC

The **major histocompatibility complex** refers to a region of the genome containing a cluster of immune response genes, including those encoding the antigen-presenting molecules, MHC class I and class II. Every vertebrate species has an MHC, although it is located in different chromosomal regions in different species. Since histocompatibility antigens were initially characterized on leucocytes, the MHC is also known as the **leucocyte antigen system** and this nomenclature is used to describe the MHC of different species, that is, the human leucocyte antigen (HLA) system, dog leucocyte antigen (DLA) system, feline leucocyte antigen (FLA) system, equine leucocyte antigen (ELA) system and so on.

The HLA and murine (H2) systems are best characterized and genetic maps showing the precise arrangement of the genes within the complex are available (**Figure 5.19**). Maps of animal MHC regions are now emerging and reveal a high level of conservation, but with some species differences. There are three regions of the MHC described in most species. The centromeric **class II region** encodes the MHC class II α and β chains. These genes are described as 'D region' genes and are named by the suffix 'D'. For example, in the dog, the key class II genes

recognized to date are *DLA-DRA1*, which is monomorphic, and the polymorphic loci *DLA-DRB1*, *DLA-DQA1* and *DLA-DQB1*. There are two further non-functional pseudogenes in this area, *DLA-DQB2* and *DLA-DRB2*. Of comparative interest is that the cat is relatively unique in having a deletion of the DQ region. The telomeric **class I region** encodes the class I genes, which in humans are known as A, B and C. In contrast, there are four canine class I genes recognized, namely *DLA-12*, *DLA-64*, *DLA-79* and *DLA-88*, with the latter probably being the most important. Sandwiched between the class I and class II regions is a series of **class III region** genes. These genes do not encode classical histocompatibility antigens, but an array of other immunologically relevant molecules including some complement factors (C2, C4 and Factor B) and the cytokine tumour necrosis factor-alpha (*TNFA*). Interestingly, the *CYP21A2* gene, which encodes 21-hydroxylase, an important autoantigen in Addison's disease (hypoadrenocorticism), is located within the MHC class III region. Their inclusion within the MHC is again probably a quirk of evolution whereby a genetic recombination event led to incorporation of this set of genes in this chromosomal region. The presence of a number of class I and class II genes within the complex probably reflects a series of gene duplication events throughout evolution, mediated by pressure to

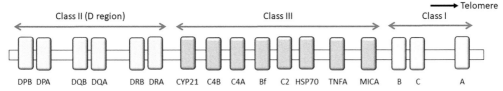

Figure 5.19 **The major histocompatibility complex.** A simplified genetic map of the human MHC is shown and the arrangement is likely to be somewhat similar in most mammalian species. The D region genes encode α and β chains of class II molecules, whereas the A, B and C genes encode class I molecules. Between these areas of the MHC are the class III genes that encode a range of immunological molecules not directly related to antigen presentation. The inclusion of this group of genes probably represents a genetic recombination event during evolution.

develop greater complexity within the immune system to allow for some degree of diversification in antigen presentation.

A unique feature of the canine (DLA) and feline (FLA) MHC is that the gene complex is split between two chromosomal areas. In the dog, class I, II and III genes are located on chromosome 12, with the remainder of the MHC on chromosome 35. Feline class I, II and III genes are on the long arm of chromosome B2, while the remainder of the complex lies on the short arm of chromosome 2. That part of the MHC on distant chromosomes includes a segment of DNA encoding some of the *TRIM* (tripartite motif) genes through to genes encoding olfactory receptors. This genetic event must have occurred before the evolutionary divergence of dogs and cats, 55 million years ago.

As we can see from **Figure 5.19**, there are multiple genes encoding MHC class I and multiple genes encoding MHC class II molecules. This is referred to as **polygeny**. The reason for this relates to our previous discussion of the structure of MHC molecules and their interaction with peptide epitopes during antigen presentation. In order to bind, the peptide needs to possess particular anchor residues to 'fit' the peptide-binding groove of the MHC molecule. If we consider MHC class I (and the same is true for MHC class II), if the gene encoding the alpha protein was monogenic and monomorphic, then every individual of the same species would express the same MHC class I molecule and the peptides presented to the immune system from a given antigen would be relatively restricted and the same for all individuals (**Figure 5.20a**). This is a risky strategy, since pathogens are adept at genetic mutation and it is feasible that a mutant virus would emerge whose antigens fail to express any peptides capable of binding to MHC, making the virus effectively 'invisible' from the adaptive immune system (**Figure 5.20b**) and a pandemic of virus infection would potentially lead to an extinction event. Clearly, a 'monogenic/monomorphic'

strategy for MHC genes is not a biologically sensible option in nature. However, selective breeding of some animals (e.g. commercial chicken breeds and some pedigree dogs) has led to generation of inbred populations that are relatively monomorphic. This can lead to increased susceptibility to infectious disease and can lead to outbreaks with high morbidity and mortality in that particular population.

This problem has been solved in evolutionary terms by allowing gene duplication to generate polygeny and since each new locus has subsequently diverged from its originator, this increases the number of different MHC molecules that can be expressed by an individual, thus increasing the peptide repertoire that can be presented (**Figure 5.20c**). In addition, each locus is highly **polymorphic** with a large number of different alleles in a population, each with their own specific peptide-binding repertoire (**Figure 5.20d**). Since the MHC is also **co-dominantly expressed**, and most individuals in an outbred population are heterozygous, this means that there is a huge diversity of MHC types within the species, each with their own peptide repertoire for antigen presentation. To give an approximate idea of this genetic diversity, for some HLA gene loci, more than 1,000 possible alleles are recognized. In the dog, the current degree of polymorphism recognized amongst class II genes is 161 *DLA-DRB1*, 30 *DLA-DQA1* and 79 *DLA-DQB1* alleles, as currently listed on the IPD-MHC Database (www.ebi.ac.uk/ipd/mhc/group/DLA/). Not all mammalian species follow this precise strategy of polygeny and polymorphism. In California sea lions (*Zalophus californianus*), for example, diversity is generated based on the use of more polygeny than polymorphism, with *DRB* comprising at least seven loci, but each with limited polymorphism.

Since the MHC class I and MHC class II genes are close together (i.e. they are in linkage disequilibrium), MHC genes tend to be inherited as a 'set', one from the paternal side

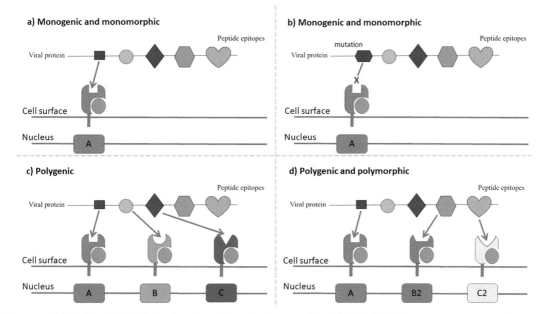

Figure 5.20 **The MHC is both polygenic and polymorphic.** (a) If the MHC were monogenic and monomorphic, only a limited repertoire of peptides that express specific anchor residues could be presented. (b) The monogenic/monomorphic strategy would mean susceptibility to pathogen mutation, whereby relatively few changes in the amino acid sequence of an antigen would render the host unable to present antigen to the adaptive immune system. (c) Polygeny increases the peptide repertoire, as each MHC molecule presents its own set of peptides. (d) Since each locus is polymorphic, there is individual variability in antigen presentation, dependent upon which MHC alleles are inherited.

and one from the maternal side. A 'set' of MHC genes is referred to as a **haplotype** and although there are large numbers of alleles, these are not inherited independently and there is a more restricted number of haplotypes, with some allelic combinations being more common than others. Therefore, in each individual human there are six MHC class I genes (an A, B and C on each chromosome) and each of these genes is expressed, such that any nucleated cell carries six different MHC class I molecules, each one of a single allotype. In the case of MHC class II genes, greater diversity is possible as different α and β chains may pair in the heterodimer to give distinct MHC class II molecules. Coupled with the knowledge that only one of a relatively large number of alleles is present at any one locus, it is possible to see that the range of MHC genes inherited by any one individual is almost unique and explains the difficulty in finding a suitable tissue match between individuals in a population, when one wishes to undertake organ transplantation.

5.8 MHC ASSOCIATIONS WITH IMMUNE FUNCTION AND DISEASE

Since the MHC is involved in positive and negative selection of T cells in the thymus and presentation of antigen to naïve T cells in secondary lymphoid tissues, it stands to reason that MHC genes will have a major influence on adaptive immunity. Since MHC genes are variable in a population, this will determine individual resistance to infection, the response to vaccination and susceptibility to immune-mediated disease.

In human medicine it was recognized many years ago that individuals who inherited particular MHC types appeared to have a greater risk of developing particular diseases associated with the immune system. There are numerous examples of HLA associations with susceptibility versus resistance to infectious disease. For example, HLA-DR2 seems to be associated to increased susceptibility to intracellular pathogens (such as leishmania and mycobacteria). In contrast, this particular HLA type is associated with resistance to autoimmune disease, including type 1 diabetes mellitus, whereas others, such as HLA-DR3 and HLA-DR4 are associated with an increased risk (particularly in those individuals that are heterozygous HLA-DR3/DR4). In such **disease association** studies, one would typically examine a population of affected individuals and a control group of clinically normal individuals. If a particular MHC allele or combination of alleles is found with higher frequency in the diseased population, then the statistical **odds ratio** can be calculated to indicate the relative risk of developing the disease, if an individual inherited that particular MHC type.

Since dogs develop a range of immune-mediated diseases, similar to those seen in humans, investigation of possible MHC–disease associations in this species has been ongoing since the 1970s. The ability to rapidly 'DLA type' large numbers of dogs by PCR and sequence-based typing methods has advanced this area of veterinary immunogenetic research. African wild dogs (*Lycaon pictus*) have very limited diversity for DLA-DQA1 and DLA-DQB1, probably as a result of population decline and which may contribute to their susceptibility to canine distemper virus infection. It is clear that selective breeding of pedigree dogs has also had a significant impact on DLA diversity. Some breeds (e.g. Labrador retrievers) seem to have retained a degree of diversity and many different DLA types are evident. However, for other breeds, the number of DLA types is dramati-

cally restricted. It is possible that this genetic bottleneck in immune response gene diversity goes some way to explaining the increased prevalence of immunological dysfunction and immune-mediated disease seen in some dog breeds. For example, an unusual DLA type is found in the Rottweiler breed, with around 25% of dogs being homozygous for this particular DLA haplotype. It is possible that this might explain the relatively poor vaccine response (e.g. to canine parvovirus and rabies) attributed to this breed, as a result of poor positive selection of T cells against this particular MHC type, or ineffective presentation of vaccine antigen, or both. In contrast, although cocker spaniels are similarly restricted in their DLA profile (albeit different DLA haplotypes compared with the Rottweiler) this seems to lead to an increased degree of hypersensitivity, with this breed being recognized for its increased prevalence of autoimmune disease.

Clear linkage has been shown between the inheritance of particular DLA haplotypes and the susceptibility or resistance to canine rheumatoid arthritis, immune-mediated haemolytic anaemia (IMHA), diabetes mellitus, lymphocytic thyroiditis, Addison's disease (hypoadrenocorticism), chronic hepatitis in Dobermanns, necrotizing meningoencephalitis, anal furunculosis, chronic superficial keratitis and leishmaniosis. To give one example, DLA types of a population of 107 German shepherd dogs affected with anal furunculosis were compared with 196 unaffected German shepherd dogs. Both groups had similar expression of the MHC class II allele *DRB1*00102*, but significantly more dogs in the disease group expressed the allele *DRB1*00101*, with a high odds ratio of 5.01 (**Figure 5.21**).

Similar immunogenetic studies are being conducted for diseases in other species. For example, MHC links are proposed for susceptibility to bovine leukaemia virus and feline coronavirus infections, Marek's disease in chickens, muscular dystrophy in sheep and

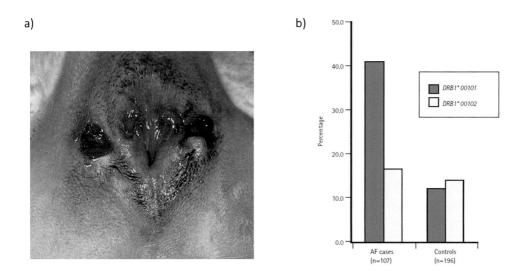

Figure 5.21 MHC disease association. (a) The perianal region of a German shepherd dog (GSD) affected with anal furunculosis. There is a region of ulceration and inflammation dorsal to the anus and bilaterally there are two ulcers that extend into the tissue as sinus tracts. GSDs have a particular predisposition for this disease. (b) This graph displays the prevalence of two MHC *DRB1* class II alleles in populations of affected and unaffected dogs. The overrepresentation of allele *DRB1*00101* in the diseased population is highly significant. Inheritance of this allele by a GSD confers an odds ratio of 5.01 for development of the disease. (Data from Kennedy LJ et al. (2007) Risk of anal furunculosis in German shepherd dogs is associated with the major histocompatibility complex. *Tissue Antigens* 71:51–56.)

allergic disease, sarcoids and equine recurrent uveitis in the horse. Another role for MHC typing in veterinary medicine relates to the **identification** of individual animals. MHC type provides an unalterable means of identifying an animal for registration purposes and a means of determining the paternity of specific offspring. Indeed, MHC typing has already been used in legal cases where the blood line of pedigree dogs was at question. Of greater practical relevance is that MHC type is not simply linked to the development of disease but, for domestic livestock, may be associated with **production traits** such as meat or milk production or resistance to infection (such as bovine tuberculosis). Selection of breeding stock on the basis of MHC type might be used to enhance health and economic output in their progeny. Examples of such associations include linkage between MHC type and resistance of cattle to *Boophilus* ticks and infection by *Theileria*, resis-

tance to intestinal endoparasitism in sheep, egg production in chickens and a range of parameters (piglet mortality, sow fertility, litter size, growth rate and carcass characteristics) in pigs. Recent intriguing studies have also shown that MHC type may be involved in the **selection of breeding partners** and be related to specific olfactory stimuli that indicate an optimal (different) MHC background in a potential mate, preferentially selecting for heterozygosity in the offspring. These studies have demonstrated that small peptides bound to MHC class I molecules are excreted in the urine. The peptides become detached from the MHC molecules and bind to receptors expressed by olfactory neurons within the vomeronasal organ. Several pheromone olfactory receptors are also encoded within the class I region of the MHC of some species, suggesting that MHC type may influence odour recognition and mate selection (even in humans).

5.9   MHC AND TRANSPLANTATION/ TISSUE GRAFT REJECTION

Transplantation of organs (e.g. kidney, liver, heart, heart and lungs) or cells (e.g. bone marrow) has had a huge impact on medical science over the past few decades. However, the success of these procedures depends upon matching donor and recipient **tissue type** (i.e. MHC alleles). Transplantation may take one of several forms. **Autografts** (e.g. transplantation of skin from one area of the body to another) and **isografts** (transplantation between genetically identical individuals; i.e. monozygotic twins) are generally very successful, as there is perfect matching between the donor and recipient. The most widely practised form of transplantation is **allografting** in which cells or tissue from a closely matched, but still genetically dissimilar, donor are transplanted into a recipient. The success of allografting is increased by obtaining the closest possible MHC match between the tissues of donor and recipient. It is for this reason that national or international registers are kept, allowing rapid deployment of matched organs to those individuals in need as they become available. Rejection of allografts may also be controlled by use of immunosuppressive drugs. The use of **xenografts** (transplantation of cells or tissue from one species to another) has been widely researched, particularly the potential for the use of pigs as donors. Despite generation of transgenic pigs to facilitate xenotransplantation, reservations about cross-species transmission of uncharacterized infectious agents (e.g. endogenous retroviruses) have slowed translation of this science into clinical practice. Since the introduction of gene editing technology, transgenic pigs have now been produced that have had endogenous retrovirus sequences removed, bringing us closer to routine use of pig-to-human organ transplants.

Although much of the basis for human transplantation surgery was first defined experimentally in porcine and canine models, this procedure has limited application in modern veterinary medicine. Canine and feline kidney transplantation is carried out in some US centres and bone marrow transplantation has been performed in individual cases. However, ethical concerns have limited such practices in most other countries. Despite these reservations, an understanding of the process of graft rejection is important, as this is an immunological event that has bearing on knowledge of fundamental aspects of immunology.

The basis for graft rejection is summarized in **Figure 5.22**. Any tissue or organ will contain a resident population of **'passenger' leucocytes** of donor origin that are very difficult to remove entirely from that tissue. After transplantation, donor leucocytes may migrate out of the donor tissue into afferent lymphatics and thus to regional lymph nodes. Either these leucocytes or recipient APCs that have taken up shed donor antigen will activate alloreactive lymphocytes within that lymphoid tissue. Activated **alloreactive T and B lymphocytes** leave the lymph node in efferent lymph and home to the site of the transplanted tissue, where they may mediate graft rejection. A variety of immunological factors, including antibody, complement, cytotoxic T cells and NK cells, are involved in this process. The alloreactive immune response may target the graft vasculature (leading to **ischaemic necrosis** of the graft) or result in **cytotoxic destruction** of grafted cells. Graft rejection may be **peracute** (in an immunologically sensitized individual), **acute** (days to weeks) or **chronic** (months to years) in nature.

In a **recipient who is highly immunosuppressed** it is also possible for the reverse scenario to occur, that is, the donor leucocytes may attack the tissue of the recipient in a process known as **graft-versus-host disease (GVHD)**. GVHD may arise when donor lymphocytes in transplanted bone marrow infiltrate a range of recipient tissues. GVHD is recognized in some immunosuppressed dogs following bone marrow transplantation, where it manifests clinically as erythematous and exudative skin lesions, mucous membrane inflammation, jaundice, diarrhoea and Coombs-positive haemolytic anaemia.

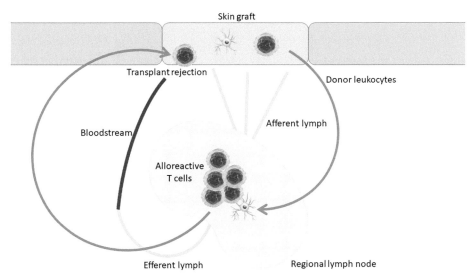

Figure 5.22 Graft rejection. Passenger leucocytes within a graft or donor histocompatibility antigens enter afferent lymphatics and travel to regional lymphoid tissue where alloreactive lymphocytes are activated. These cells home to the site of the graft and mediate graft rejection by attacking the vascular supply to the graft tissue (causing ischaemic necrosis) or infiltrating the graft, leading to cytotoxic destruction of graft cells.

KEY POINTS

- Major histocompatibility complex (MHC) molecules interact with the T cell receptor (TCR) in the 'immunological synapse'.
- MHC class I molecules comprise an α chain and β_2 microglobulin and are found on all nucleated cells of the body.
- MHC class II molecules comprise an α and β chain and are mostly expressed by antigen presenting cells.
- MHC molecules have a peptide-binding groove, which allows attachment of peptides (eight to ten amino acids in length for MHC class I and 13–25 amino acids in length for MHC class II), provided they express the correct anchor residues.
- Endogenous (cytoplasmic) antigens are degraded by the proteasome and transported into the endoplasmic reticulum by TAP transporters, where they encounter (and can bind to) nascent MHC class I molecules.
- MHC class I-peptide complexes travel via the Golgi to the cell surface where they can interact with the TCR on CD8+ T cells.
- MHC class II molecules are synthesized and bind to an invariant chain that acts as a chaperone protein and also blocks binding of endogeneous peptides in the endoplasmic reticulum.
- MHC class II molecules travel via the Golgi apparatus to enter a specialist vesicular compartment termed the MIIC compartment, where the invariant chain is degraded, leaving the CLIP in the peptide-binding groove.
- Antigen is taken up via phagocytosis, pinocytosis or receptor-mediated endocytosis and enters the endosomal pathway, where antigen is degraded by the action of lysosomal enzymes.

- The endosomes fuse with the MIIC, allowing foreign peptides to displace the CLIP, whereupon the MHC class II-peptide complexes traffic to the cell surface where they can interact with the TCR on CD4$^+$ T cells.
- Cross-presentation is the presentation of peptides derived from extracellular antigens via MHC I molecules and allows activation of CD8$^+$ T cells by dendritic cells.
- CD1 molecules present lipid antigens to T cells and NKT cells.
- Dendritic cells are important professional antigen presenting cells (APCs) that express both MHC class I and class II molecules and traffic to secondary lymphoid tissues where they can activate naïve T cells.
- Macrophages and B lymphocytes are important APCs and present antigen via MHC class II in order to receive T cell help.
- Some cell types can be stimulated by cytokines to become non-professional APCs and present antigen.
- The major histocompatibility complex comprises class I, II and III gene regions.
- The MHC is polygenic and MHC genes are the most highly polymorphic in the genome.
- Due to polymorphism, MHC haplotype is almost unique in individuals of an outbred population.
- MHC haplotypes are typically inherited as a set of alleles.
- There are associations between particular MHC alleles and the susceptibility or resistance of immune-mediated or infectious disease in man and other animals.
- In veterinary medicine MHC typing may be used for identification purposes and can be associated with production traits in livestock.
- Major histocompatibility antigens mediate graft rejection via the activation of alloreactive T and B lymphocytes in the recipient.
- Graft-vs-host disease (GVHD) may occur in highly immunosuppressed recipients of graft tissue containing viable donor leucocytes.

CYTOKINES, CHEMOKINES AND THEIR RECEPTORS

OBJECTIVES

At the end of this chapter, you should be able to:

- Understand that the immune system requires cytokines in order for cells to communicate with each other.
- Appreciate that cytokines can act via an autocrine, paracrine and/or endocrine manner.
- Understand that there are families of chemokines that are categorized depending upon the organization of cysteine residues, forming disulphide bonds within the protein.
- Describe how chemokines are responsible for chemotaxis, ensuring that leucocytes are targeted and migrate to the site of infection.
- Understand that chemokine receptors are G-protein coupled receptors that convert an extracellular signal to an intracellular response, leading to changes in cellular function.
- Explain how colony-stimulating factors affect production of white blood cells by the bone marrow.
- Describe the role of type I interferons in the antiviral response.
- Explain why interferon-gamma is important for macrophage function.
- Appreciate the diversity of interleukins.
- Describe the JAK/STAT pathway of signalling from interleukin receptors.
- Appreciate how a knowledge of cytokines and cytokine receptors has led to advances in veterinary clinical practice.
- Describe the role of TNF family members and their receptors in the inflammatory response and inducing apoptosis.

6.1 INTRODUCTION

This chapter examines immune cell communication, between each other and with other cells of the host. Cytokines and chemokines are 'immunological hormones' that act in an autocrine, paracrine or endocrine manner (**Figure 6.1**). These molecules are important in orchestrating immune responses, ensuring that cells migrate to where they are needed and perform their function as required. Cytokine and chemokine receptors are necessary for transmitting an extracellular signal into an intracellular response, whereby the biological activity of the target cell is altered in a way that benefits the host. Some pathogens can subvert the

DOI: 10.1201/9781003310969-6

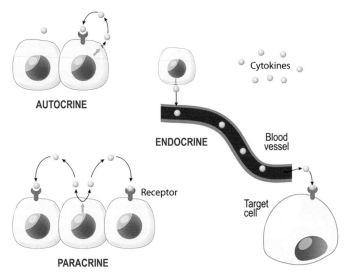

Figure 6.1 **How immune cells communicate.** A cytokine is a soluble messenger protein released by one cell to bind to a specific cytokine receptor expressed by the same cell (autocrine), a nearby cell (paracrine) or a distant cell (endocrine). Binding of cytokine to receptor activates intracellular signalling pathways, and this results in functional changes to the target cell.

immune system by producing 'fake' cytokines or cytokine receptors, which benefits the pathogen to the detriment of the host.

6.2   CHEMOKINES AND CHEMOKINE RECEPTORS

Chemokines are a family of small chemoattractant molecules that are involved in recruitment of leucocytes from blood and directing their migration in tissues, targeting immune cells to the site of infection. Chemokines are grouped according to their structure, in particular, the arrangement of cysteine residues that form disulphide bonds generating their tertiary structure (**Figure 6.2**). Additionally, they are given the suffix 'L' or 'R' indicating their role as a ligand or a receptor, respectively. The CCL family is the largest and contains various chemokines that are important in migration of cells of the monocyte lineage (i.e. macrophages and dendritic cells). These include CCL2, also known as monocyte chemotactic protein-1 (MCP-1); CCL3, also known as macrophage inflammatory protein-1 alpha (MIP-1α); and CCL4, also known as MIP-1β. In addition, there are a number of CCL molecules, known as **eotaxins**, that are involved in recruitment of eosinophils to the sites of helminth parasite infection and in pathological recruitment of eosinophils to tissues in allergic disease (**Figure 6.3**).

Chemokine receptors belong to the G-protein coupled receptor family. Binding of the ligand generates an intracellular response, such that an attached guanosine diphosphate (GDP) is phosphorylated, leading to release of the Gsα subunit, which subsequently interacts with membrane-bound adenylate cyclase, the enzyme that converts ATP to cyclic AMP. The cAMP second messenger activates protein kinase A, which can phosphorylate a number of cytoplasmic proteins, including enzymes and transcription factors, leading to a change in the biological response of the target cell (**Figure 6.4**).

6.3   CYTOKINES AND CYTOKINE RECEPTORS

Cytokines are highly potent molecules that function to orchestrate the biological responses of immune cells and other cells in the body, both locally (autocrine and paracrine effects)

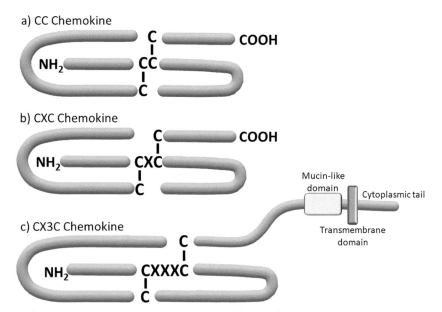

Figure 6.2 **The family of chemokines are categorized according to their structure.** The way that cysteine residues are positioned to form disulphide bonds is used to place chemokines into various families.

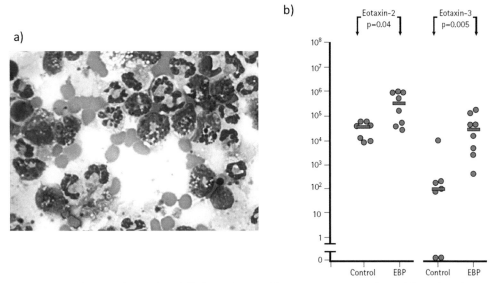

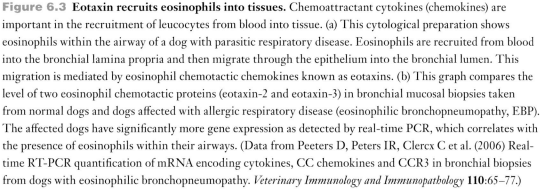

Figure 6.3 **Eotaxin recruits eosinophils into tissues.** Chemoattractant cytokines (chemokines) are important in the recruitment of leucocytes from blood into tissue. (a) This cytological preparation shows eosinophils within the airway of a dog with parasitic respiratory disease. Eosinophils are recruited from blood into the bronchial lamina propria and then migrate through the epithelium into the bronchial lumen. This migration is mediated by eosinophil chemotactic chemokines known as eotaxins. (b) This graph compares the level of two eosinophil chemotactic proteins (eotaxin-2 and eotaxin-3) in bronchial mucosal biopsies taken from normal dogs and dogs affected with allergic respiratory disease (eosinophilic bronchopneumopathy, EBP). The affected dogs have significantly more gene expression as detected by real-time PCR, which correlates with the presence of eosinophils within their airways. (Data from Peeters D, Peters IR, Clercx C et al. (2006) Real-time RT-PCR quantification of mRNA encoding cytokines, CC chemokines and CCR3 in bronchial biopsies from dogs with eosinophilic bronchopneumopathy. *Veterinary Immunology and Immunopathology* **110**:65–77.)

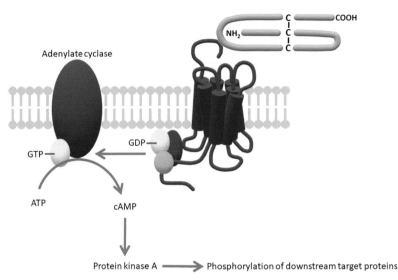

Figure 6.4 Chemokines exert their effect through G-protein coupled chemokine receptors. Ligand binding generates a conformational change in the receptor following conversion of GDP to GTP. The activated Gsα subunit (yellow) stimulates adenylate cyclase to generate the second messenger cAMP, which activates protein kinase A and subsequent phosphorylation of cytoplasmic proteins leads to a change in cellular biological function.

or at a distance (endocrine effect). The binding of a cytokine to its receptor generates signals that induce the target cell to perform a specific action. Cytokines are therefore one of the key mechanisms of **communication between cells** of the immune system. Any one cell may produce multiple cytokines, and multiple cytokines may act on any one target cell, providing it expresses the appropriate receptors. Any one cytokine may have multiple actions (pleiotropy) depending on the immunological circumstances and the nature of the target cell to which it binds. Cytokines often work in a **network**, whereby a series of cytokines has a concerted effect with a common end goal or, alternatively, where some cytokines antagonize the effects of others. Several **families of cytokines** are recognized, and many of these molecules will be considered further as we progress through the rest of this book. Some cytokines are important in bone marrow haematopoiesis and in the development of T and B lymphocytes, others are involved in the activation or suppression of immune cells and some are considered pro-inflammatory in their effects.

6.3.1   Colony-Stimulating Factors

This family of cytokines are typically released by cells (particularly macrophages) at the site of infection and have an endocrine effect on bone marrow function, stimulating production and release of white blood cells. Granulocyte colony-stimulating factor (G-CSF) and granulocyte macrophage colony-stimulating factor (GM-CSF) are responsible for the neutrophilia and monocytosis seen during a bacterial infection. A modified version of G-CSF is commercially available (pegbovigrastim; Imrestor™, Elanco) for administration to dairy cows prior to calving, designed to boost circulating neutrophil numbers during the peri-parturient period, potentially reducing the risk of clinical mastitis.

6.3.2   Interferons and Their Receptors

There are three main families of interferons: type I, type II and type III. Type I interferons (including IFN-α, IFN-β and IFN-ω) and their signalling pathway were described in Chapter 3 as part of the innate immune response against viruses. Large quantities of type I interferons are produced by plasmacytoid dendritic cells that express a number of PRRs to detect viral nucleic acid (e.g. TLR-3, TLR-7 and TLR-9). Interferon-ω (Virbagen® Omega, Virbac) is commercially available for antiviral therapy of companion animals. It may slow down viral replication and reduce the viral load and has been used to treat cats with persistent virus infections such as FeLV, FIV and FIP. Type III interferons (IFN-λ) have a similar function to type I interferons, but their receptors and antiviral activity are restricted to epithelial cells.

Interferon-gamma (IFN-γ) is a type II interferon and is primarily produced by T cells (particularly CD4+ T helper type 1 cells) and NK cells. Its major action is on macrophages, which are stimulated to enhance their antimicrobial responses (e.g. increase nitric oxide production via upregulation of the induced nitric oxide synthase; iNOS) and upregulate antigen processing and presentation. This is particularly important in the immune response to vesicular pathogens (e.g. mycobacteria) that can survive inside endosomes following phagocytosis. Although recombinant IFN-γ has been used experimentally, this has not yet translated into veterinary clinical practice, where (theoretically) there might be a benefit, such as in the case of leishmaniosis, where a relative deficiency in production of this cytokine is associated with a worse prognosis.

6.3.3   Interleukins and Their Receptors

This is a large family of cytokines, which is constantly expanding as more members are discovered. Some interleukins are involved in the local inflammatory response, such as IL-1β and IL-6, which also stimulate the acute phase response (see Chapter 3). Interleukin 8 is misclassified as an 'interleukin' as it is actually a chemokine (CXCL8). Some interleukins, such as IL-3 and IL-7, are important for the development of lymphocytes, and their receptors are expressed in primary lymphoid tissues on stem cells and lymphoid precursors. Other cytokines are produced primarily by lymphocytes and have an important autocrine and paracrine function in

secondary lymphoid tissues and at the site of infection. We will focus on specific interleukins as we progress through the book and focus on different aspects of the adaptive immune response, but in this section, we will discuss a few examples to illustrate important concepts of this cytokine family.

Interleukin 2 (IL-2) is produced by naïve T cells in secondary lymphoid tissues in response to antigenic stimulation. These cells additionally upregulate their expression of IL-2 receptor proteins, such that the cytokine can act in an autocrine and paracrine manner (**Figure 6.5**). Ligation of the IL-2R is the driving force for proliferation during the 'clonal expansion' phase in secondary lymphoid tissues, leading to generation of large numbers of antigen-specific T cell clones (see **Figure 1.7**).

Oncept IL-2 (Boehringer Ingelheim) is an immunotherapeutic for feline fibrosarcoma. It consists of a canarypox vector, engineered to carry the feline IL-2 coding sequence. It is licensed as an adjunct to surgery or radiotherapy in cats affected with fibrosarcoma. Specifically, it can be injected into the local tissues, following resection of the tumour, which reduces the risk of local recurrence and/or metastasis. The likely mechanism of action is that, following expression of recombinant IL-2 at the injection site, this enhances activation of tumour-specific T cells, which aid in immune-mediated destruction of residual cancer cells.

Canine X-linked severe combined immune deficiency (SCID) can occur in Basset hounds and Cardigan Welsh Corgi dogs. In both cases, a mutation (although different in the two breeds) in the gene encoding the IL-2Rγ chain leads to a truncated, dysfunctional receptor, incapable of responding to IL-2. Thus, the T cells in affected dogs are unable to undergo clonal expansion in response to foreign antigen, and these dogs have a severely restricted adaptive immune response. The resultant immunodeficiency typically leads to death from infection in affected male puppies (see also Chapter 16).

One other cytokine/receptor interaction that is relevant to veterinary medicine is that

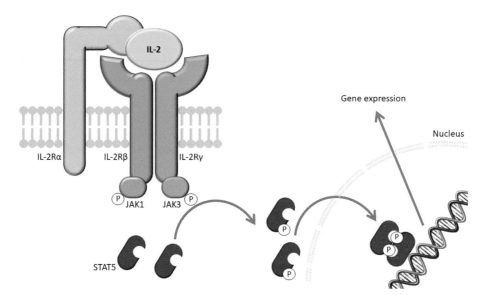

Figure 6.5 Interleukin 2 receptor signalling. The IL-2 receptor consists of three proteins that interact for binding to the cytokine. The IL-2Rα chain (also known as CD25) is upregulated on naïve T cells in response to antigenic stimulation and comes together with the β and γ chains to create the high-affinity IL-2 receptor. Cytokine binding stimulates the Janus kinases (JAKs) to become phosphorylated, which in turn phosphorylate signal transducer and activator of transcription proteins (STATs). These subsequently form dimers and migrate into the nucleus where they act as transcription factors, activating specific immune response genes through binding to their promoter elements.

between IL-31 and the IL-31R. Interleukin 31 can be secreted by lymphocytes (particularly CD4+ T helper type 2 cells) and acts on its cognate receptor expressed on sensory nerves. This particular cytokine seems to play an important role in driving the 'itch/scratch cycle' in dogs affected with allergic skin disease. Two commercial products are available that interfere with this mechanism. The first works by blocking the phosphorylation of specific JAKs (oclacitinib; Apoquel™, Zoetis), thus preventing intracellular signalling following binding of IL-31 to its receptor. The other is a monoclonal antibody (lokivetmab, Cytopoint™, Zoetis) that binds to and neutralizes the cytokine, so that it is unable to bind to its receptor. Both of these products illustrate how our knowledge of cytokines and their receptors potentially allows these to be manipulated, when necessary. However, since many cytokines have pleiotropic

effects, administration of agonists (e.g. recombinant cytokines) or antagonists (neutralizing antibodies or receptor blockers) into this network can potentially have unwanted or unexpected consequences.

6.3.4 Tumour Necrosis Factor Family and Their Receptors

In contrast with previous cytokines, members of this family are relatively cytotoxic, capable of inducing apoptosis in a selective manner. This is particularly important in the immune response against infected cells and those that have undergone malignant transformation (as indicated by their name). Tumour necrosis factor (TNF, also known as TNF-alpha) is one of the most important cytokines of this family and is a pro-inflammatory cytokine secreted by macrophages, NK cells and some T cells. It is particularly important for stimulating vascular

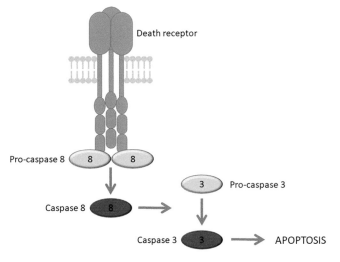

Figure 6.6 TNF receptor family members are 'death receptors'. TNF receptors, such as FAS form a trimer which associates with pro-caspase 8. Upon ligand binding, pro-caspase 8 is converted to its active state, which is released to further activate downstream 'executioner' caspases (such as caspase 3) and the cell subsequently undergoes apoptosis.

endothelial cells to upregulate their cell-adhesion molecules, thereby recruiting circulating leucocytes to the site of infection. Overproduction of TNF and other inflammatory cytokines such as IL-1 and IL-6 can be detrimental (the so-called **cytokine storm**) and may lead to toxic shock syndrome. Polymorphisms in the *TNF* gene promoter might influence the amount of this cytokine produced under certain circumstances, which could be detrimental in some individuals, if an exaggerated response is elicited.

Members of the TNF receptor family can activate the extrinsic pathway of apoptosis, whereby an extracellular signal triggers the caspase cascade (**Figure 6.6**). They are therefore termed **death receptors** and while some function by binding to a secreted cytokine ligand (e.g. lymphotoxin, released by CD8⁺ cytotoxic T cells), others interact with membrane-bound ligands (e.g. FAS ligand interaction with the death receptor FAS). There are a number of TNF-related apoptosis inducing ligands (TRAIL) that can interact with functional or decoy receptors (the latter protecting healthy cells from undergoing apoptosis accidentally).

Although monoclonal anti-TNF antibody therapy has been used in human medicine since 1998 (e.g. for the treatment of rheumatoid arthritis), this strategy has not yet translated into veterinary medicine. For this to happen, this would likely require species-specific monoclonal antibodies to be developed, owing to the lack of cross-reactivity with the human TNF antibodies and a requirement for 'caninizing' the immunoglobulin molecule.

KEY POINTS

- Cytokines are an important means of communication between cells of the immune system and between immune cells and other host cells (via autocrine, paracrine and endocrine effects).
- A cytokine binds to its specific receptor, leading to altered function of the target cell.

- Chemokines recruit leucocytes from blood into tissue and facilitate chemotaxis towards the site of infection.
- Chemokines are grouped and named according to their structure.
- Chemokine receptors are G-protein coupled receptors.
- Colony-stimulating factors are produced at the site of infection and have an endocrine effect on the bone marrow, increasing production and release of white blood cells.
- Interferons alpha and beta are released as part of the innate immune response against viruses and act to slow down their replication.
- Interferon gamma is released by T cells and functions to stimulate macrophages, which is particularly important in the response against vesicular pathogens.
- Interleukin 2 is an important cytokine for clonal expansion of T cells in response to antigenic stimulation in secondary lymphoid organs.
- The interleukin 2 receptor (and other interleukin receptors) signals via the JAK/STAT pathway.
- Immunomodulators can be used in clinical practice that act as agonists (recombinant cytokines) or antagonists (JAK inhibitors and monoclonal antibodies).
- TNF is an important inflammatory cytokine.
- TNF receptor family members typically act as death receptors, inducing apoptosis in cells targeted by their ligands, which may be secreted or cell-surface expressed.

OBJECTIVES

At the end of this chapter, you should be able to:

- List the major molecules expressed on the surface of T lymphocytes and describe their function.
- Explain the three signals required for activation of naïve T cells.
- Understand the concept of clonal proliferation and differentiation as it applies to T cells.
- Describe the functional differences between CD4+ and CD8+ T cells.
- Discuss the functional differences between different subsets of CD4+ T cells.
- Describe how target cells are recognized and killed by CD8+ cytotoxic T cells.
- Define what is meant by T cell memory.

7.1   INTRODUCTION

Following positive and negative selection in the thymus, naïve T cells are released into the circulation to perform their role in immune surveillance. They travel via the blood and lymphatic system in search of foreign antigen in secondary lymphoid tissues. Activation of antigen-specific T lymphocytes is pivotal to the generation of an adaptive immune response. In this chapter we consider the nature of the T cell receptor (TCR), T cell activation and the roles of different T lymphocyte effector subsets in the response to infection.

7.2   THE T CELL RECEPTOR AND T CELL ACTIVATION

T lymphocytes are distinguished by their expression of a **TCR** and associated **CD3 complex**, as well as one of two major co-receptors, either CD4 (expressed by **helper T lymphocytes**; Th) or CD8 (expressed by **cytotoxic T lymphocytes**; CTLs).

The TCR of conventional T cells consists of an α and β chain, each of which has outermost variable domains and constant region domains closest to the cell membrane, into which the proteins anchor. This core receptor molecule is surrounded by a number of signal transduction molecules (δ, ε, γ and ζ chains) that collectively form the **CD3 complex** and which express 'immunoreceptor tyrosine-based activation motifs' (ITAMs) that become phosphorylated when in an activated state. The CD4 molecule consists of a single elongated chain, while CD8 may be composed of a shorter α and β chain or an alternative form consisting of two α chains (the CD8 homodimer) (**Figure 7.1**). Another population of T cells expresses the TCRγδ.

DOI: 10.1201/9781003310969-7

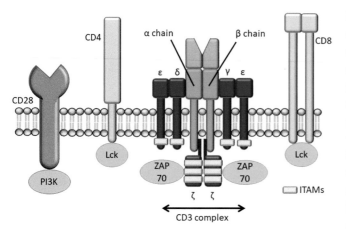

Figure 7.1 **The T cell receptor associates with the CD3 complex, co-receptors and several signalling molecules.** The TCR recognizes MHC-peptide complexes and requires the CD3 molecules to translate the recognition of peptide into an intracellular signal. The various elements of the CD3 complex have intracellular 'immunoreceptor tyrosine-based activation motifs' (ITAMs), which, when phosphorylated by the Lck molecule associated with the co-receptor (either CD4 or CD8), will allow binding of ZAP-70. CD28 is an important co-stimulatory molecule and associates with PI3 kinase. Note that naïve T cells will express either CD4 or CD8, not both.

These cells were discussed in Chapter 3 and will not be further considered here.

One key concept in immunology is that the TCR of **CD4+ T cells** is only allowed to view peptides presented by **MHC class II molecules**, whereas the TCR of **CD8+ T cells** is only allowed to view peptides presented by **MHC class I molecules**. This specificity is determined by binding specificity of these co-receptors (CD4 vs CD8) to their respective MHC molecules. This allows the TCR to engage with contact residues on the peptide as well as with the surface of the surrounding MHC molecule, a phenomenon known as **MHC restriction**. Thus, this represents the opposite side of the 'immunological synapse' that was discussed in Chapter 5 (see **Figures 5.7, 5.11**).

The interaction between the TCR (on the T cell) and MHC (on an APC) occurs within secondary lymphoid tissues, typically within the paracortex of a lymph node. The APC (generally a dendritic cell) has carried antigen from the tissues and has processed it for presentation on MHC molecules. T cells within the recirculating pool continually enter the paracortex from the blood (via high endothelial venules). As the dendritic cells take up residence in the paracortex, presenting antigen, large numbers of T cells will approach them and screen the peptides to see whether their TCR recognizes any of those presented to them. Using video microscopy it has been estimated that up to 500 different T cells may approach any one dendritic cell each hour.

Naïve T cells require two specific signals to become fully activated to start proliferating and a third signal is required to drive their differentiation (**Figure 7.2**). The first of these signals (**signal 1**) is delivered via binding of the TCR (and co-receptor; CD4/CD8) to the MHC-peptide complex. The second signal (**signal 2**) requires an interaction between the T cell co-stimulatory molecule, **CD28** and either **CD80** or **CD86** on the surface of the APC. CD80/CD86 can be thought of as **warning signals** and are upregulated on APCs (particularly dendritic cells) that have come from an environment where these cells have sensed (using their PRRs) the presence of microbes (via their PAMPs) or cellular distress or death (via DAMPs). Without signal 2, ligation of the TCR is insufficient on its own to stimulate activation, with presentation of 'harmless' antigen (no PAMPs/DAMPs) typically leading to a form of tolerance, known as **anergy** (see Chapter 14).

When the TCR and CD28 have been engaged, this allows an intracellular signal to be

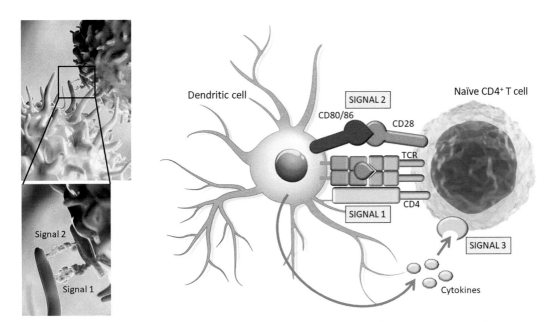

Figure 7.2 Activation of naïve T cells requires multiple signals. T cell activation requires three signalling events: the recognition of peptide–MHC by the TCR (signal 1), the interaction of a 'danger signal' and co-stimulatory molecules (signal 2) and the delivery of APC-derived cytokines to a T cell cytokine receptor (signal 3).

generated, initially by phosphorylation of the ITAMs and subsequently recruitment of additional signalling proteins (such as ZAP-70, IP3 kinase and phospholipase c gamma; PLC-γ), Together, these generate an **increase in intracellular Ca^{2+}** as a second messenger. The Ca^{2+} binds to and allows interaction between two important intracellular proteins, **calmodulin and calcineurin**. These are capable of de-phosphorylating the **nuclear factor of T cells (NF-AT)**, which is the main transcription factor that upregulates gene expression of IL-2 and the IL-2Rα chain (**Figure 7.3**). It is these molecules that stimulate cell proliferation and clonal expansion of antigen-reactive T cells in the secondary lymphoid tissues. The mechanism of action of the immunosuppressive drugs ciclosporin and tacrolimus is to inhibit calcineurin, thus effectively blocking the T cell signalling cascade and preventing activation of NF-AT.

A third signal (**signal 3**) is required to enable proliferating T cells to differentiate into their final effector phenotype. This takes the form of specific **cytokines** that are released by the APC or other innate immune cells, which act in a paracrine manner, binding to the respective cytokine receptor expressed by the T cell. A range of different cytokines can fulfil this role and the specific cytokine environment present during this stage of T cell differentiation is determined by the nature of the stimulating antigen and the type of immune response required (see more on this later). Only when all three signals have been received can T cells become fully operational. Most of the cells will become '**effector**' T cells, which participate in the immune response, although some will become a specialized population of '**memory**' **T cells**, which can be thought of as 'reserve troops', retaining the immunological memory of this activation event (**Figure 7.4**).

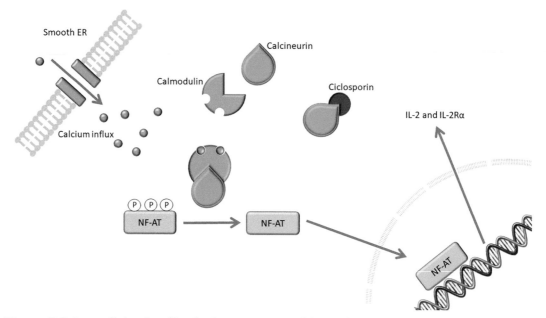

Figure 7.3 Intracellular signalling leads to activation of the nuclear factor of activated T cells (NF-AT), an important transcription factor. Following cell signalling events at the T cell surface, the intracellular calcium concentration rises. This facilitates formation of a calmodulin/calcineurin complex that dephosphorylates NF-AT to an active state. NF-AT migrates into the nucleus, where it acts as a transcription factor, stimulating transcription of key genes including IL-2 and the IL-2Rα chain. Ciclosporin is a calcineurin inhibitor that is used clinically as an immunosuppressive drug, via its action in blocking T cell activation.

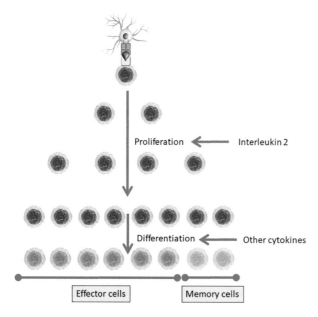

Figure 7.4 T cell clonal proliferation and differentiation. Activation of the T cell initiates clonal proliferation to produce large numbers of antigen-specific effector T cells and a smaller population of memory T cells.

7.3 THE ROLE OF CD4+ T LYMPHOCYTES

CD4$^+$ T cells are also termed 'helper' T cells (Th) and their major function is in providing help (via production of cytokines) for the generation of antibody production (**humoral immunity**) or enhancing macrophage function (**cell-mediated immunity**; CMI). In addition, as will become evident in Chapter 15, these cells also have a key role to play in **regulation and suppression** of the immune response. There are several subsets of CD4$^+$ T cells (helper and regulatory) that have been recognized, representing different differentiation pathways of naïve CD4$^+$ T cells to an effector phenotype that best suits the needs of the immune system (**Figure 7.5**).

Following a period of proliferation, CD4$^+$ T cells initially remain uncommitted (often referred to as Th0) and are dependent upon cytokine signals (typically from the dendritic cells but also from other cells of innate immunity) to determine which subtype of effector cell best suits that particular situation. The strongest influence would appear to be the **nature of the antigen** and the way in which this interacts with the APC (via PAMPs and PRRs), which in turn modifies how that APC will subsequently interact with the naïve CD4$^+$ T cells when it reaches the lymphoid tissues. The most important determining factors are the cytokines released by the APC to provide signal 3. The cytokine environment in which the antigen-specific T cells find themselves will determine which receptors are active and which intracellular signals (STATs) are entering the nucleus (**Figure 7.6**). Other influences in this decision-making process may include the nature of the APC (e.g. dendritic cell, macrophage or B cell),

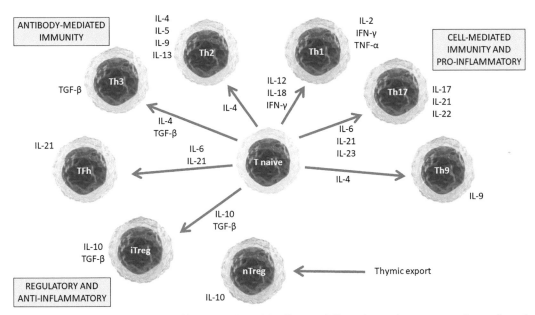

Figure 7.5 Subsets of CD4+ T cells. Naïve CD4+ T cells may differentiate to become one of a number of different effector T cells, dependent upon the cytokine environment (signal 3; indicated adjacent to the arrows). Effector T cell subsets are characterized by their production of a specific cytokine signature (indicated adjacent to each cell type) and may function in cell-mediated/pro-inflammatory, humoral or regulatory roles. Induced T regulatory (iTreg) cells are also generated from naïve CD4+ T cell precursors, while natural T regulatory (nTreg) cells are exported directly from the thymus.

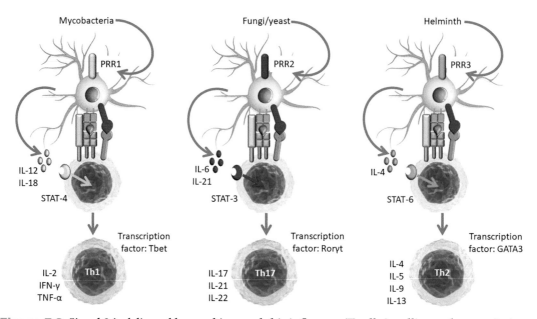

Figure 7.6 **Signal 3 is delivered by cytokines and this influences T cell signalling and transcription factor usage.** The decision of the naïve CD4$^+$ T cell to differentiate to one of the effector subtypes is determined by the cytokine signals that the cell receives from the APC. These are modified according to what the APC has 'sensed' at the site of antigen acquisition. Differentiation to a Th1 cell requires IL-12 or IL-18 and signalling through their receptors via STAT-4, which induces usage of the transcription factor Tbet to drive production of the signature cytokines characteristic of this cell type. Other cytokine receptors signal through other STAT combinations to commit T cells to differentiating to other defined subsets as shown.

the stage of the immune response and the tissue environment in which the immune response is being generated (as determined by regional environmental hormones or cytokines).

The first of the T cell subsets to be characterized were the **T helper 1 (Th1)** and **T helper 2 (Th2)** cells (**Figure 7.7**). These cells are difficult to distinguish on the basis of their cell surface molecular phenotype. However, they are functionally distinct based on the fact that these two subsets mediate (help) entirely different immunological processes by virtue of the panel of cytokines that each one produces. The **Th1** cells preferentially secrete **IL-2** and **IFN-γ** and this latter cytokine stimulates classical M1 macrophages that are effective in killing intracellular pathogens, driving a **cell-mediated immune response**. Th1 cells are less able to provide help for B cells, although they can pro-

vide limited help for production of the specific subclass of IgG involved in cytotoxic responses. In contrast, **Th2** cells preferentially produce **IL-4, IL-5, IL-9 and IL-13** and by virtue of this cytokine profile play a major role in providing help for class-switching by B cells, which secrete antibody (generally of the IgG, IgA or IgE classes) in the **humoral immune response**. In addition, Th2 cell cytokines induce mast cell proliferation, alternative activation of macrophages into M2 macrophages and recruitment/activation of eosinophils. These latter features are typically seen in allergic inflammation, the response to helminth parasites and also play a role in tissue repair. Th1 and Th2 cells can be selectively recruited to sites of infection, as they may express particular adhesion molecules and are attracted by specific chemokines for which they bear receptors. In addition to these

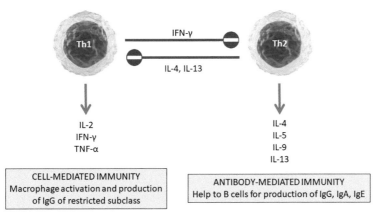

Figure 7.7 Th1 and Th2 CD4+ T lymphocytes. These two subsets of CD4+ T cells are functionally distinct based on the cytokine signature that each one produces. Th1 cells secrete IL-2 and IFN-γ and stimulate cell-mediated immune responses. Th2 cells produce IL-4, IL-5, IL-9 and IL-13 and preferentially provide help for B cell differentiation, class-switching and antibody-mediated immunity, leading to the production of IgG, IgA or IgE antibodies. These T cell subsets are mutually antagonistic, as IFN-γ is inhibitory towards Th2 cells, and IL-4 and IL-13 are inhibitory of Th1 cell function.

two cell subsets mediating distinctly different effector immune responses, they also tend to be **mutually antagonistic**, such that when one subset is activated the other is less so. This '**cross-regulation**' again relates to the cytokines that each cell produces. IFN-γ produced by Th1 cells inhibits the function of Th2 cells and conversely the IL-4 and IL-13 produced by Th2 cells suppresses the action of Th1 cells. This effect means that immune responses can have a relatively **polarized** outcome, with either CMI or humoral effects dominating in the immune response to a particular pathogen. This phenomenon is referred to as '**immune deviation**' and there are several good examples of immune responses that are polarized in this manner. However, in reality, for most immune responses, elements of both types of immunity are important and a balanced response is favoured. A specific example of these effects is shown in **Figure 7.8**. A dog infected by the sand fly-transmitted intracellular protozoan pathogen *Leishmania infantum* (which infects and replicates within macrophages) will require production of a strong cell-mediated immune response with production of IFN-γ to contain the pathogen and recover from infection. In this instance the protective immune response must be mediated by Th1 cells. In contrast, systemic migration of helminth parasites such as *Toxocara canis* (roundworm) and *Ancylostoma caninum* (hookworm) will require the production of IgE antibody, which enhances anti-parasite defences mediated by mast cells and eosinophils. The protective immune response in this instance must be mediated by Th2 cells.

Other CD4+ T cell subsets play more specialist roles in adaptive immune responses. CD4+ **Th3 cells** are preferentially produced in the mucosa-associated lymphoid tissues and secrete TGF-β, an important cytokine for B cell class-switching to IgA for mucosal immunity (see Chapters 8 and 9). CD4+ **Th17 cells** are characterized by their secretion of the cytokines IL-17, IL-21 and IL-22. These cells mediate an inflammatory response in various infectious diseases (especially at mucosal surfaces) as well as in some autoimmune and neoplastic diseases. Exposure of epithelial cells to IL-17 induces the secretion of chemokines that are effective

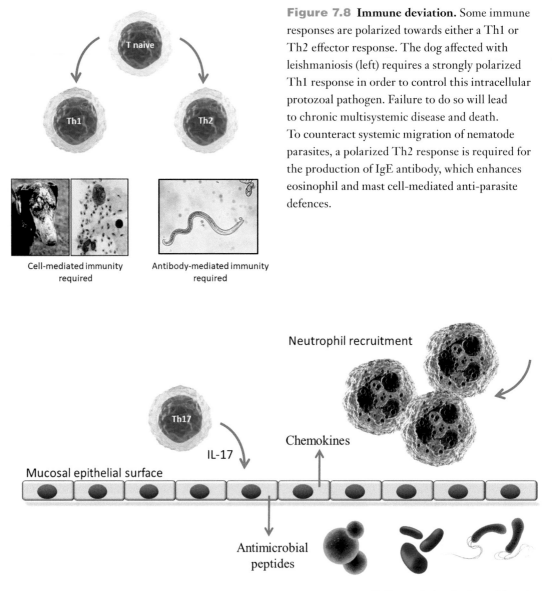

Figure 7.8 Immune deviation. Some immune responses are polarized towards either a Th1 or Th2 effector response. The dog affected with leishmaniosis (left) requires a strongly polarized Th1 response in order to control this intracellular protozoal pathogen. Failure to do so will lead to chronic multisystemic disease and death. To counteract systemic migration of nematode parasites, a polarized Th2 response is required for the production of IgE antibody, which enhances eosinophil and mast cell-mediated anti-parasite defences.

Cell-mediated immunity required

Antibody-mediated immunity required

Neutrophil recruitment

Chemokines

IL-17

Mucosal epithelial surface

Antimicrobial peptides

Pathogenic bacteria and fungi

Figure 7.9 Th17 cells help to target neutrophils to the site of infection. CD4+ Th17 cells are particularly effective at mucosal surfaces, where there are certain types of pathogenic bacteria or fungi. They release IL-17 and IL-22, which stimulate the epithelial cells to secrete antimicrobial peptides into the lumen and to produce chemokines to recruit neutrophils to the site of infection.

in enhancing neutrophil recruitment, whereas both IL-17 and IL-22 stimulate the secretion of antimicrobial peptides (**Figure 7.9**). The **Th9 cell** (characterized by dominant production of IL-9) has a role in defence against helminth infections and in the chronic phase of allergic disease. **T follicular helper cells** (TFh) reside in lymphoid follicles and are important in the formation and maintenance of germinal centres (see Chapters 2 and 9) and the regulation of B cell differentiation to plasma cells and memory B cells (see Chapter 9). Induced and natural **CD4+ T regulatory (Treg) cells** function to suppress the immune response in a targeted manner, via production of anti-inflammatory and regulatory cytokines including IL-10. These cells and responses will be considered further in Chapter 14.

7.4   THE ROLE OF CD8+ T LYMPHOCYTES

The main role of CD8+ T cells is to seek out and destroy virus-infected cells. Their cytotoxic effect is an **MHC class I-restricted** phenomenon that is part of the adaptive immune response. CD8+ T cells are activated by professional APCs in similar manner to that described for CD4+ cells, except their TCR recognizes antigenic peptide expressed by MHC class I and the CD8 molecule acts as the co-receptor. While direct interaction between a CD8+ T cell and an APC is sufficient to activate the cell, more effective activation occurs when the APC has previously encountered a CD4+ Th1 cell specific for the same antigen. The interaction with the Th1 cell leads to increased expression of peptide-loaded MHC class I and production of IL-12. This allows more effective stimulation of the naïve CD8+ T cells. Further paracrine signalling to the CD8+ cell by Th1 derived cytokines (particularly IL-2 and IFN-γ) also contributes to their activation (**Figure 7.10**).

Activated CD8+ T cells leave the lymphoid tissues and travel via the blood to home to sites of inflammation, where they seek out and destroy their targets (typically virus-infected cells). Recognition is an MHC-restricted process, through the MHC class I-peptide: TCR complex, with additional interactions between adhesion molecules expressed by the target cell and cytotoxic T cell. Following recognition, the next stage in the sequence is **adhesion** (**Figure 7.11**). At this point, the target and

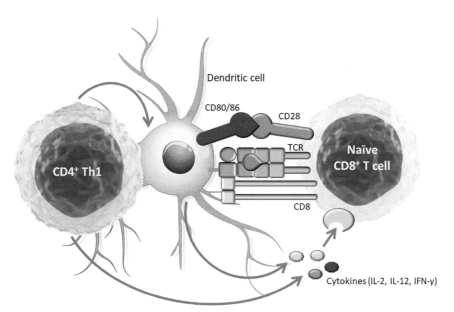

Figure 7.10 Activation of naïve CD8+ T cells. Although CD8+ T cells are activated in a similar manner to CD4+ T cells in terms of signal 1 (via TCR) and signal 2 (via CD28), the presence of CD4+ effector T cells (particularly Th1 cells) against the same antigen can enhance this process by activating the dendritic cell to be a more effective APC (signals, 1, 2 and 3) and by providing additional IL-2 that can act in a paracrine manner.

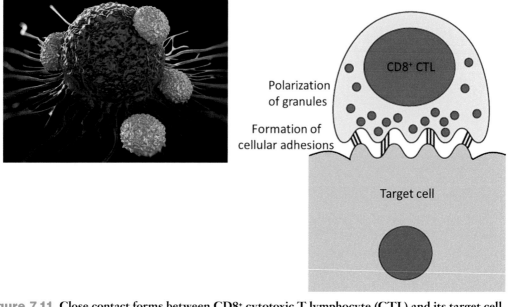

Figure 7.11 **Close contact forms between CD8⁺ cytotoxic T lymphocyte (CTL) and its target cell.** After recognition, the cytotoxic T cell forms close contact with the target cell. This involves interdigitation of the cell surface and the formation of adhesions between the cells in order to create a contained microenvironment. The cytotoxic cell cytoplasmic granules polarize to the surface of the cell in apposition to its target.

cytotoxic T cell forms a close association, with membrane interdigitations and intercellular **adhesions**, creating a defined space between the two cells. The cytoplasmic granules within the cytotoxic T cell polarize towards the surface of the target cell in advance of degranulation.

Adhesion is followed by **cytotoxicity**, which is mediated by one or more different mechanisms. The first of these involves degranulation and release of the contents of **intracellular granules** into the confined space between the two cells. The close adhesion prevents these cytotoxic molecules from diffusing away and causing collateral damage to nearby healthy cells. The granule contents include the molecule **perforin**, which polymerizes and inserts into the membrane of the target cell to form a channel somewhat similar to the MAC of the complement pathway (in fact, there is sequence homology between these molecules). Although perforin may establish an osmotic imbalance (akin to the MAC), the more important effect

is that this channel provides a route by which other granule contents (enzymes known as **granzymes**) enter the cytoplasm of the target cell (**Figure 7.12**). These enzymes may cause direct damage but also, importantly, they activate intracellular pathways that lead to **apoptosis** ('programmed cell death' or 'cell suicide') of the target cell. Granzymes activate the **intrinsic pathway of apoptosis** in which damaged cells activate pro-apoptotic Bax and Bak proteins that cause mitochondria to release cytochrome c, leading to formation of the 'apoptosome' (a multiprotein complex). The apoptosome activates the 'initiator caspase' (caspase-9), which in turn activates the 'effector caspases' (particularly caspase-3) that are responsible for the molecular effects of apoptosis, via the action of the caspase-3 activated DNase (CAD) molecule (**Figure 7.13**).

The cytotoxic T cell may also secrete cytotoxic cytokines such as lymphotoxin (a member of the TNF family), which binds to receptors on

Perforin and granzymes released
onto surface of infected cell

Perforin forms a pore in the
target cell membrane

Entry of granzymes into
cytoplasm of infected cell

Perforin pore and entry of granzymes

Figure 7.12 Degranulation of CD8⁺ cytotoxic T lymphocyte and release of perforin and granzymes.
Degranulation of the CTL onto the surface of the target cell releases perforin and granzymes. The perforin forms a pore in the target-cell membrane allowing entry of granzymes into the cytoplasm, where they initiate the caspase cascade, leading to apoptosis.

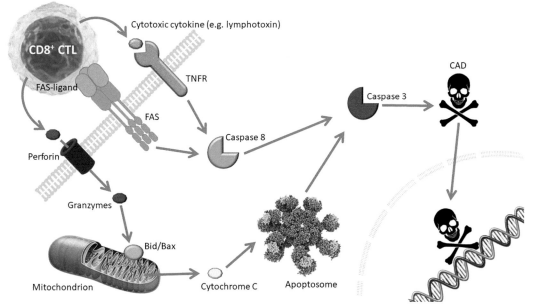

Figure 7.13 Cytotoxicity. Several mechanisms are used by the cytotoxic T cell to destroy the target. The cytoplasmic granule contents are released into the space between the two cells. One of these molecules (perforin) polymerizes and inserts into the target-cell plasma membrane to form a channel. Granzymes enter the target cell through this channel and cause direct damage in addition to activating the intrinsic pathway of apoptosis in the target cell. Apoptosis is also induced via the extrinsic pathway following signalling through a TNF receptor and following interaction between FAS-ligand and the molecule FAS.

the target cell and provides further signalling events, leading to apoptosis via the **extrinsic pathway**. The extrinsic pathway of apoptosis involves activation of initiator caspases-8 and -10, leading to effector caspase induction. The same pathway is triggered following interaction between the molecules **FAS** (expressed by the target cell) and **FAS ligand** (expressed by the CTL), which also triggers apoptosis. Co-stimulatory cytokines such as IFN-γ may amplify these effects.

Apoptosis is a carefully regulated process, characterized by blebbing of the cell membrane, loss of cytoplasmic organelles and clumping of the nuclear chromatin, leading to nuclear fragmentation. The final stages of apoptosis are the generation of 'apoptotic bodies' (cellular remnants), which can be phagocytosed by macrophages (**Figure 7.14**). Once the target cell is destroyed, the cytotoxic cell can detach and move on to locate and kill another target. In this manner, cytotoxic cells are **'multi-hit'** cells that may destroy a number of targets. A clinical example of the value of the cytotoxic response is provided by the benign canine skin tumour known as histiocytoma (**Figure 7.15**). These are common skin tumours seen in young dogs. The small nodular lesions often grow rapidly, but in a matter of weeks undergo spontaneous regression. The regression of the tumour is a direct consequence of active infiltration by CD8+ cytotoxic T cells into the lesion. This is one of the rare examples of an immune response to neoplastic cells being able to resolve the tumour.

7.5 MEMORY T CELLS

A key feature of the adaptive immune response is that it retains the memory of past antigen experience, so that on re-exposure to the same antigen, a more potent **'secondary'** immune response is generated. Both T and B **memory lymphocytes** exist, but they are poorly characterized. Immunological memory can be **long-lived** (sometimes for the entire lifetime in humans), as epitomized by observations made many years ago of indigenous populations living

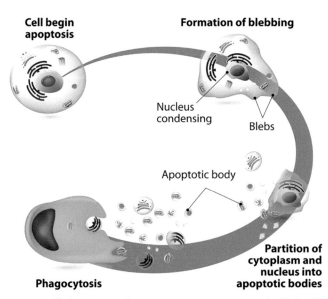

Figure 7.14 Apoptosis. As cells begin to undergo apoptosis (programmed cell death) the nuclear material is seen to condense and the cell membrane undergoes blebbing. Subsequently, nuclear fragmentation and formation of apoptotic bodies leads to phagocytosis by macrophages.

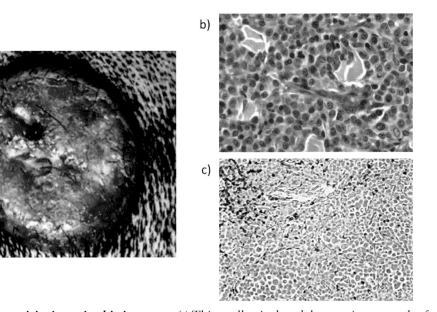

Figure 7.15 Cytotoxicity in canine histiocytoma. (a) This small, raised, nodular mass is an example of a common canine cutaneous tumour, the histiocytoma. (b) This is a tumour of Langerhans cells that most often occurs in young dogs and undergoes spontaneous regression over a period of weeks. (c) This section of a histiocytoma is labelled to show an infiltrate of T cells expressing the CD3 complex. These are CD8[+] cytotoxic T cells and are responsible for destruction of neoplastic cells. This lesion provides one of the few examples of the immune system successfully destroying a tumour.

on remote islands. When such islands were first visited by European explorers, pathogens (e.g. measles virus) were introduced to these populations and disease outbreaks occurred. After departure of the Europeans the virus failed to become endemic. Many decades later, when the virus was reintroduced by subsequent visitors, those individuals who had been alive at the time of the first exposure were resistant to infection, whereas others who had been born subsequently became ill. The principle of immunological memory underlies the process of vaccination, which we shall consider further in Chapter 12.

Despite firm evidence for the existence of memory T cells, it has proven to be difficult to identify and characterize these cells. In some species, different isoforms of the leucocyte common antigen (CD45) are expressed, comparing naïve and memory T cells, with the truncated CD45R0 form associated with the memory

population. It is also possible that memory cells have selective recirculation pathways and homing receptor expression. Such cells may also display greater sensitivity to antigenic stimulation and have functional differences (e.g. in terms of cytokine production) relative to naïve T cells.

A contentious issue regarding memory lymphocyte biology is the longevity of such cells. Although immunological memory can last for many years, the evidence indicates that individual memory cells have a much shorter lifespan, indicating that they require some form of periodic **restimulation** to maintain their presence. This is probably mediated by cytokines, in particular IL-7 (which is also important in early development of lymphocytes) and IL-15. Antigen may also be required for restimulation and several models have been proposed to account for the long-term availability of such antigen. One proposal is that small **reservoirs**

of antigen may persist (perhaps associated with dendritic cells in lymphoid tissue) and periodically presented to the memory lymphocyte population. An alternative theory is that environmental antigens may express **cross-reactive epitopes** that permit restimulation of memory lymphocytes. The required antigen may also be provided by **re-exposure** to the infectious agent or via interventional booster vaccination.

KEY POINTS

- Conventional T cells express an αβ TCR and CD3 complex with either CD4 or CD8 used as the co-receptor.
- Antigen presentation by MHC class II activates CD4+ T cells; presentation by MHC class I activates CD8+ T cells.
- T cell activation requires signal 1 (MHC/peptide–TCR interaction), signal 2 (co-stimulatory molecular interaction) and signal 3 (APC-derived cytokines).
- Activated T cells undergo clonal proliferation and differentiation to effector and memory T cells, coordinated by signal 3.
- CD4+ T cell subsets include Th1, Th2, Th9 and Th17 cells.
- Th1 cells produce IL-2 and IFN-γ and are important for macrophage activation as part of cell-mediated immunity.
- Th2 cells produce IL-4, IL-5, IL-9 and IL-13 and facilitate humoral immunity.
- Naïve CD4+ T cells give rise to Th1, Th2, Th9 or Th17 cells depending on the nature of the simulating antigen and signalling by the APC that presents it.
- Th1 and Th2 cells are mutually antagonistic, allowing polarized immune responses (immune deviation) in some instances.
- CD8+ T cells recognize target cells in an MHC class I-restricted manner.
- Cytotoxicity involves adhesion, induction of apoptosis and detachment.
- Cytotoxicity involves degranulation, perforin insertion and entry of granzymes into the target cell's cytoplasm, which triggers the caspase cascade.
- Target-cell apoptosis is induced by intracellular granule contents, cytokine–cytokine receptor interaction and FAS–FAS ligand interaction.
- Memory T cells are poorly characterized, but are likely to require periodical restimulation with antigen to provide long lived immunity.

ANTIBODY STRUCTURE AND FUNCTION

OBJECTIVES

At the end of this chapter, you should be able to:

- Describe the basic structure of an immunoglobulin molecule.
- Describe how an antibody binds to an antigenic epitope.
- Define 'affinity' and 'avidity' as they relate to antigen binding by antibody.
- Describe the basic biological functions of antibody, following binding to its target antigen.
- List the five classes of immunoglobulin and describe their structure and function.

8.1 INTRODUCTION

Unlike T lymphocytes, that require antigen to be internalized, processed and presented to them by other cells, B lymphocytes express antigen receptors capable of recognizing surface epitopes of antigen in the extracellular fluid. When activated, B lymphocytes repurpose their surface antigen receptor to become a secreted molecule, known as **antibody**. Antibodies can be thought of as molecular 'heat seeking missiles' that can be fired out of lymphoid tissues to target, neutralize and destroy any pathogen expressing the antigen that stimulated their production. Thus, whereas the main 'immunological output' from CD4$^+$ T lymphocytes are cytokines, the main output from B lymphocytes is antibody.

An antigen may be defined as any substance capable of binding to a lymphocyte antigen receptor, typically stimulating an immune response. The immune system is usually 'tol-erant' towards self-antigens (as we shall see in Chapter 14), although in some instances, auto-antibodies can be produced that may lead to pathology and autoimmune disease. The majority of foreign antigens are proteins and include those from infectious agents (e.g. viruses, bacteria, fungi, protozoa or helminths) or environmental substances (e.g. dietary proteins and pollens). The epitopes associated with these antigens may come from a peptide component (termed **linear epitopes**) or more commonly, a region on the surface of the protein brought about through its secondary, tertiary or quaternary structure (termed **conformational epitopes**). It is also possible for non-proteinaceous molecules to act as antigens (e.g. polysaccharides), although (as we shall see in the next chapter) the antibody response to these is not so potent or effective. It is also possible for inorganic substances and small organic molecules, including some chemicals and drugs, to stimulate an antibody response, although they can

DOI: 10.1201/9781003310969-8

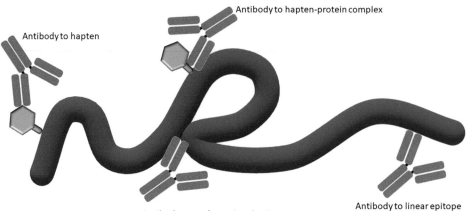

Figure 8.1 Linear epitopes, conformational epitopes and haptens. This small chemical group is not immunogenic, unless it is conjugated to a large carrier protein. The combined hapten–carrier may trigger an immune response (represented by antibody) to the hapten alone, the carrier protein alone or a novel antigen formed of both the hapten and carrier.

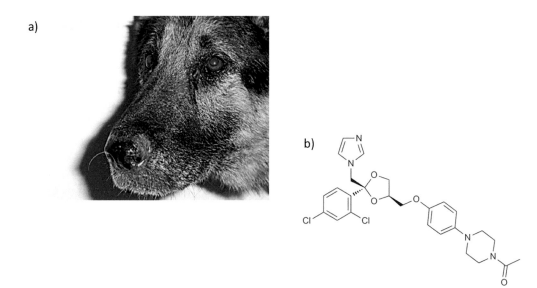

Figure 8.2 Drug reaction. (a) This German shepherd dog was being treated with the systemic antifungal drug ketoconazole (structure shown in (b)) and subsequently developed lesions affecting the planum nasale and periorbital skin. In this instance the drug may be acting as a hapten by binding to dermal proteins and triggering a local immune response. Such reactions may spontaneously resolve once administration of the drug is halted.

only do so when bound to a larger host **carrier protein** and are referred to as **haptens** (**Figure 8.1**). The clinical relevance of this phenomenon lies in the fact that some drugs may bind host proteins and inappropriately stimulate a **drug reaction** (**Figure 8.2**).

In this chapter, we take a closer look at the structure of antibodies, how they interact with antigen and what happens as a consequence of antibody-binding to its target.

8.2 ANTIBODY STRUCTURE

The fraction of blood that remains fluid after clotting represents the **serum** (i.e. plasma without fibrinogen); it may be electrophoretically separated into constituent proteins including albumin and the alpha, beta and gamma globulins (**Figure 8.3**). The gamma globulin fraction consists chiefly of **immunoglobulins**, also known as **antibodies**.

The basic molecular structure of an immunoglobulin molecule is well characterized (**Figure 8.4**). It consists of four glycosylated protein chains, held together by interchain disulphide bonds in a Y-shaped conformation. Two of the chains are of higher molecular mass (**heavy chains**, approximately 50 kD) and two chains are smaller in size (**light chains**, approximately 25 kD). In terms of structure and amino acid sequence, the two heavy chains in any one

immunoglobulin are identical to each other, as are the two light chains. This means that the two 'halves' of the molecule are essentially mirror images of each other. Although we commonly depict immunoglobulins diagrammatically as linear structures, these molecules have a complex tertiary structure. Each chain is formed of a series of **domains**, which have a roughly globular structure that is created by the presence of intrachain disulphide bonds. A further basic feature of an immunoglobulin molecule is that the structure and amino acid sequence towards the C-terminal end of the protein are relatively uniform (conserved) from one immunoglobulin isotype to another, while the structure towards the N-terminal end has considerable variability between immunoglobulins. Within each of the four chains this gives rise to the presence of an N-terminal **variable region** and a series of **constant domains** towards the C terminus. Each light chain is composed of two domains, a variable domain of the light chain (V_L) and a constant domain of the light chain (C_L). Similarly,

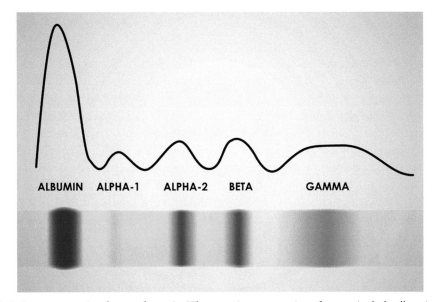

Figure 8.3 Serum protein electrophoresis. The constituent proteins of serum include albumin and the alpha, beta and gamma globulins. The gamma globulin fraction contains the antibody molecules (aka immunoglobulins).

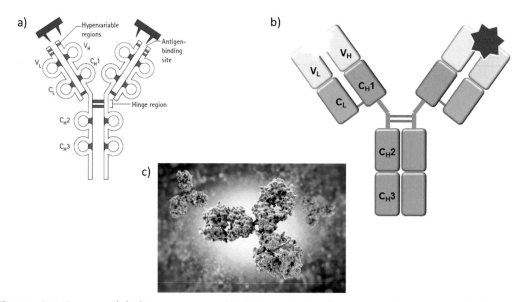

Figure 8.4 **Immunoglobulin structure.** (a, b) The basic Y-shaped structure of an immunoglobulin is composed of two identical heavy chains and two identical light chains linked together by interchain disulphide bonds. Intrachain disulphide bonds confer a globular domain structure to each chain. The N-terminal domains (V_L and V_H) show greatest variation in amino acid sequence between immunoglobulins and represent the site of antigen binding, particularly at the hypervariable regions. The C-terminal domains (V_H, C_{H1} to C_{H3}) have a more conserved amino acid sequence and undertake functions including complement fixation and receptor binding. The hinge region between C_{H1} and C_{H2} allows mobility of the short arms of the molecule. (c) 3D rendering of an immunoglobulin molecule.

each heavy chain is comprised of one variable domain of the heavy chain (V_H) and three or four constant domains named C_{H1}, C_{H2}, C_{H3} and C_{H4}. Most immunoglobulins have a distinct region between the C_{H1} and C_{H2} domains, involving interchain disulphide bonds known as the **hinge region**. This confers on the molecule the ability of the short arms of the Y-shaped structure to move through approximately 180°. The variable domains contain regions in which there is the greatest degree of variation in amino acid sequence between different immunoglobulins, which have a major impact on antigen binding. These sub-areas are known as the **hypervariable loops** or **complementarity determining regions** (CDRs) (**Figure 8.5**).

The globular domains of the immunoglobulin molecule have distinct functional attributes. The structure formed by the variable regions (V_L and V_H) is that portion of the molecule that binds to an antigenic epitope (the **antigen-binding site**; **Fab**). The C_{H2} domains are involved in activation of the complement pathway (see Chapter 3) and the C_{H3} domains are important in binding of the antibody molecule to cellular immunoglobulin receptors called **Fc receptors**.

Historically, the structure of the immunoglobulin molecule was characterized by studying the various fragments, following incubation with particular proteolytic enzymes (**Figure 8.6**). The enzyme **papain** cleaves the molecule on the

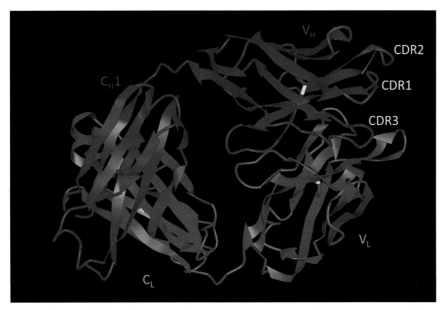

Figure 8.5 The hypervariable loops (CDRs) of the variable region. The three hypervariable loops (CDRs) are indicated on the variable heavy chain domain. Molecular model generated in iCn3D, from PDB ID: 2HKH.

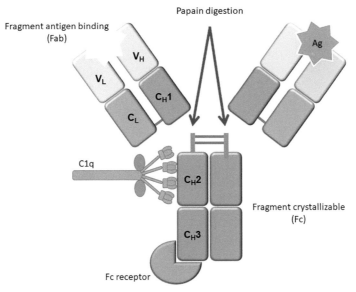

Figure 8.6 Papain digestion fragments of immunoglobulin. Digestion of the immunoglobulin molecule with the enzyme papain liberates two identical fragments that retain the ability to bind to antigen (Fab) and a third fragment that retains the biological properties of the immunoglobulin molecule in terms of its interaction with host proteins (such as the complement protein C1q and the Fc receptor expressed by phagocytic cells).

N-terminal side of the hinge region, creating three fragments: the joined C-terminal heavy chains (the 'body' of the Y shape), known as the **Fc region** (for 'fragment crystallizable'), and the joined N-terminal heavy chains and light chains (the 'arms' of the Y shape), each known as a **Fab** fragment ('fragment antigen-binding').

As we can see, the two main regions of the immunoglobulin molecule (Fab and Fc) serve different purposes, with the Fab (N-terminal heavy and light chain variable regions) responsible for antigen specificity and the Fc (heavy chain C-terminal region) responsible for the biological activity of the antibody after binding to its target. For the latter, there are different heavy chains that can be utilized by the immune system to modify the antibody molecule, so that it best suits the type of infection (virus, bacteria or parasite), with five different immunoglobulin isotypes available for selection (IgD, IgM, IgG, IgA and IgE). We will now focus more on these two key attributes of antigen binding and biological activity and compare/contrast the structure and function of the different types of antibody.

8.3 ANTIBODY BINDING TO ANTIGEN

The interaction between antibodies and their target antigen specifically involves the recognition of an antigenic epitope by the N-terminal variable (Fab) regions of the immunoglobulin molecule. In three dimensions, the V_L and V_H domains form a **cleft or groove**, lined by the hypervariable regions (or CDRs), into which the epitope fits or, alternatively, these domains flatten out to form a planar area of interaction that permits a larger area of contact with the antigen. This interaction is exquisitely precise and there are specific points of interaction (**contact residues**) between the epitope and antigen-binding site (**Figure 8.7**). Some antigenic epitopes will have perfect match for

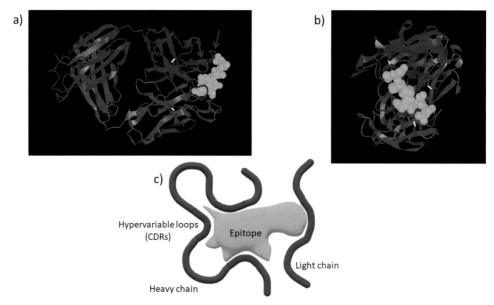

Figure 8.7 Antigen–antibody binding. (a) Side view (with red arrows indicating the three hypervariable loops) and (b) surface view of antibody binding to an influenza haemagglutinin epitope (green). Note how the epitope sits between the variable domains of the heavy and light chains. (c) There is a close association between the hypervariable loops and the epitope. Molecular models generated in iCn3D, from PDB ID: 1FRG.

the antigen-binding site of immunoglobulin and interact like a '**lock and key**' to produce a **high-affinity** binding. Antibody affinity refers to the strength of binding of one Fab of an antibody to one antigenic epitope (**Figure 8.8**). In this situation the antigen is held tightly in place by interactions involving the formation of van der Waal's forces, hydrogen bonds, electrostatic forces and hydrophobic forces. Other antigenic epitopes will interact with antigen-binding sites with **low affinity** or display no interaction at all. A second term used to describe the strength of interaction between antigen and antibody is avidity. **Avidity** refers to the strength of overall binding of the two molecules, such that a low-avidity interaction might involve the binding of one Fab within an immunoglobulin to one epitope on an antigen, whereas an interaction of higher avidity would involve multiple Fab–epitope binding (**Figure 8.9**).

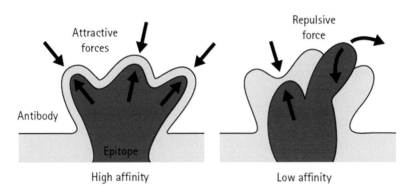

Figure 8.8 Antigen antibody binding affinity. Affinity is the sum of attractive and repulsive forces between the antigenic epitope and the Fab antigen-binding site of the antibody. High- and low-affinity interactions are demonstrated.

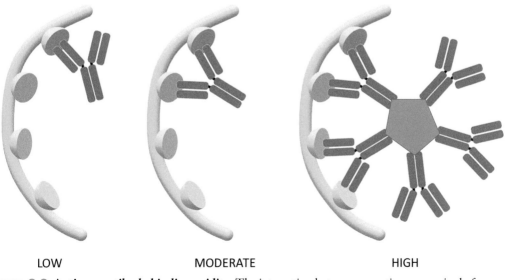

Figure 8.9 Antigen–antibody binding avidity. The interaction between an antigen comprised of repeating units of an epitope and specific antibodies is shown. The greater the number of interactions the higher is the avidity of binding.

8.4 BIOLOGICAL PROPERTIES OF ANTIBODY

Antibody binding to its target antigen (and the pathogen expressing it) can have various biological consequences. Simply binding to surface epitopes that are required for attachment to host-cell receptors can interfere with antigen function. This is particularly important for **virus-neutralizing (VN) antibody**, which targets viruses in the extracellular fluid during their transmission from host to host and cell to cell within infected tissues. Viruses use their surface antigens (capsid or spike proteins) to attach to specific host-cell receptors (this determines their species specificity and tissue tropism). Production of VN antibody blocks this interaction and the virus is effectively neutralized (**Figure 8.10**). Neutralizing antibody (particularly IgA) can also prevent microbial adhesins and microbial toxins from causing harm to the host cells they come into contact with.

Antibodies (IgM and IgG in particular) can enhance innate immune mechanisms. Phagocy-tosis is an important host defence against bacteria. Receptor-mediated endocytosis is more effective than non-specific uptake of microbes and PRRs as well as complement receptors enhance this innate immune mechanism. Phagocytosis can also be enhanced through production of **opsonizing antibody** (IgG) since antibody-coated microbes can be detected by Fc-gamma receptors expressed on the surface of phagocytic cells such as neutrophils and macrophages. The interaction between the antibody (bound to its target antigen) and the Fc receptors allows effective uptake of the microbe, where it can be killed and digested (**Figure 8.11**).

Fc-gamma receptors are also expressed on the surface of natural killer (NK) cells, except these cells are not phagocytic and instead degranulate when bound to cell-surface antibody (IgG). This mechanism of **antibody-dependent cellular cytotoxicity (ADCC)** is particularly effective against enveloped viruses that exit the infected cell by budding (**Figure 8.12**). These viruses export their surface glycoproteins (spike proteins) to the cell surface in advance of budding, where they can be bound by antibody.

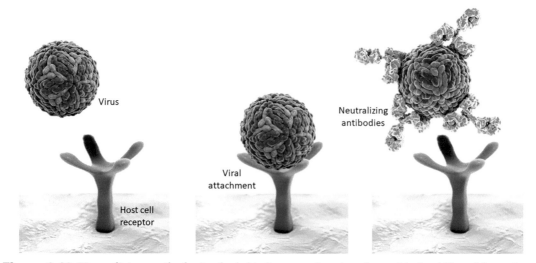

Figure 8.10 Neutralizing antibody. Antibody binding to antigen interferes with the ability of the antigen to attach to its target-cell receptor. This is a particularly important adaptive immune defence against viral attachment, but also prevents attachment of microbial adhesins and exotoxins.

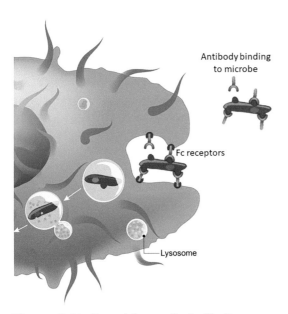

Figure 8.11 **Opsonizing antibody.** Binding of antibody (particularly IgG) via the Fab can be detected by host-cell Fc receptors (Fc-gamma), expressed on the surface of macrophages and neutrophils. This enhances their ability to phagocytose the bound antigen/pathogen.

Detection of immune complexes on the surface of these infected cells triggers NK cell degranulation and (in a similar manner to CD8+ CTLs) this induces apoptosis. ADCC can be exploited for cancer immunotherapy, whereby antibodies directed against cell surface antigens on cancer cells can be used to target the immune system for their destruction (see Chapter 20).

Another example whereby the adaptive immune system enhances the function of cells of innate immunity is via production of specialist types of antibody, such as IgE. This is preferentially produced in the presence of antigen from helminth parasites and is utilized by mast cells and basophils (and to some extent eosinophils) as they express the high-affinity IgE receptor (Fc-epsilon receptor; FcεR). Thus, these innate immune cells adsorb antigen-specific IgE (produced by the adaptive immune system) and become 'sensitized' to the presence of the parasite. Binding and cross-linking of the IgE molecules on the mast cell surface is a

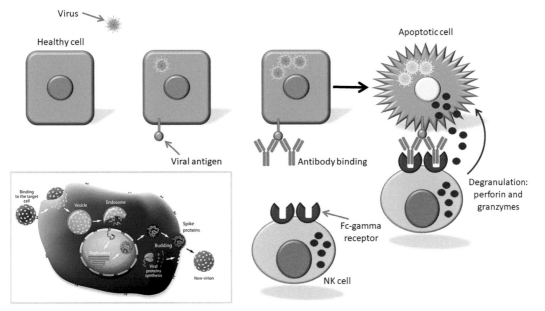

Figure 8.12 **Antibody-dependent cellular cytotoxicity (ADCC).** Natural killer (NK) cells express Fc-gamma receptors that allow them to recognize host cells coated in antibody. This is particularly prominent in virus-infected cells, where the virus exits by budding and viral proteins are expressed on the host cell surface (see inset).

potent trigger for degranulation and release of toxic, digestive enzymes and pro-inflammatory molecules onto the surface of the pathogen (**Figure 8.13**).

We have already seen (in Chapter 3) how the complement system provides innate immune defence against microbes. This innate immune mechanism can also be enhanced by production of antibody. Deposition of IgM or several molecules of IgG can react with the adaptor molecule C1q to initiate the classical complement pathway, leading to formation of the membrane attack complex (MAC) and release of pro-inflammatory complement by-products (such as C3a and C5a) (**Figure 8.14**).

8.5 IMMUNOGLOBULIN CLASSES

There are five different classes (isotypes) of antibody, determined by which immunoglobulin heavy chain is used. These are named by the Greek letters α (alpha), δ (delta), ε (epsilon), γ (gamma) and μ (mu). There are also two distinct forms of light chain named κ (kappa) and λ (lambda). Given that one Y-shaped immunoglobulin is composed of two identical heavy and two identical light chains, a single immunoglob-

ulin must therefore consist of a pair of one of the five types of heavy chains, coupled to a pair of either κ or λ light chains. Five **immunoglobulin (Ig) classes** are defined by usage of these heavy chain molecules: **IgA** (α chain), **IgD** (δ chain), **IgE** (ε chain), **IgG** (γ chain) and **IgM** (μ chain) (**Figure 8.15**). For IgG and IgA, duplication of the genes encoding the heavy chains means that there are additional **subclasses** of immunoglobulin that have subtle differences between their amino acid sequence and the structure of their constant region.

The best characterized immunoglobulins and immunoglobulin genes are those of humans and experimental rodents. However, readers should be aware that while extrapolations may be made to domestic animal species, distinct species differences do exist. Domestic animal species share the same basic five immunoglobulin classes, but there is variation in the range of IgG and IgA subclasses and some unique species-specific modifications in immunoglobulin structure exist (Table 8.1). For example, four IgG subclasses (IgG1–IgG4) are recognized in humans and dogs, while there are seven in the horse (IgG1–IgG7). Humans have two IgA subclasses encoded by distinct IgA heavy chain

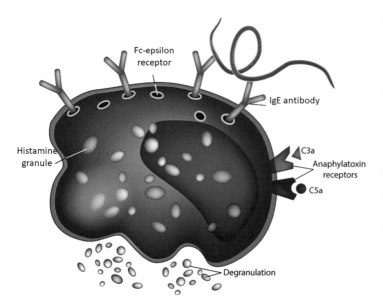

Figure 8.13 Antibody sensitization of mast cells. In addition to expressing receptors that react to the presence of complement by-products (part of the innate immune response), mast cells also express specialist Fc receptors (Fc-epsilon) that allows them to adsorb IgE antibodies. Thus, these cells can exploit the adaptive immune response to become antigen-specific and this enhances their parasite targeting ability.

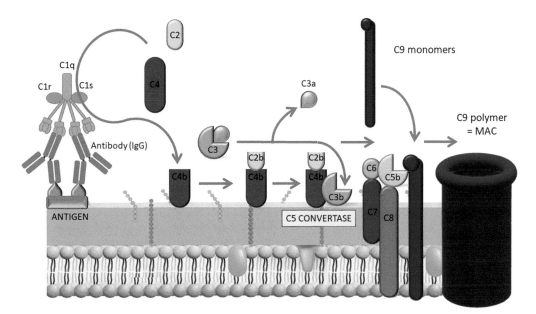

Figure 8.14 Complement-fixing antibody. Binding of antibody (IgM or IgG) to its antigen on the surface of a pathogen allows attachment of the complement protein C1q, which triggers the classical pathway of complement activation and formation of the membrane attack complex (MAC).

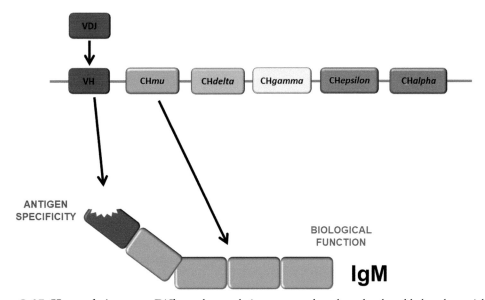

Figure 8.15 Heavy chain genes. Different heavy chain genes can be selected to be added to the variable region (generated during V(D)J recombination) to create immunoglobulin heavy chains of different classes, but with the same antigen specificity. In this example, CH*mu* is selected to generate the IgM heavy chain.

(*IGHA*) genes and rabbits have 13 such genes, most of which produce functional protein. In contrast, other domestic animals have a single *IGHA* gene, but genetic variation in the hinge region of the IgA heavy chain gene has been identified in pigs, sheep and dogs, which may translate into functionally distinct IgA proteins. In addition to the standard four-chain IgG,

Table 8.1 IgG, IgA and IgM in Man and Animals

SPECIES	IGG SUBCLASSES	SERUM IGG CONCENTRATION (MG/ML)	IGA SUBCLASSES[1]	SERUM IGA CONCENTRATION (MG/ML)	SERUM IGM CONCENTRATION[2] (MG/ML)
Man	G1, G2, G3, G4	7.5–22.0	A1, A2	0.5–3.4	0.2–2.8
Dog	G1, G2, G3, G4	10.0–20.0	Four allelic variants in hinge region	0.2–1.6	1.0–2.0
Cat	G1, G2, G3	5.0–20.0	Not recognized	0.3–2.0	0.2–1.5
Horse	G1, G2, G3, G4, G5, G6, G7	11.5–21.0	Not recognized	1.0–4.0	1.0–3.0
Cow	G1, G2, G3	17–27	Not recognized	0.1–0.5	2.5–4.0
Sheep	G1, G2, G3	18–24	Three allelic variants in hinge region	0.1–1.0	0.8–1.8
Pig	G1, G2a, G2b, G3, G4	9.0–24.0	Two allelic variants in hinge region	0.5–1.2	1.9–3.9
Mouse	G1, G2a, G2b, G3	2.0–5.0	Two allelic variants in hinge region	1.0–3.2	0.8–6.5

[1] Two *IGHA* genes are recognized in man, but only one in other species.

[2] Subclasses of IgM are not recognized.

camelid species produce homodimeric IgG made up of two heavy chains and no light chain. Chickens have a major immunoglobulin known as IgY, a four-chain molecule not dissimilar to mammalian IgG in structure. The heavy chain of IgY (υ or upsilon chain) has one variable and four constant domains, but lacks a hinge region. Some avian species also have a truncated version of IgY with heavy chains of only two constant domains. As this molecule does not have an Fc region, it is known as IgY(ΔFc) and is of unknown function.

8.5.1 IgD

The IgD molecule is a single Y-shaped immunoglobulin unit comprising two δ heavy chains and with an antigen-binding valence of two. IgD has limited distribution in the body and is restricted to being expressed on the surface of naïve B lymphocytes, where it functions as an antigen receptor (**Figures 8.16, 8.17**). There are numerous species differences in the structure of this immunoglobulin. The heavy chain of mouse IgD has two constant domains and an extended hinge region, making it susceptible to proteolysis. The delta chain of domestic animals, man and primates has three constant domains. In most of these species (except the pig) the hinge region is also long relative to that of other immunoglobulin classes, providing flexibility in terms of antigen detection by the B cells.

8.5.2 IgM

The IgM molecule exists as both a monomer and a pentamer. In its monomeric form, it is expressed alongside IgD as a cell surface immunoglobulin receptor (**Figure 8.18**). However, upon activation, the transmembrane domain is spliced out and the heavy chains associate with a joining (J) chain at a ratio of 5:1, with additional disulphide bonds between the C-terminal domains further stabilizing the molecule (**Figure 8.19**). Each of the component units comprises paired μ heavy chains and each of these carries an additional C-terminal C_{H4} domain. The IgM molecule **lacks a distinct hinge region**, but there

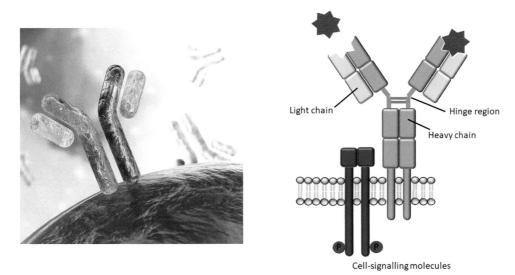

Figure 8.16 **Immunoglobulin D (IgD).** IgD is not typically secreted as antibody and instead acts as one of the surface immunoglobulins (sIg) on naïve B lymphocytes. It functions as an antigen receptor and associates with signalling molecules, triggering activation of naïve B cells in response to antigenic stimulation.

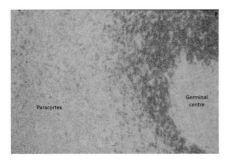

Figure 8.17 **IgD is expressed on the surface of naïve B cells.** Immunohistochemistry of a reactive lymph node stained using an anti-IgD antibody. The naïve B cells are present in the mantle zone surrounding the germinal centre of a secondary follicle.

is some mobility at the level of C_{H2} to C_{H3}. The large molecular mass of IgM means that this immunoglobulin is mainly found in the plasma, as it does not readily diffuse between vascular endothelium to enter the tissue fluid. For this reason, IgM plays an important role in the immune response to infections of the blood (e.g. bacteraemia). IgM is readily able to fix complement (see Chapter 3) and because of its antigen-binding valence of 10, it is able to attach to and draw together multiple particulate antigens in a process known as **agglutination** (**Figure 8.20**). IgM is of greatest importance in the **primary immune response** (see Chapter 9).

8.5.3 IgG

The IgG molecule consists of a single Y-shaped unit with a molecular structure as described for a 'standard' immunoglobulin. IgG consists of paired γ heavy chains with either paired κ or λ light chains. The molecule has two N-terminal antigen-binding sites and therefore has an antigen-binding **valence** of two. IgG is the dominant form of immunoglobulin found in the serum, and where IgG subclasses are defined, the relative serum concentrations of these may vary. The molecular mass of IgG (approximately 150 kD) means that it is readily able to leave the circulation and enter the extracellular

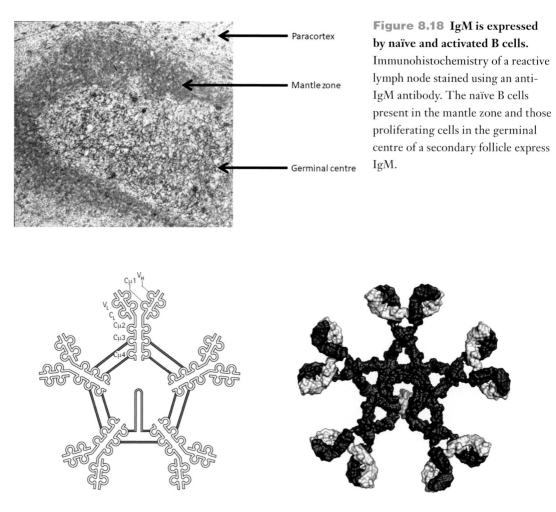

Figure 8.18 IgM is expressed by naïve and activated B cells. Immunohistochemistry of a reactive lymph node stained using an anti-IgM antibody. The naïve B cells present in the mantle zone and those proliferating cells in the germinal centre of a secondary follicle express IgM.

Figure 8.19 Structure of IgM. IgM is a pentamer of five basic immunoglobulin units linked together by a joining chain and additional disulphide bonds between the C_{H3} and additional C_{H4} domains present in the μ heavy chain.

fluid when there is increased permeability of the vascular endothelium (for example in an inflammatory response), but some IgG is normally found in this environment and may be actively transported there from the blood. The diverse functional properties of IgG include neutralization (of viruses and microbial toxins), complement fixation, opsonization and ADCC. IgG is the dominant immunoglobulin in the **secondary immune response** (see Chapter 9).

8.5.4 IgA

The IgA molecule may be found as a **monomer** consisting of a single Y-shaped immuno-globulin (utilizing α heavy chains) or as a **dimer** of two such units linked together by a J chain (**Figure 8.21**). These variants have an antigen-binding valence of 2 or 4, respectively. The occurrence of monomeric versus dimeric IgA is dependent on the species and anatomical location. Relatively small quantities of IgA may be found within the circulation as a monomer in humans and as a dimer in most domestic animal species. The highest concentration of IgA is found in the secretions bathing the **mucosal surfaces** of the body (i.e. the gastrointestinal, respiratory and urogenital tracts, the eye and the mammary glands) or in secretions related

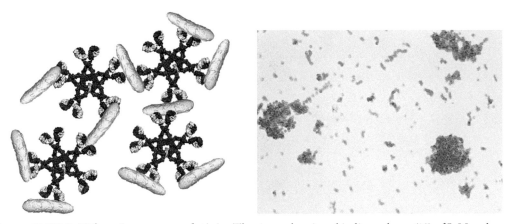

Figure 8.20 IgM functions as an agglutinin. The size and antigen binding valence (10) of IgM makes this molecule particularly effective at binding and drawing together particulate antigens in the process of agglutination. A clinical example of this phenomenon occurs in immune-mediated haemolytic anaemia (IMHA) in which IgM autoantibodies cause aggregation and destruction of erythrocytes within the circulating blood. IgM molecules bind to erythrocytes to form an agglutinate. A drop of blood from a dog with IMHA shows the presence of large agglutinates of red blood cells.

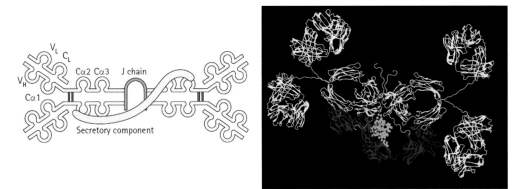

Figure 8.21 Structure of the IgA dimer. IgA consists of two Y-shaped immunoglobulin units with α heavy chains linked by a J chain (green). Where secreted onto mucosal surfaces the molecule also carries a secretory component (blue) that wraps around the C-terminal ends of the immunoglobulin units to protect the IgA from proteolytic degradation. (Molecular model generated in iCn3D, from PDB ID: 3CHN.)

to these surfaces (e.g. bile, tears, colostrum or milk). This distribution reflects the fact that the majority of IgA in the body is produced at mucosal surfaces and plays a key role in protecting mucosal epithelial cells from attachment by viruses and bacterial adhesins or in neutralizing harmful toxins produced by pathogens at mucosal sites.

This secreted form of IgA is dimeric and in those species with distinct IgA subclasses there are regional preferences for their secretion at mucosal surfaces. As mucosal surfaces are rich in proteolytic enzymes, secreted dimeric IgA has one further modification that protects it from enzymatic degradation. This modification is the presence of a **secretory component**, which wraps around the C-terminal ends of the subunits and is attached to the C_{H2} domains by disulphide bonds. The pathway of mucosal IgA secretion is shown in **Figure 8.22**.

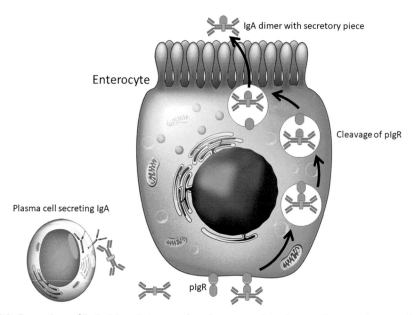

Figure 8.22 Secretion of IgA. Most IgA is produced at mucosal surfaces and secreted across the lining epithelium as a dimeric molecule. During this process, the secreted IgA molecule acquires a protective secretory component that is formed from a portion of the polymeric immunoglobulin receptor (pIgR) that mediates its transport across the epithelial barrier.

The dimeric IgA molecule is secreted by a **plasma cell** within the lamina propria underlying the mucosal epithelium. This molecule attaches to the **polymeric immunoglobulin receptor (pIgR)** on the basal surface of the epithelial cell (in the intestine the epithelial cells at the base of the crypts express pIgR) and the complex of pIgR and IgA dimer is internalized by the epithelial cell. By a process of transcytosis, the IgA moves towards the luminal surface during which time the pIgR is cleaved, releasing the IgA into the vesicle, but with a remnant of the receptor, **the secretory component**, still attached. Finally, the IgA is exocytosed and released into the lumen where it can carry out its neutralization functions. The pIgR can also transport IgM into mucosal secretions, but under normal circumstances the bulk of mucosal plasma cells secrete IgA, which occupies most of the receptors. However, in IgA deficiency, one may see a compensatory increase of secretory IgM in mucosal fluids.

8.5.5 IgE

The IgE molecule is a single Y-shaped immunoglobulin unit composed of two ε heavy chains and has a valence of two. Similar to IgM, the IgE molecule has an additional C_{H4} **domain** and an **indistinct hinge** region. IgE is found at low concentration free in the circulation of humans, but at proportionally higher concentration in the blood of most domestic animal species. This is thought to relate to the relatively greater level of parasitism of domestic animals, as one of the major beneficial roles of IgE is as a participant in the immune response to endoparasites such as helminths. The IgE produced by plasma cells becomes attached to specific Fcε receptors on the surface of tissue mast cells (**Figure 8.23**) and circulating basophils. This sensitizes the mast cells to the presence of the parasite (see Figure 8.13) and cross-linkage of the IgE on the cell surface triggers degranulation and release of histamine as well as other toxic and pro-inflammatory molecules. IgE is sometimes produced

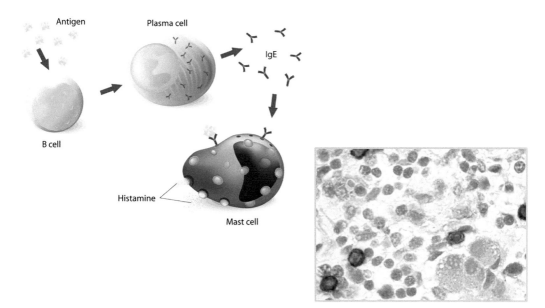

Figure 8.23 IgE sensitizes mast cells. Plasma cells produce IgE, typically against helminth parasites. This is adsorbed onto the surface of mast cells as they express the high affinity IgE receptor (FcεR). Inset: Section of intestinal mucosa from a foal labelled using immunohistochemistry to show the presence of IgE (red colour).

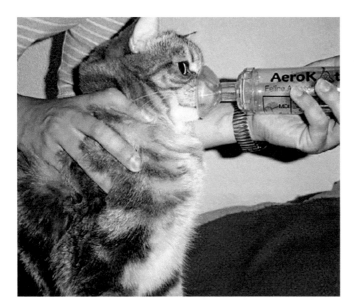

Figure 8.24 IgE is involved in allergic disease. This cat is receiving inhaled medication (a bronchodilator drug) through a purpose-designed delivery system for the treatment of asthma. Asthma is an immune-mediated disease involving the excessive production of IgE antibodies against environmental antigens.

inappropriately and is involved in the immunological phenomenon known as **type I hypersensitivity** (see Chapter 16). Related to this mechanism is the key role of IgE in mediating a range of **allergic diseases** (e.g. asthma, atopic dermatitis, flea allergy dermatitis) of major significance in human and animal populations (see Chapter 17) (**Figure 8.24**).

KEY POINTS

- Most antigens are foreign to the body and can contain linear and conformational epitopes.
- Haptens are small (often inorganic chemicals/drugs) molecules that can stimulate antibody responses when bound to host proteins (such as plasma albumin).
- Antibodies are Y-shaped molecules composed of two identical heavy and two identical light chains.
- The Fab region of the immunoglobulin molecule (consisting of the paired V_H and V_L chains) determines its antigen specificity.
- The Fc region of the immunoglobulin molecule (consisting of C_H2-C_H3/C_H4 domains) determines the biological consequences after antibody binding to its target antigen.
- Antibody binding can neutralize pathogens or their toxins, bind cellular receptors (opsonization, ADCC and mast cell sensitization) and activate the complement system.
- The five classes of antibody (IgD, IgM, IgG, IgA and IgE) have a distinct structure, anatomical distribution and function.
- IgD is expressed on naïve B cells and functions as an antigen receptor.
- IgM in monomeric form acts as an antigen receptor and in pentameric form is secreted during the primary immune response.
- IgM antibody is an agglutinin and is a potent activator of the classical pathway of the complement system.
- IgG is secreted as a monomer and is an adaptable type of antibody in terms of its ability to contribute to neutralization, opsonization, ADCC and complement activation.
- IgA is important antibody for mucosal immunity and is present in secretions where it acts predominantly as neutralizing antibody.
- IgE is produced in response to helminth parasites and enhances mast cell responses.
- IgE can be involved in allergic reactions.

OBJECTIVES

At the end of this chapter, you should be able to:

- Describe the difference in how T and B lymphocytes recognize antigen.
- Distinguish between a T cell-independent and a T cell-dependent antigen.
- Describe the three signals required for B cell activation.
- Summarize the events that occur within the secondary lymphoid follicles during B cell activation.
- Understand the mechanism of the immunoglobulin class switch.
- Describe the differences between primary and secondary humoral immune responses.
- Relate the secondary immune response to B cell memory.
- Describe how monoclonal antibodies are produced.

9.1   INTRODUCTION

There are two types of B lymphocytes, termed B1 and B2. B1 lymphocytes are thought to arise from stem cell precursors in the fetal liver or abdominal omentum in early life, after which time the population is maintained by self-renewal through cell division. There is, however, debate as to whether B1 and B2 cells derive from unique precursors or have a common precursor from which particular selection processes give rise to the two lineages. B1 lymphocytes have a restricted receptor diversity with limited V region repertoire and produce low affinity **polyreactive immunoglobulins** (IgM and IgA) with a broad antigenic specificity. These cells are recognized in a number of species, including pigs and ruminants. In rodents, they are found in the abdominal and thoracic cavities, while a large proportion of IgA plasma cells in the intestinal tract are of B1 origin. B1 cells and the antibodies they secrete may be considered part of the innate immune defence of these mucosal surfaces.

Conventional B2 lymphocytes recognize antigen in the extracellular fluid using their surface immunoglobulin as the B cell receptor (BCR) and when fully activated (often with T cell help), they transform into plasma cells and secrete antibody, which disseminates around the body (**Figure 9.1**). This chapter considers the development and maturation of these B cells in secondary lymphoid tissues and the effector and memory functions they perform as part of adaptive immunity.

DOI: 10.1201/9781003310969-9

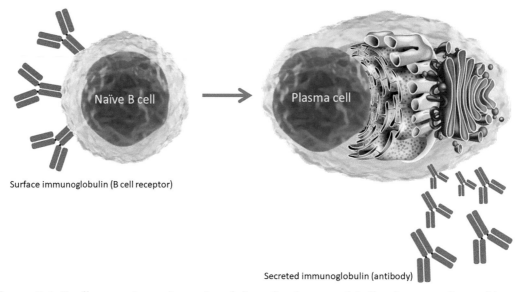

Surface immunoglobulin (B cell receptor)

Secreted immunoglobulin (antibody)

Figure 9.1 **B cells recognize antigen using their surface immunoglobulin, then reconfigure this molecule for secretion as antibody.** Naïve B cells express surface immunoglobulin (IgM and IgD), which functions as the B cell receptor (BCR). When activated, the B cells proliferate and differentiate to plasma cells and secrete the immunoglobulin (IgM, IgG, IgA or IgE) as antibody.

9.2 ANTIGEN RECOGNITION BY B CELLS

Lymphocytes are able to respond to antigenic stimulation, but B cells and T cells do so in an entirely distinct fashion. As we discussed in Chapter 7, the TCR recognizes small peptide fragments from processed antigen, presented in association with MHC molecules. In contrast, the **BCR** (surface immunoglobulin; sIg) interacts with a much larger area of an **intact antigenic epitope** that generally has **conformational or planar structure**. Consequently, the B cell has no requirement for antigen processing and may directly recognize antigen, usually on the surface of a pathogen in the extracellular fluid. Antigen may sometimes be deposited on the surface of follicular dendritic cells, but the epitopes remain intact and are bound by the BCR without any requirement for prior processing. Although the surface immunoglobulin (IgD and IgM on naïve B cells) is responsible for recognition of antigen, similar to the situation with the TCR and the CD3 com-

plex, the recognition signal requires additional cell surface molecules. Signal transduction for the BCR is undertaken by a complex of CD79a and CD79b molecules, alongside co-receptors CD19 and CD21, which have a somewhat analogous signalling capability to the T cell co-stimulatory molecule, CD28 (**Figure 9.2**).

B cell recognition of antigen may be **T cell-independent** (**TI**) or **T cell-dependent** (**TD**). T cell-independent antigens are limited to those molecules composed of polymers of simple repeating structural units (such as bacterial polysaccharide). Such TI antigens are able to cross-link BCRs and trigger stimulatory signals for B cell activation in the absence of T cell help (**Figure 9.3**). However, this type of antigen stimulation is relatively weak, with transient low levels of IgM antibody and no memory cells produced. In contrast, the majority of antigens recognized by B cells are proteins, which are only able to fully activate these cells with assistance from T helper cells (i.e. they are T cell dependent).

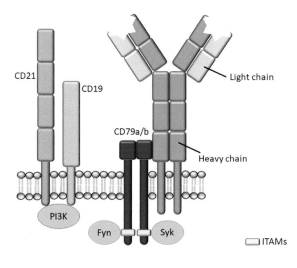

Figure 9.2 The BCR requires additional cell surface molecules for signal transduction. The surface immunoglobulin associates with CD79a and CD79b (also known as Igα and Igβ), which express 'immunoreceptor tyrosine-based activation motifs' (ITAMs). These can associate with various tyrosine kinases (e.g. Syk, which is the B cell equivalent of the T cell ZAP-70). The PI3 kinase associated with CD19 and CD21 acts in the same manner as that associated with the CD28 molecule found on T cells.

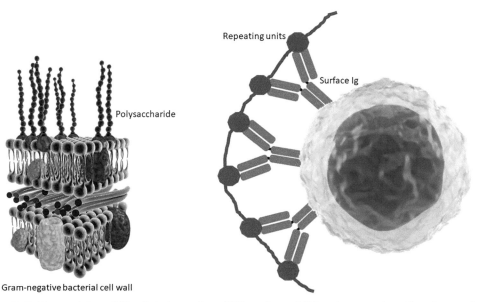

Figure 9.3 Recognition of T cell-independent (TI) antigen. A TI antigen consists of a polymer of repeating subunits (such as bacterial polysaccharide) that are able to cross-link BCRs and activate the B cell in the absence of T cell help.

9.3 ACTIVATION OF B LYMPHOCYTES

In the bone marrow, the earliest form of B cells (the **pre-B cell**) is a precursor that produces the μ heavy chain within the cytoplasm. In the next stage of development, complete IgM monomers are synthesized and the **immature B cell** displays these on its plasma membrane. The immature B cell leaves the bone marrow and undergoes a final maturation step in secondary lymphoid tissues such as the spleen. Mature naïve B lymphocytes, characterized by the co-expression of surface **IgM and IgD** molecules, become part of the recirculating B cell pool.

Activation of a naïve B cell is not dissimilar to the process of T cell activation described in Chapter 7. The B cell requires three specific signals in order to become fully activated (**Figure 9.4**):

- **Signal 1** for B cell activation is the **recognition of antigen by the BCR**.
- **Signal 2** comprises **intermolecular interactions** between surface molecules on that B cell and an antigen-specific T helper cell. One such interaction results from the B cell acting as an APC, internalizing antigen and presenting peptide fragments associated with MHC class II molecules. These can be recognized by the TCR of a CD4$^+$ T helper cell. This allows further interaction between the two cells, involving the B cell surface molecule **CD40** binding to the **CD40 ligand** expressed by the T helper cell.

- **Signal 3** for B cell activation takes the form of a **co-stimulatory cytokine** released by the T helper cell that binds to cytokine receptors on the surface of the B cell. A range of cytokines may act in this manner (depending on the nature of the antigen, the subset of T helper cell and the type of immune response required) including IL-4, IL-5, TGF-β and IFN-γ.

The delivery of all three signals permits activation of the B cell, which transforms morphologically to become a **lymphoblast**. These cells upregulate expression of surface MHC class II and undergo a unique phenomenon, the '**immunoglobulin class switch**', whereby its surface membrane IgM and IgD are replaced with a receptor of a single immunoglobulin class (IgG, IgA or IgE). In addition, the activated B cell undergoes **clonal proliferation and differentiation**, generating large numbers of antigen-specific B lymphocytes (**Figure 9.5**).

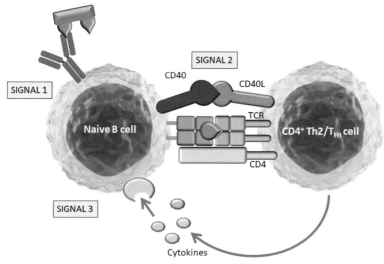

Figure 9.4 Activation of the B lymphocyte. B cell activation requires three signalling events. Signal 1 for the B cell is the recognition of the antigen by the BCR. The antigen is internalized and the B cell presents peptide epitopes in association with MHC class II to recruit T cell help. Signal 2 requires cell:cell contact between the B cell and the CD4$^+$ Th2/T$_{FH}$ cell via CD40 engagement by CD40 ligand. Signal 3 occurs following delivery of cytokines (including IL-4 and IL-5) derived from the Th2/T$_{FH}$ cell binding to cytokine receptors on the surface of the B cell.

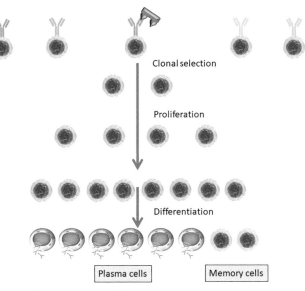

Figure 9.5 B cell clonal proliferation and differentiation. Activation of the naïve B cell leads to clonal division, which is further assisted by T helper cell-derived cytokines. The terminal stage of differentiation for a B cell is the effector B cell or plasma cell that is the source of antibody (initially IgM, but subsequently IgG, IgA or IgE, following class-switching). These plasma cells may be short or long lived. A population of long-lived B memory cells is also produced.

The effector stage of B cell activation involves transformation of these B cells into **plasma cells**, a process that requires the input of further cytokine signalling (e.g. IL-6, IL-11). Plasma cells will synthesize and secrete immunoglobulin of the same antigen specificity as the parent B cell and of the same immunoglobulin class determined by class switching. In addition to effector cells, some activated B cells differentiate to become **memory B lymphocytes**.

At this point, it is possible to consider all the stages of the adaptive immune response that lead to antibody production (**Figure 9.6**) and involve the triad of dendritic cell, T cell and B cell. The precise location of these cellular interactions within lymphoid tissues has been defined by studies in which antigen-specific T and B lymphocytes were labelled and tracked as the adaptive immune response was generated. The **interaction between dendritic cells and CD4⁺ T cells** occurs within the lymph node **paracortex**, as cells from the recirculating pool of naïve and memory T cells enter through HEVs. The activated T helper cells then migrate to the edge of **primary follicles** where they may recognize the antigenic peptide presented by MHC class II on the antigen-specific B cell and, in return, provide co-stimulatory signals for the B cell's activation and differentiation.

The activated B cells undergo clonal proliferation within the follicle, which transforms to become a **secondary follicle**, with a germinal centre and surrounding mantle zone. The **germinal centre** consists of a **dark zone** and a **light zone**, the former being the site of the most active proliferation and containing 'centroblasts'. The B cells that form by clonal proliferation migrate from the dark to the light zone, where they are known as 'centrocytes' and are able to re-encounter antigen associated with **follicular dendritic cells** and also encounter antigen-specific T_{FH} cells (**Figure 9.7**). It is within the light zone that there is a further opportunity for 'fine tuning' of the

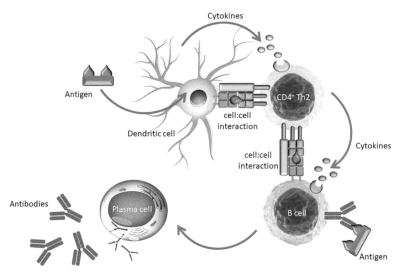

Figure 9.6 Generation of the adaptive immune response. This diagram summarizes the key stages in the generation of the adaptive immune response. Antigen is taken up by APCs, processed and presented by MHC class II molecules. The naïve CD4+ T cell is activated by the APC following the delivery of signals 1–3. Activated T helper cells subsequently provide appropriate stimulation for the activation of antigen-reactive B cells, via provision of signals 2 and 3. B cell activation leads to terminal differentiation to an antibody-producing plasma cell.

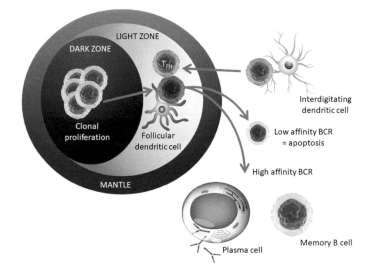

Figure 9.7 Anatomical location of T and B cell activation. Antigen-laden dendritic cells enter the lymph node paracortex where they activate antigen-specific T cells from the recirculating pool that have entered the paracortex via HEVs. Antigen-specific T cells migrate to the edge of a primary follicle where they encounter antigen-specific B cells from the recirculating B cell pool. Activation of the B cell leads to formation of the germinal centre. Active B cell proliferation occurs within the dark zone of the germinal centre. B cells thus formed migrate into the light zone where they encounter T_{FH} cells and antigen present on the surface of follicular dendritic cells (FDC). B cells that have formed low-affinity BCRs fail to recognize antigen, do not receive T cell help and die by apoptosis. B cells with high-affinity receptors are preferentially selected and undergo terminal differentiation to plasma cells or memory B cells.

B cell response, in a process known as **affinity maturation**. During clonal proliferation, each generation of dividing B cells may acquire additional somatic mutations in their BCR gene segments, leading to subtle modifications in the antigen-specificity of the expressed surface immunoglobulin. If these mutations create a receptor with **lower affinity** for antigen, the B cell will fail to effectively bind to antigen and they undergo **apoptosis**. The elimination of B cells carrying low-affinity receptors (**negative selection**) occurs on a large scale, with almost one in every two developing B cells being eliminated. Similar to the thymus, evidence of this may be seen on microscopic examination of a germinal centre in which there is significant macrophage phagocytosis of apoptotic cell debris. This process also selects against B cells that inadvertently become self-reactive and could potentially produce autoantibodies. In contrast, those B cells with **higher affinity receptors** will effectively capture antigen and receive T cell help, thus allowing them to progress and become plasma cells or memory B cells. Selected B cells migrate from the follicles to take up residence in the lymph node medullary cords, splenic red pulp, bone marrow or mucosal lamina propria, where they differentiate into plasma cells and secrete their antibodies.

Early in the immune response, when antigen is abundant, a greater number of B cells may be positively selected in this manner; however, in the late stages of the adaptive response only those B cells with the highest-affinity receptors will be allowed to mature. Thus, as the immune response progresses, antibodies are produced with increasing affinity for their target antigen, making the response more effective. Differentiation to a memory cell phenotype typically occurs towards the end of the immune response as the pathogen is eliminated. These cells re-enter immune surveillance, alongside the naïve lymphocytes and can become re-activated upon subsequent exposure to the same pathogen in the **secondary response**.

9.4 THE IMMUNOGLOBULIN CLASS SWITCH

When a B cell is first activated, it adjusts production of surface immunoglobulin (IgD and monomeric IgM) to synthesis of IgM, without its transmembrane domain. In parallel, synthesis of the joining (J) chain protein allows polymerization of IgM into a pentameric molecule that can be secreted as antibody. IgM is the first antibody produced in the immune response and is a major feature of the **primary response**. With T cell help and, under direction of other cytokine signals, the heavy chain gene used for synthesis of the immunoglobulin molecule can be switched to produce different classes of antibody, best suited to the type of infection present. Thus, **immunoglobulin class switching** is the process whereby the most appropriate type of antibody (IgG, IgA or IgE) is selected. Therefore, V(D)J recombination in primary lymphoid tissues (and subsequent somatic mutation in secondary lymphoid tissues) establishes the antigen-specificity of the antibody response and class-switching in secondary lymphoid tissues determines the biological response, once the antibody has bound to its target.

The class switch is achieved by the ability of the B cell to undertake **multiple DNA rearrangements**. In naïve B cells the heavy chain variable region was established by VDJ recombination when the cell was in the bone marrow. The VH segment sits upstream of the heavy chain constant genes (CHμ, CHδ, CHγ, CHε and CHα). A primary transcript, produced during transcription, and alternative RNA splicing allows the formation of both IgM and IgD (**Figure 9.8**). During **class switching**, a **second DNA rearrangement** removes various sections of intervening DNA, leading to the formation of the final class (isotype) of immunoglobulin. This comes about through recombination between '**switch regions**', which are sequences of intronic DNA that lie upstream of each constant region gene (except for Cδ). The μ chain

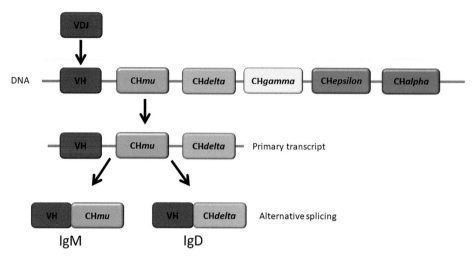

Figure 9.8 Naïve B cells synthesize both IgD and monomeric IgM as surface immunoglobulin.
Naïve B cells have undergone VDJ recombination to determine the variable region of the immunoglobulin heavy chain. Transcription of the gene leads to both CH*mu* and CH*delta* elements in the primary RNA transcript. By a process of alternative splicing, two mRNA species are produced, one encoding IgM and the other encoding IgD. These then form immunoglobulin monomers (2× heavy chains combining with 2× light chains) and are expressed on the cell surface, where they combine with other molecules of the BCR complex to function as the antigen receptor.

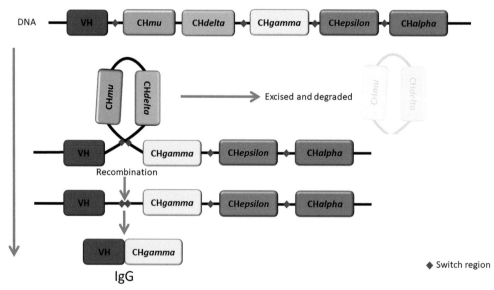

Figure 9.9 Class switching from IgM to IgG. Following antigenic stimulation of the B cell, with additional T cell help (both cell:cell contact and provision of cytokines), DNA rearrangement takes place at the immunoglobulin heavy chain locus. The CH$_\mu$ and CHδ gene segments are looped out and deleted and the switch region preceding the CH$_\mu$ is brought into apposition with the switch region upstream of the CHγ gene segment. This allows the B cell to start synthesizing IgG.

switch region (Sμ) recombines with the switch region of the selected immunoglobulin class with looping out and deletion of intervening sequences in a process that probably involves enzymes similar to those of the V(D)J recombinase complex (**Figures 9.9, 9.10**).

In the case of IgA, class switching may be T dependent or T independent in nature.

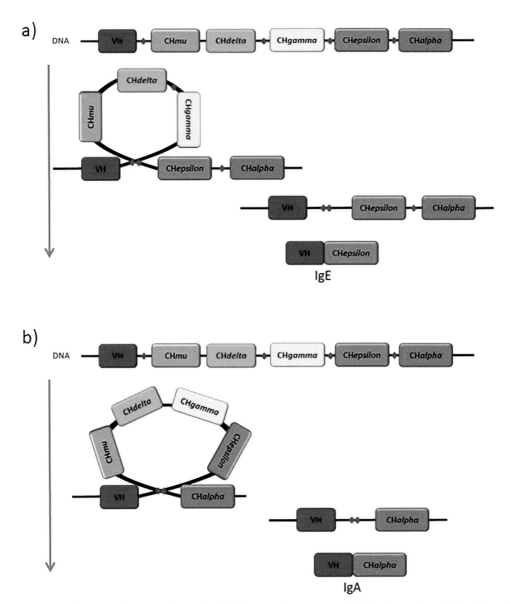

Figure 9.10 Class switching to IgE or IgA. Following interaction with T helper cells, additional cytokine signals can stimulate class-switching to more specialist types of antibody, such as (a) IgE or (b) IgA. In this instance, larger segments of the immunoglobulin heavy chain locus are excised to bring either CHε or CHα in apposition with the VH segment, allowing transcription and translation to proceed for secretion of IgE or IgA antibodies.

T-independent IgA class switching is induced by binding of the cytokines BAFF (B cell activating factor), APRIL (a proliferation-inducing ligand) and transforming growth factor (TGF)-β to receptors on the surface of B cells in the mucosa-associated lymphoid tissue. These stimulatory cytokines are produced by mucosal dendritic cells and epithelial cells.

9.5   KINETICS OF THE ANTIBODY RESPONSE

Exposure to **T independent antigen**, leads to a relatively rapid response from antigen-specific B cells and specific antibody (IgM) may be detected in the serum within **two to seven days**. However, the response to TI antigens is relatively weak (low antibody titre) and transient, typically only lasting three to four weeks. There is no class-switching in response to TI antigens; since they are not proteins they cannot be processed and presented by MHC class II molecules in order to solicit T cell help. Furthermore, they cannot stimulate production of memory lymphocytes, so repeated exposure does not lead to any enhanced response over time (**Figure 9.11**).

Following exposure to **T cell-dependent antigens** it generally takes somewhere between **four and ten days** before antibody can be detected in the serum. This is because the response requires sequential stages of antigen presentation, T cell activation, B cell activation and the generation of an effector population of plasma cells (see **Figure 9.6**), which all must be coordinated in the secondary lymphoid tissues. This phenomenon is readily demonstrated in a classical immunological experiment that describes the kinetics of a **primary and secondary immune responses** (**Figure 9.12**). In this experiment a naïve animal is injected with foreign antigen on day 1 in order to induce the primary response. The development of the antibody response is monitored by daily blood sampling and testing for the **titre** (concentration) of antigen-specific antibody within the serum. For the first few days after injection, no serum antibody is detectable (**the lag phase**),

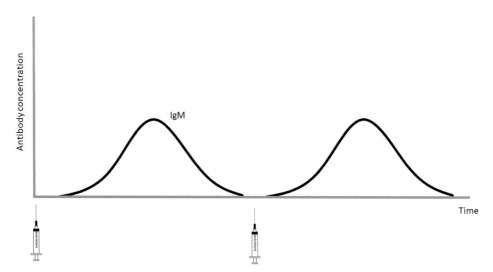

Figure 9.11 Antibody response to TI antigen. Serum antibody (IgM) can be detected relatively quickly after exposure to a TI antigen, but the titre is low and the response is transient. Repeated exposure to the same antigen results in a similar response each time.

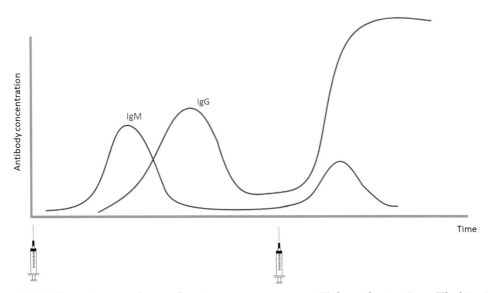

Figure 9.12 The primary and secondary immune response to a T-dependent antigen. The kinetics of serum antibody production following primary and secondary immunization with a non-replicating antigen are shown. The secondary response has a shorter lag phase, is of greater magnitude, mostly comprises class-switched immunoglobulin (IgG) of greater affinity and persists for a longer period of time. Memory B cells mediate this secondary response.

but after five to seven days, a low-titre of IgM antibody appears. Over the following few days, as the immunoglobulin class switch proceeds, this will be joined by detectable IgG antibody that gradually replaces the IgM as the dominant class of immunoglobulin. This humoral response peaks at around two to three weeks after initial exposure and the concentration of serum antibody will gradually decline, as most immunoglobulins have a half-life of around 10–14 days.

Once the primary immune response has subsided, the presence of a population of **memory B cells** is readily demonstrated by repeating the injection with the same antigen. This induces a serum antibody response with strikingly different kinetics:

- There will be a **much shorter lag phase** before antibody becomes evident within the serum.

- The **antibody titre will be significantly greater** than was seen in the primary response and this higher titre will be **achieved relatively quickly**.
- The antibody will **primarily be IgG** (or whichever class of immunoglobulin was selected during the primary response).
- The antibody will persist within the serum for a much longer period of time with a **plateau of high titre** before **gradual decline** (often months, sometimes years later).

The secondary immune response is considerably more potent than that following primary exposure. The antibodies produced will also be of much **higher affinity** than in the primary response as the proliferating B cells within germinal centres will continue to undergo **affinity maturation**. The secondary response is mediated by the memory B cells that were generated within the primary activation event. These cells are poorly understood, but are thought to be a relatively **long-lived population** that might undergo periodic low-level division upon restimulation by antigen held by follicular dendritic cells. Memory B cells may persist for the lifetime of the individual animal. Additionally, there are populations of **long-lived plasma cells** that can survive for several years and continue to produce antibodies. These cells may preferentially reside in the bone marrow and mucosal tissues. The induction of memory B cells and long-lived plasma cells and the persistence of serum antibody derived from the latter population are fundamental to vaccination strategies, which will be discussed further in Chapter 12.

9.6   MONOCLONAL ANTIBODIES

Any discussion of B cell biology would not be complete without considering the production and use of **monoclonal antibodies**, a major immunological breakthrough that has revolutionized the practice of medicine and led to the award of the 1984 Nobel Prize to Jerne, Kohler and Milstein, who first described this technique. As described previously, immunization of an animal with a complex antigen bearing many epitopes will lead to generation of a **polyclonal** immune response, since numerous lymphocyte clones will respond to the different antigenic epitopes. Serum from such an animal will contain antibodies of many different specificities and affinity for the antigen as a whole, which is referred to as a **polyclonal antiserum**. In contrast, a monoclonal antibody is derived from a single B cell clone and therefore represents a single species of high-affinity antibody of a single immunoglobulin class, reactive against a defined antigenic epitope. Monoclonal antibodies therefore provide a much more refined means of serological testing and these

immunological reagents have found wide application in diagnostic tests and, more recently, as therapeutics. Because of the exquisite specificity of such antibodies, when injected they will target and bind only the antigen against which they have been selected to interact. Such antibodies may be used to block receptors for infectious agents, to target toxins or drugs to tumours and to block immunological pathways by neutralizing cytokines or specific cell surface molecules.

Although monoclonal antibodies are produced *in vitro*, their generation often requires the use of an experimental animal. Historically, mice have been most widely used for the initial steps in monoclonal antibody production, but it is possible to use animals of any species. The original method for generating a monoclonal antibody is as follows (**Figure 9.13**). A mouse is injected with the target antigen and will make a polyclonal immune response. The spleen is subsequently removed and the lymphocytes isolated and placed into cell culture. This lymphoid population will contain some B cells that are specific for the antigen and others that are not. The second component required to produce a monoclonal antibody is a non-secreting **myeloma cell line**. The key stage is the process of cellular '**fusion**', whereby individual B lymphocytes from the immunized mouse are fused with a myeloma cell in the presence of a chemical, **polyethylene glycol** (PEG). When fusion is successful, a **hybridoma cell** is produced that has the properties of both parent cells (i.e. it produces antibody and it is immortalized). There follows a period of selective culture, whereby cells that have not fused are eliminated. This is achieved using a combination of chemicals: **hypoxanthine, aminopterin and thymidine** (HAT). Unfused myeloma cells die in the presence of HAT as they do not possess the necessary enzymes to metabolize these compounds. Unfused mouse lymphocytes will die simply because they are not immortal.

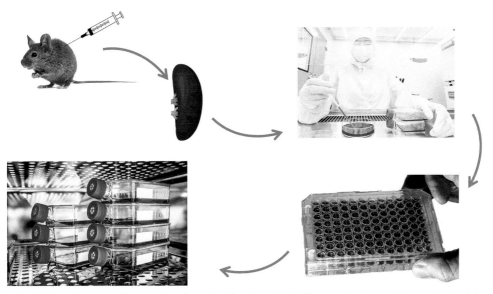

Figure 9.13 **Production of monoclonal antibodies from hybridomas.** A mouse is immunized with antigen so that reactive B cells are activated in that animal. The spleen is removed and a suspension of cultured splenic lymphocytes is fused with a myeloma cell line by addition of polyethylene glycol (PEG). The cells are plated out in a medium containing hypoxanthine, aminopterin and thymidine (HAT). Unfused splenocytes die by attrition and unfused myeloma cells die as they lack appropriate enzymes to metabolize HAT, while hybrid cells will survive. Wells containing hybridomas are diluted to a point where a single cell is present within each well. The cells are left to proliferate and, if a clone is identified that produces an antibody of the required specificity, it is selected for bulk culture and production of monoclonal antibody.

Hybridoma cells, however, will survive and replicate because they possess properties of both parent cells.

At this stage, the surviving cells are titrated so that individual cells are placed into single wells of culture plates. These need to be screened to identify those clones that are producing antibody of the desired specificity (i.e. against the immunogen). A small sample of medium is collected and tested in an **immunoassay** (most commonly an ELISA). The cells from positive wells will be selected and **bulk cultured** to produce large quantities of monoclonal antibody, which can be harvested and purified for use in research, diagnostics or therapy. In order for monoclonal antibodies to be used therapeutically, the protein sequence of the antibody has to be exchanged from mouse to the target species, apart from the complementarity-determining regions that make up the antigen binding sites. This process of 'caninization' or 'felinization' has been used successfully to generate therapeutic monoclonal antibodies for the treatment of atopic dermatitis in dogs and of chronic pain in dogs and cats.

More recently, advances in molecular techniques (e.g. generation of antibody phage-display libraries) means that synthetic monoclonal antibodies can now be produced, without requiring experimental animals. These synthetic monoclonal antibodies are not usually full immunoglobulin molecules and are often **single chain Fv,** consisting of fused heavy and light chain variable regions (**Figure 9.14**).

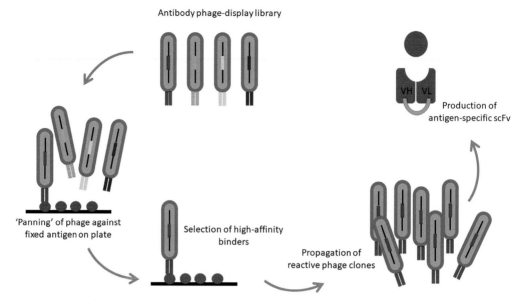

Figure 9.14 Synthetic monoclonal single chain Fv. Antibody phage-display libraries are now available that express antibody fragments on their surface, representing the antibody repertoire of a human. The target antigen is immobilized on a plate and the phage-display library is allowed to interact ('panning'). Only those phages expressing a high affinity receptor to the target antigen will bind. The bound phages are recovered and propagated, and after screening, suitable clones are selected for production of monoclonal recombinant single chain Fv (scFv) protein, which represents a dimer of VH and VL elements of an antibody molecule.

KEY POINTS

- The BCR often recognizes a large conformational determinant of antigen, typically on the surface of a pathogen in the extracellular fluid.
- T cell-independent antigens are polymers of repeating units that cross-link BCRs and directly activate B cells.
- T cell-dependent antigens require T cell help for B cell activation.
- Naïve B cells co-express surface membrane IgD and IgM.
- B cell activation requires signal 1 (BCR antigen recognition), signal 2 (interaction with T helper cell co-stimulatory molecules) and signal 3 (T helper cell-derived cytokines).
- Activated B cells undergo clonal proliferation and differentiation to form effector B cells (plasma cells) and memory B cells.
- B cell clonal proliferation occurs within the germinal centre of secondary follicles.
- Affinity maturation of the BCR during clonal proliferation leads to selection of those B cells bearing high-affinity receptors.
- Activated B cells initially secrete pentameric IgM antibody.
- T cell help leads to B cell DNA rearrangement that underlies the immunoglobulin class switch from IgD/IgM to IgG, IgA or IgE.

- The secondary humoral immune response has a short lag phase and a more rapid production of a high titre of class-switched antibody, which persists for a longer period of time.
- The secondary humoral immune response is mediated by long-lived memory B cells.
- Monoclonal antibodies are of a single specificity high-affinity antibody generated *in vitro* by fusing immunized splenocytes with a myeloma cell line.
- Monoclonal antibodies have wide application in research, diagnosis and immunotherapy.

IMMUNODIAGNOSTICS

Serology, Immunoassays and Measurement of Cell-Mediated Immunity

---

OBJECTIVES

At the end of this chapter, you should be able to:

- Define 'serology' and understand the role of serological testing in clinical veterinary medicine.
- Describe the production of an antiserum.
- Briefly describe the different types of immunoassays employed for diagnostic purposes.
- Understand why it would be useful to test the function of lymphoid or phagocytic cells in a clinical situation.
- Describe the principle of the lymphocyte stimulation test.
- Describe the principle of evaluation of lymphocyte or NK cell cytotoxic function.
- Describe how the function of phagocytic cells may be evaluated.

10.1 INTRODUCTION

Serology is the *in vitro* study/analysis of **antibodies in serum, plasma or other biological fluids** and is commonly undertaken in clinical medicine. Practising veterinarians will regularly employ serological testing as part of their diagnostic approach to animals suspected of suffering from infectious or immune-mediated disease or sometimes to determine whether an animal has responded appropriately to vaccination (or requires a booster vaccine). Other immunoassays use specific antibodies for analysis of clinical samples to detect the presence of antigens associated with infection by a specific pathogen or to measure a biomarker (e.g. hormone or cytokine).

When dealing with an infectious disease, it is important to appreciate the stage of infection, which will dictate whether the best diagnostic approach is to look for **antigen at the site of infection** or to look for **antibodies in the blood** (**Figure 10.1**). More refined serological tests may be used to quantify the amount of antibody present and by repeat blood sampling of a patient (acute and convalescent serum), one can determine whether the amount of antibody might be increasing (as is the case in an active infection). Some serological tests are designed to **detect either IgM or IgG** antibody and the relative proportions of these may indicate the **stage of the infection**, with the former being indicative of recent exposure.

The level of specific antibody present in a sample is defined as the **antibody titre**. A titre is the reciprocal of the highest serum dilution giving an unequivocally positive reaction in a serological test (**Figure 10.2**). In some assays,

DOI: 10.1201/9781003310969-10

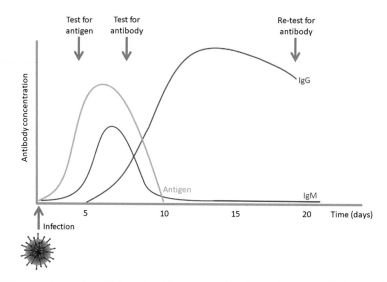

Figure 10.1 Diagnostic testing. If the animal is presented in the early stages of infection (e.g. a puppy affected with canine parvovirus), samples should be taken from the site of infection (i.e. a faecal sample in this example) and a diagnostic test employed to detect the presence of antigen. An antibody test on a blood sample would be (false) negative during the lag phase. Later in the course of infection, antibodies can start to be detected in the blood, with IgM the first produced, typically followed by a rise in IgG (i.e. IgM > IgG during the early stages). After recovery (convalescence), a high titre of IgG antibodies is usually present, which persists for a period of time before going into decline.

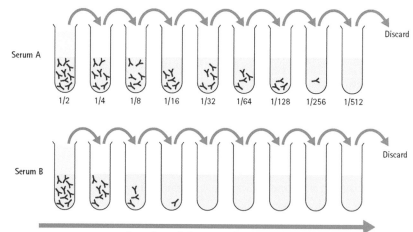

Figure 10.2 Antibody titre. Serum samples are taken from two animals, A and B. Both animals have been exposed to the same antigen, but the serum from animal A contains more antibodies than that of animal B. A series of test tubes is set up for each serum sample with a fixed volume of diluent in each one. The serum samples are diluted 1 in 2 repeatedly from one tube to the next. This process is repeated along the line of test tubes to produce a 'doubling dilution series' in which the final tube has a 1/512 dilution of the original sample. The contents of each tube are now tested in the immunoassay. For the serum from animal A the test will be positive up to and including the tube containing the 1/256 dilution of serum. For serum from animal B the test will be positive up to and including the tube containing the 1/16 dilution of serum. The titre of each sample is expressed as the reciprocal of the last positive serum dilution (i.e. a titre of 256 for animal A and 16 for animal B).

the antigens are large and particulate, allowing direct visualization of the reaction, when the two are mixed and form immune complexes, but in many tests both components (antigen and antibody) are too small to see and determination of their interaction requires an **indicator system**. This is often achieved by modifying one of the antibodies in the reaction by chemically adding a label/tag, such as an enzyme, fluorochrome, inorganic particle or radioligand.

An **antiserum** is derived from blood collected from an animal (most commonly a mouse, rabbit, sheep or goat) that has been immunized with a particular antigen of interest and therefore contains antibodies to that particular antigen. Where the immunizing antigen carries multiple epitopes, it is likely that the immunized animal will generate antibodies against several of these and such an antiserum is said to be '**polyclonal**' in nature. It is also possible to generate antibody of a single specificity and the production of **monoclonal** antibodies was discussed in Chapter 9. The challenge of identifying the presence of antibody binding to antigen in these assays has been overcome by the production of secondary antibody reagents. This is based on the concept that immunoglobulin proteins differ between species and an immunoglobulin from one species is immunogenic in another. Thus, when canine IgG is injected into a rabbit, this is seen as a foreign antigen and the rabbit makes antibodies against it. These antisera can then be chemically labelled and used to detect the presence of canine IgG binding to a specific antigen in a diagnostic immunoassay for dogs (**Figure 10.3**).

Cell-mediated immunity may also be studied *in vitro* by the application of an array of techniques. Components of cell-mediated immunity may be evaluated and enumerated within tissues by use of antibody-labelled cells as is the case in immunofluorescence microscopy or immunohistochemistry, or within fluid suspensions by the use of flow cytometry. Assessing the function of lymphoid and phagocytic cells can also be achieved *in vitro*, although these cellular

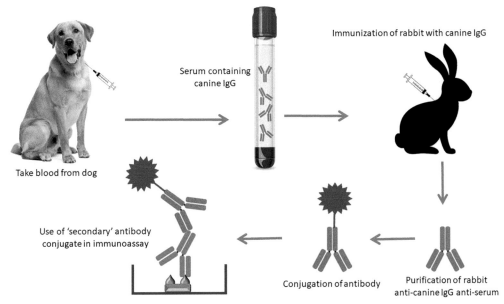

Figure 10.3 Production of antiserum. To produce an antiserum that is able to bind to canine IgG, blood is taken from a dog, the IgG is purified from the serum and injected into a rabbit. Once the rabbit has mounted an immune response to the foreign dog protein, anti-serum taken from the rabbit will contain antibodies specific for canine IgG. These antibodies can be labelled (e.g. with an enzyme, fluorochrome or radiolabel etc.) to generate an antibody conjugate that can be used to detect canine IgG binding in an immunoassay.

assays are more challenging from a logistical and technical perspective than immunoassays designed to detect the presence of antigen or antibody in clinical samples.

10.2 SEROLOGICAL TESTS AND IMMUNOASSAYS

A range of different serological tests and immunoassays are currently used for diagnostic purposes in veterinary medicine. Many of these tests are used by specialist diagnostic laboratories, but there are commercially available immunoassays designed for use in practice laboratories or patient-side. Immunoassays vary in their complexity; some are very simple and based on principles that were first defined over a century ago, whereas others have been developed relatively recently and their increased sophistication generally means that they have greater sensitivity

(reduced risk of false negatives) and specificity (reduced risk of false positives).

10.2.1 Agglutination

Agglutination is a relatively simple procedure that may be performed when the antigen of interest is **particulate** (e.g. RBCs, bacteria) or when soluble antigen is coated onto particles such as latex beads. Particulate antigen is mixed in optimum proportions with antibody and if the antigen bears appropriate epitopes, the antibody binds these and cross-links different antigens to form an **agglutinate**. The agglutinate may be visible to the naked eye or require light microscopic examination. A number of commonly employed veterinary immunodiagnostic procedures are based on the phenomenon of agglutination and blood-typing is shown as one example in **Figure 10.4**. The microscopic agglutination test (MAT) is the 'gold standard'

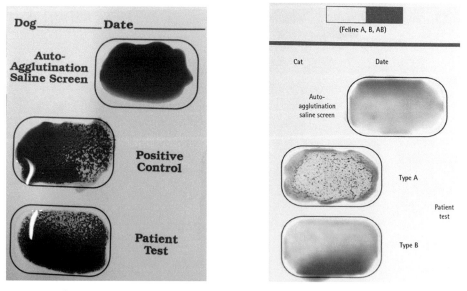

Figure 10.4 Agglutination. There are commercially available test kits to determine the blood group of companion animals, designed for rapid in-practice use where an emergency blood transfusion might be required. The 'wells' on the card are first activated by the addition of a diluent. A drop of blood from the animal is added to each well and mixed with a spatula. The top well contains no antiserum and serves as a (negative) control to ensure that the erythrocytes in the blood sample do not spontaneously agglutinate. Other wells are impregnated with antiserum, able to detect specific blood group antigens. The canine test includes a well with antiserum against a common red cell antigen ('Positive Control'). The 'Patient Test' well has antiserum against the DEA1.1 antigen, which is positive in this case. In the feline test, the cat's red cells express A, but not B, as seen by agglutination of red cells in the middle well after incubation.

diagnostic assay for serological diagnosis of leptospirosis and is based on mixing serum from the patient with different leptospira serovars and observing whether an agglutination reaction can be visualized by light microscopy (**Figure 10.5**).

10.2.2 Precipitation

This test relies on the ability of antibodies to precipitate antigen within solution or in an agar gel matrix, where it is known as the agar gel immunodiffusion (AGID) test. Although now mostly a historical technique, the method is still used in veterinary diagnostics for some infectious diseases, such as for equine infectious anaemia, where it is known as the Coggins test (**Figure 10.6**). The formation of precipitates depends on the two reactants (antigen and antibody) being present at optimum concentration, where they are able to form a classical 'lattice-like' structure.

10.2.3 Complement Fixation Test (CFT)

The CFT is used in the diagnosis of infectious disease (e.g. to determine whether cattle may have been exposed to *Brucella* antigens). Antigen is added to a tube containing serum from the test animal and if the animal is seropositive, an antigen–antibody complex will form. A source of complement (e.g. fresh plasma) is added to the tube and where an antigen–antibody complex is present, the complement will be activated and integrated into the complex. If the animal is seronegative, the complement will remain intact as the free antigen does not activate it. As all of these interactions are invisible to the eye, a secondary indicator system is employed made up of a second antigen–antibody complex. This system utilizes erythrocytes (often sheep red blood cells) that are pre-coated with anti-erythrocyte antiserum. The antibody-coated erythrocytes are added to the CFT tube. In a test from a seronegative animal, the free complement within the tube will be activated by the indicator system and result in haemolysis of the erythrocytes. Lack of haemolysis in this test indicates a positive reaction (**Figure 10.7**).

10.2.4 Haemagglutination Inhibition

The HAI test is commonly used in virology and relies on the ability of some viruses (e.g. influenza virus expressing haemagglutinin) to cause

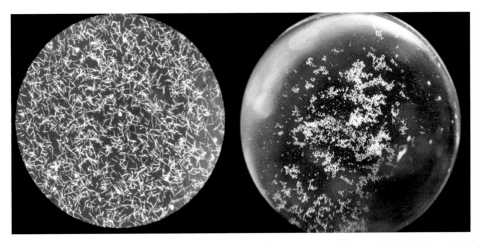

Figure 10.5 Microscopic agglutination test (MAT). This assay is used for serological diagnosis of leptospirosis, whereby different leptospira serovars are mixed with serum and the presence of agglutination visualized by light microscopy. The image on the left is of a well containing leptospires in the absence of serum. The well on the right shows agglutination in the presence of reactive serum antibodies. (Images courtesy of George Souter, Animal and Plant Health Agency and Collette Taylor, Royal Veterinary College.)

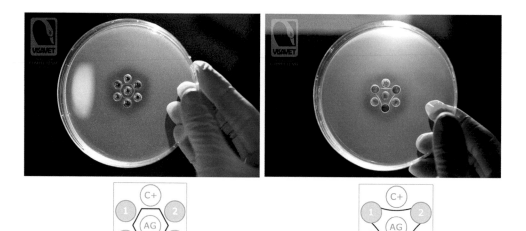

Figure 10.6 The Coggins test. This immunoprecipitation assay is used to detect serum antibodies against equine infectious anaemia virus. In a positive Coggins test (left), an immunoprecipitation occurs due to the binding of the antigen (central well) and the antibodies in the positive controls (C⁺) and the samples (1, 2 and 3). In the picture, the three samples (1, 2 and 3) are positive to EIA and therefore a line can be seen in all wells, giving the appearance of a hexagon. In a negative Coggins test (right), an immunoprecipitation occurs due to the binding of the antigen (central well) and the antibodies in the positive controls (C⁺), but the immunoprecipitation does not happen in the samples (1, 2 and 3). In the picture, the three samples (1, 2 and 3) are negative to EIA and therefore a line can only be seen in the wells containing positive controls, giving the appearance of a triangle. (Image courtesy of Fátima Cruz López, SEVISEQ. VISAVET Health Surveillance Centre. Universidad Complutense Madrid.)

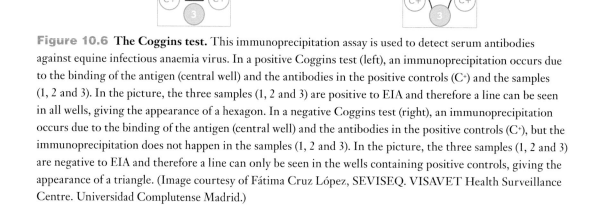

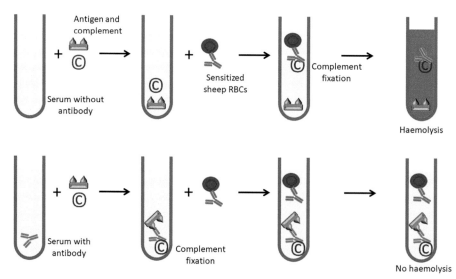

Figure 10.7 Complement fixation test. During the first stage of the test, serum is incubated with the antigen of interest and an immune complex will form where that animal is seropositive. A source of complement is subsequently added, but will only be utilized where an antigen–antibody complex is present. Finally, an indicator system of antibody-coated erythrocytes is added. Where the animal is seronegative, complement is still available to lyse the erythrocytes, providing a visual read-out for the test.

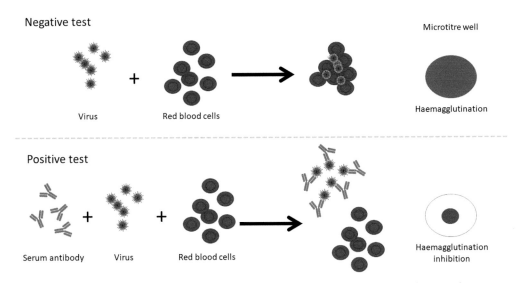

Figure 10.8 Haemagglutination inhibition. The ability of some viruses to induce haemagglutination can be exploited in a diagnostic test for the presence of serum antibodies. In the absence of antibodies, when the virus is mixed with red blood cells in a round-bottom microtitre well, agglutination can be observed as a 'shield' of red cells at the bottom of the well. If antibodies are present they will bind to the added virus and the red blood cells will remain in suspension, eventually forming a 'button' at the very bottom of the well (i.e. haemagglutination inhibition).

agglutination of erythrocytes from particular animal species. If a patient has serum antibody specific for the virus, pre-incubation of virus particles with antibody will inhibit the ability of the virus to subsequently cause haemagglutination when incubated with erythrocytes (**Figure 10.8**).

10.2.5   Immunofluoresence Antibody (IFA) Test

The IFA test has wide application in the diagnosis of a number of infectious diseases of animals. In the direct IFA, a **fluorochrome-labelled antibody** against a target antigen is used to 'probe' tissue sections for the presence of a pathogen (**Figure 10.9a**). In the indirect IFA, the infectious agent (either alone or associated with cultured cells in which it has been propagated) is applied to a microscope slide and serum from the patient is added to it. This primary interaction requires a secondary detection system. A fluorochrome-labelled secondary antibody, designed to detect immunoglobulin (usually IgG) of the species of interest, is subsequently layered over that region of the slide. When the slide is examined under a fluorescence microscope, the fluorochrome becomes excited and the reaction becomes visible (**Figure 10.9b**). A commonly used fluorochrome for this purpose is fluorescein isothiocyanate (FITC), which emits an apple-green fluorescence under ultraviolet light.

10.2.6   Virus Neutralization (VN) Assay

The VN test is also used to detect the presence of serum antibody specific for a virus. Test serum is incubated with live virus particles that are subsequently incubated with a monolayer of cells capable of being infected by that virus. The 'read out' for the test is the observation of a **cytopathic effect** (cell damage and destruction) in the cell monolayer. In a positive test, serum antibody will neutralize the virus, thereby preventing infection *in vitro*

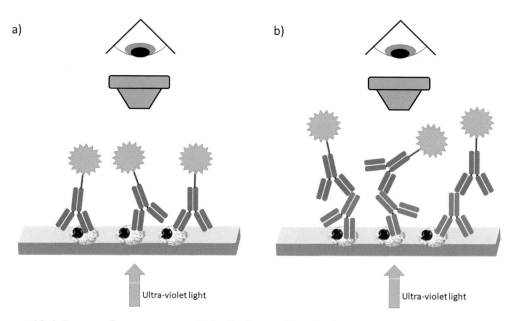

Figure 10.9 Immunofluorescence antibody (IFA) test. (a) In the direct IFA, a labelled antibody against the target antigen is used to probe tissues for the presence of a pathogen. (b) In the indirect IFA, a source of antigen is applied to a microscope slide then serum from the animal is added and allowed to bind. Unbound serum antibody is washed off and a fluorochrome-labelled secondary antibody is applied. Microscope slides are visualized using a fluorescence microscope.

and preserving the integrity of the target cells. For those viruses that do not cause an obvious cytopathic effect (such as rabies virus) an indicator system must be employed (in this case a fluorescent labelled antibody) to allow visualization of virus infection (or not) of the cells. This modified assay is known as the fluorescent antibody virus neutralization (FAVN) test (**Figure 10.10**).

10.2.7 Enzyme-Linked Immunosorbent Assay (ELISA)

The ELISA is probably the most common format of immunoassay currently used in veterinary diagnostic testing and forms the foundation from which newer modalities (e.g. immunochromatography, western blotting) were subsequently developed. There are two fundamental types of ELISA. The first (**indirect ELISA**) is designed to **detect the presence of serum antibody** specific for a particular antigen as evidence that the animal has been previously exposed to that antigen and has made

an immune response. The second (**sandwich ELISA**) is designed to **detect the presence of antigen in a clinical sample**, as evidence that antigen from an infectious agent is present in an animal with an active infection.

In a laboratory setting, the ELISA is most often performed in 96-well microtitre plates with flat-bottomed wells. The indirect ELISA technique is shown in **Figure 10.11**. The antigen of interest is first coated onto the bottom of the wells of the plate and a second unrelated protein (often bovine serum albumin or skimmed milk protein) is used to block any areas that are not coated by antigen. Appropriately diluted serum from the patient is added to each test well and positive and negative control sera may also be applied to control wells. If the test serum contains relevant antibodies, these will bind to the antigen in the primary phase of the reaction. As the antigen and antibody are not visible to the eye, a secondary detection system is required to determine whether such binding has occurred. In this instance, after

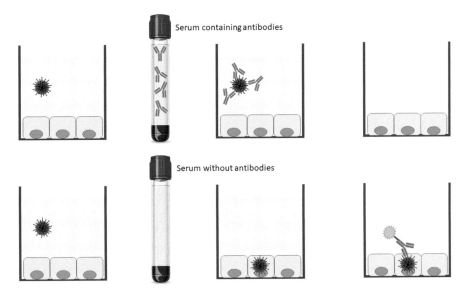

Figure 10.10 Fluorescent antibody virus neutralization (FAVN) test. Cells are cultured and exposed to virus in the presence of serum from the patient. Where virus-neutralizing antibody is present, the cells remain uninfected. In the absence of neutralizing antibodies, the virus infects the cells. The presence of intracellular virus antigen is detected using the direct IFA technique. The absence of fluorescence indicates a positive result in the test. This test is typically used to determine the antibody response following rabies vaccination.

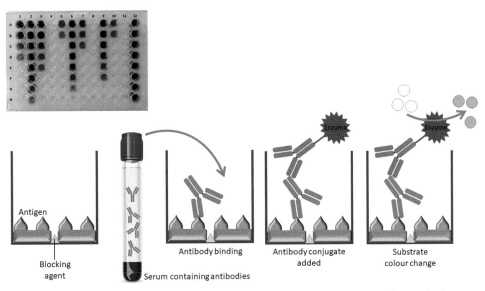

Figure 10.11 Indirect ELISA for detection of serum antibody. Antigen is coated onto the bottom of the wells of a 96-well microtitre plate. A blocking agent (e.g. skimmed milk protein) is added to cover up the remainder of the plastic surface. After washing, serum is diluted and added to the well and incubated to allow any antibodies to bind. After further washing, the secondary antibody conjugate is added and incubated. After final wash steps, the substrate is added, which changes colour to indicate the presence of bound immune complexes. Inset (top left) an ELISA plate following completion of the assay. Serial dilutions of test sera (columns 1–10) are shown; column 11 is a negative control and column 12 is a positive control. The sera show variable antibody titres against the target antigen.

unbound antibody is washed from the wells, an appropriately diluted antiserum, specific for the immunoglobulin of interest (e.g. rabbit anti-canine IgG when assaying dog serum samples—see **Figure 10.3**) is added. This antibody will have been chemically conjugated to an enzyme. A range of enzymes can be used, but the most commonly employed are **alkaline phosphatase** or **horseradish peroxidase**. If the patient serum contains antibodies specific to the antigen of interest, the enzyme-linked antiserum will detect and bind to that patient immunoglobulin (which is in turn bound to the antigen). After washing to remove any unbound antibody conjugate, an appropriate **substrate** is added to each well to initiate a **colour change**, which is measured spectrophotometrically at a particular time point when the enzyme–substrate reaction may be deliberately stopped. The same colour change would occur in wells incubated with positive control serum, but no colour should develop in negative control wells. The optimum technical conditions for performance of an ELISA (e.g. dilutions of antigen, serum and antiserum, time and temperature of incubations) must first be determined experimentally.

ELISAs may also be used for the detection of antigen within a biological fluid sample. The antigen may be derived from an infectious agent or be a host molecule (e.g. cytokine). This form of ELISA is known as a **sandwich ELISA (Figure 10.12)**. In this technique, the wells of the microtitre plate are first coated with an antibody (antiserum) specific for the antigen of interest. The test sample is subsequently added to the well and if antigen is present, it will be 'captured' and bound by the antiserum. Unbound antigen is then washed from the well. As this primary interaction remains invisible, a secondary detection system is required. In this ELISA, a second antibody, also specific for the antigen of interest, but usually designed to interact with a different epitope, will bind to any antigen that has previously been captured by the primary antibody. The detection antibody is a conjugate, with an enzyme and after this incubation and further washing steps, the appropriate substrate is added to elicit a colour change in the positive wells.

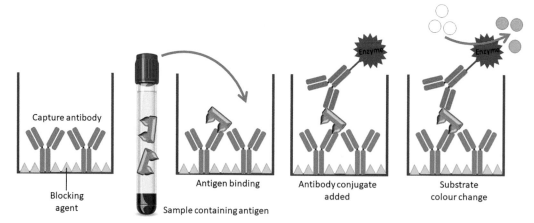

Figure 10.12 Sandwich ELISA for detection of antigen in a biological sample. Capture antibody is applied to the well of a 96-well microtitre plate and blocked. The diluted sample potentially containing antigen is added and allowed to bind. After washing, a second antigen-specific antibody (although usually to a different epitope) in the form of an antibody-enzyme conjugate is added and after removal of unbound antibodies the substrate is added. A colour change in the well indicates a positive result.

10.2.8 Lateral Flow Test/Rapid Immunomigration Assay

An alternative method of visualizing antigen/antibody interaction is utilized by the rapid immunomigration (RIM) assay (also known as the lateral flow test; LFT). In this case the antibodies are often labelled with gold particles to allow the accumulation of antibodies at the test line to be observed. Historically, the most common use for this type of immunoassay is probably the home pregnancy test, which is designed to detect the presence of human chorionic gonadotrophin in the urine during early pregnancy (**Figure 10.13**). However, in recent times, we have become accustomed to using this type of test for COVID-19.

10.2.9 Other Types of Immunoassay

A further refinement of the ELISA is the process of western blotting, which provides specific information concerning the specific proteins to which patient antibodies bind. This has particular relevance in infectious disease and can be used to distinguish between vaccinal antibody and antibody produced through field exposure to pathogen. In western blotting the antigenic mixture under question is electrophoretically separated by molecular weight in an acrylamide gel and the separated proteins are subsequently transferred ('blotted') onto a nitrocellulose membrane. The membrane may be cut into small strips, and each strip is incubated (after a blocking step) with appropriately diluted patient serum. Serum antibody binds selectively to specific antigenic determinants, and this binding is visualized by the use of a secondary antiserum that is conjugated to an enzyme, as for an ELISA. The reaction is finally demonstrated via either a colour change or emission of light (chemiluminescence).

Some other diagnostic laboratory immunoassays require specialist equipment and

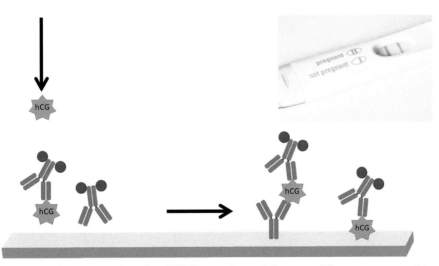

Figure 10.13 **Rapid immunomigration (RIM) assay/lateral flow test.** The pregnancy test kit utilizes RIM technology. Specific antibodies labelled with gold particles are present at one end of the test device. Urine is applied and if human chorionic gonadotrophin (hCG) is present, it will bind with some of the antibodies. As the liquid moves along the membrane, immune complexes are captured by other antibodies at the test (pregnant) line, indicating a positive result. Excess antibodies are captured and accumulate at the control (non-pregnant) line where hCG is embedded onto the test membrane. A similar process is used to detect antigen associated with COVID-19 virus in throat/nasal swabs.

expertise. The Immulite® system uses a bead-based method, somewhat similar to the sandwich ELISA, but with a chemiluminescence reaction at the end (instead of a colour change) and the reactivity is measured using a high sensitivity photon counter. There are Immulite® assays available for various biomarkers, but this is mainly used for measurement of hormones (e.g. cortisol, thyroxine) in veterinary medicine. Antibodies may also be labelled with a radioligand and antibody-binding to antigen determined by measuring the amount of radiation present in any immune complexes formed.

10.2.10 Immunoassays Designed for Use in the Veterinary Clinical Practice Setting

Various immunoassays have been developed and sold commercially as rapid, in-house, diagnostic test kits that can be used in a veterinary practice setting. One variant of this type of test, which has been adapted for several purposes is the SNAP® technology developed by IDEXX Laboratories (**Figure 10.14**). SNAP test kits

are available for a variety of infectious disease diagnostics, including feline leukaemia virus (detection of FeLV antigen) and feline immunodeficiency virus (detection of FIV antibodies). In addition, other antigens can be detected using this system, such as pancreatic lipase, whereby elevated levels of this protein in the blood are indicative of acute pancreatitis. The RIM/lateral flow technique is another widely used test system in point-of-care diagnostic test kits that can be used for diagnosis of a variety of infectious diseases (**Figure 10.15**).

One important practice-based application of the ELISA is in serological testing to ensure that dogs are protected by vaccination. There is a strong correlation between the presence of serum antibodies specific for canine distemper virus (CDV), canine adenovirus (CAV) and canine parvovirus (CPV) and protection from these infectious diseases. There are currently two commercial tests that can rapidly determine whether a dog is protected or whether it may require a booster vaccination (**Figure 10.16**). The first test kit (Titrechek®, Zoetis) utilizes

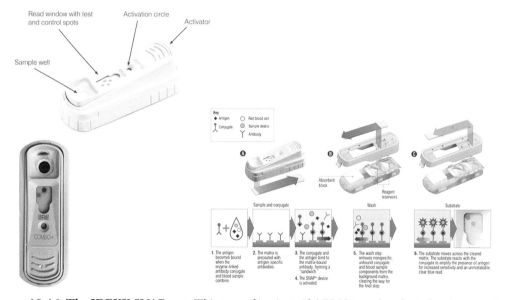

Figure 10.14 The IDEXX SNAP test. This type of 'patient side' ELISA can be adapted to detect various antigens or antibodies. (Reprinted from O'Connor (2015), SNAP Assay Technology. *Topics in Companion Animal Medicine*, 30(4): 132–138. Copyright (2015), with permission from Elsevier.)

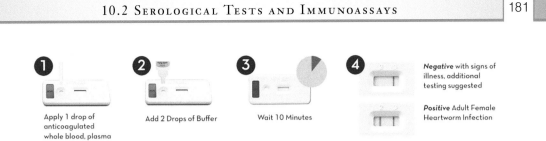

Figure 10.15 The Witness® test. This type of lateral flow test can be adapted to detect various antigens or antibodies. In this example, the test is designed to detect *Dirofilaria immitis* (heartworm) antigen in a whole blood sample. (Copyright Zoetis 2022, reproduced with permission.

a)

b)

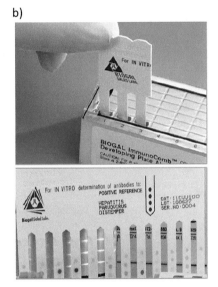

Figure 10.16 ELISA test kits for detection of serum antibody to vaccine antigen. (a) Titrechek® (Zoetis) utilizes a well-based ELISA to test for the presence or absence of CDV and CPV antibodies. Shown is a CDV test. Well 1 is a positive control (blue) and well 2 is a negative control (clear). Wells 3 to 7 test serum samples from individual dogs. Only dog 5 is seronegative and requires revaccination. (b) The second test kit (Vaccicheck®, Biogal Laboratories) uses a system in which viral antigens are impregnated into spots on the teeth of a comb. The assay is performed by sequentially dipping the comb into a series of wells containing diluted dog serum, enzyme-linked antiserum and substrate. Each 'tooth' presents four dots—the top dot represents the positive control and beneath this are the reactions for CAV, CPV and CDV. A grey colouration represents a positive reaction that can be scored semi-quantitatively against a colour scale.

a well-based ELISA to test for the presence or absence of CDV and CPV antibodies. The second test kit (Vaccicheck®, Biogal Laboratories) uses a system in which viral antigens are impregnated into different spots on the teeth of a comb. This 'dot-ELISA' is performed by sequentially dipping the comb into a series of wells containing diluted dog serum, enzyme-linked antiserum and substrate. Each 'tooth' presents four dots—the top dot represents the positive control and beneath this are the reactions for CAV, CPV and CDV. These test kits provide a major advance in rapidly determining, in the practice, the vaccine requirements of individual dogs.

10.3   IMMUNOPHENOTYPING

Cytological and histological examination of cells and tissues by trained clinical and anatomic pathologists can provide information in terms of the cell types present. However, identification of cells in terms of their precise phenotype/lineage can sometimes be problematic, particularly within a pathological lesion and/or where the cells are not morphologically recognizable (as might be the case in a neoplastic lesion). When lymphocytes are present in a cytological preparation (e.g. blood smear) or histological section of tissue, it is not possible, by light microscopy alone, to determine whether any one individual cell might be a T or a B cell, let alone whether they are one of the many phenotypes of T cells. In the case of a tissue section, the microanatomical location is suggestive of identity (i.e. most follicular lymphocytes are B cells and most paracortical or PALS lymphocytes are T cells). In order to identify more precisely the phenotype of cells in suspension or tissue section, immunologists have developed a range of tools and techniques, based on the use of antibodies against specific cell-surface markers, that provide additional information on the cell-types present, when analyzing cellular material obtained from clinical patients. These molecules are classified by the **cluster of differentiation** (CD) numbering system. We have already come across various 'CD' molecules in earlier chapters, but this an ever increasing list, as more cell surface proteins are identified and characterized (www.hcdm.org/).

The ability to discriminate between different lymphocyte populations is based on the fact that different cell types express unique molecules on their cell surface and that these molecules can be identified by the use of polyclonal antisera or monoclonal antibodies. For example, only T cells express a surface **T cell receptor** (TCR) and only B cells carry surface immunoglobulin as the **B cell receptor** (BCR). Other molecules largely restricted to T cells include CD3, CD4 and CD8, and the molecules restricted to B cells include CD19 and CD21. In reality, there are now several hundred surface membrane molecules expressed by different cell types and defined by specific antisera.

Antisera or monoclonal antibodies specific for surface membrane molecules may be used to tag specific subpopulations of cells while held in liquid suspension. At the simplest level, the antiserum or monoclonal antibody is chemically conjugated to a fluorochrome and when added to the cell suspension it will selectively bind to those cells expressing the cell-surface marker. Labelled cells may then be viewed with a fluorescence microscope that emits light of a wavelength able to excite the fluorochrome (see **Figure 10.8**). In a more refined procedure, labelled cells may be passed through a **flow cytometer**, which separates the cell suspension into droplets containing individual cells that are 'interrogated' by a laser beam to determine whether fluorescence is emitted or not. Each fluorescent cell is counted by the machine, which will determine the proportion of labelled cells within the suspension (**Figure 10.17**). Flow cytometers are able to measure light emitted by multiple fluorochromes conjugated to multiple antisera or monoclonal antibodies, and this enables individual cells to be characterized for their expression of a number of surface membrane or cytoplasmic molecules. Some flow cytometers are also able to sort differentially labelled cells into separate test tubes, which permits subsequent experimental studies of the function of those subpopulations.

A similar immunolabelling process may be used to identify cells within tissue sections (known as **immunofluorescence** microscopy). Fluorochrome-labelled antisera or monoclonal antibodies may be applied to tissue sections on a microscope slide and will selectively

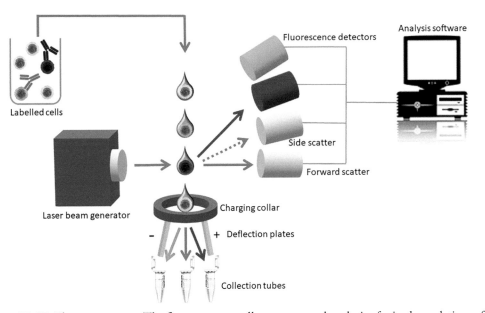

Figure 10.17 Flow cytometry. The flow cytometer allows automated analysis of mixed populations of cells in suspension. In this example, the cells have been incubated with two separate monoclonal antibodies, one conjugated to a fluorochrome that emits green fluorescence (fluorescein) and another to a different fluorochrome that emits red fluorescence (phycoerythrin). The population contains some cells that label with each reagent and some cells that label with neither. The cells are taken up into the machine and are funneled into a single cell flow. Each cell is then 'interrogated' by laser beams and will emit light if labelled by the relevant antibody. The forward scatter (size), side scatter (granularity) and fluorescence signature is recorded and the data can be analyzed by computer, once a representative number (e.g. 10,000) of cells have been analyzed. Some flow cytometers (fluorescence-activated cell sorters) subsequently allow the different populations to become electrically charged as a means of purifying the different populations, so that they may then be further studied.

bind to target molecules expressed by specific cell types. When viewed with a fluorescence microscope, these cells appear coloured on a black background (the colour determined by the type of fluorochrome used) (**Figure 10.18**). As for flow cytometry, several different antisera or monoclonal antibodies conjugated to different fluorochromes may be applied to any one tissue section. An alternative to this technique is **immunohistochemistry** (**Figure 10.19**). In this method the antiserum or monoclonal antibody is conjugated to an enzyme such that when substrate is added to the section in the presence of a chromogen, there is colour (often brown, blue or red) deposited at the position the cell

occupies in the tissue. Immunohistochemistry allows better understanding of the relationship of labelled cells to the tissue, as the background may be counterstained and the section examined with a standard light microscope.

10.4 LYMPHOCYTE STIMULATION ASSAYS

The lymphocyte stimulation assay is designed to determine whether there are lymphocytes in a sample (generally blood, although lymphoid tissue can also be disaggregated to provide a suitable cell suspension) that are able to respond to antigenic stimulation *in vitro* (**Figure 10.20**).

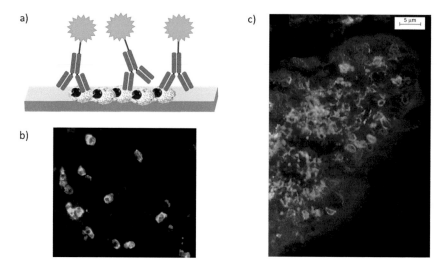

Figure 10.18 Immunofluorescence. (a) This technique permits the identification of specific cells or molecules within a histological section of tissue. The section is overlaid with an antiserum or monoclonal antibody specific for the antigen of interest. The antibody is conjugated to a fluorochrome such that when the section is examined under an appropriate wavelength of light there will be localized emission of fluorescent light. (b) This section of dog tissue is labelled with a fluorescein-conjugated antiserum specific for IgA. Examination under an ultraviolet microscope reveals the presence of plasma cells containing cytoplasmic IgA within the tissue. (c) This section of a canine intestinal villus is labelled with two fluorescent antibody markers that show the location of T cells expressing the αβ TCR (green) within the lamina propria, and T cells expressing the γδ TCR (red) within the epithelium. (From German AJ, Hall EJ, Moore PF et al. (1999) Analysis of the distribution of lymphocytes expressing the αβ and γδ T cell receptors and expression of mucosal addressin cell adhesion molecule-1 in the canine intestine. *Journal of Comparative Pathology* 201:249–263, with permission.)

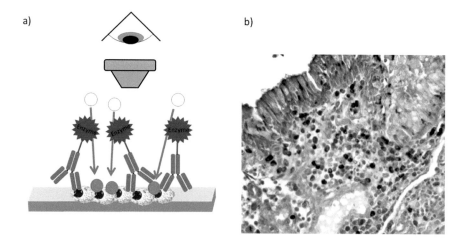

Figure 10.19 Immunohistochemistry. (a) This technique is as described in for immunofluorescence except that the antiserum is conjugated to an enzyme rather than a fluorochrome. Addition of an appropriate substrate and a chromogen leads to deposition of colour at the site of antibody binding. In this technique, the surrounding tissue can be counterstained to evaluate the morphology and the section examined with a standard light microscope. (b) A section of bronchial mucosa from a dog with allergic respiratory disease labelled for IgA by immunohistochemistry. Numerous plasma cells within the lamina propria have cytoplasmic IgA, and secreted IgA is present on the surface of the overlying epithelial cells.

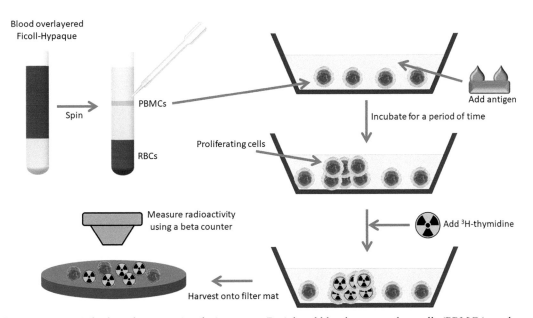

Figure 10.20 **The lymphocyte stimulation assay.** Peripheral blood mononuclear cells (PBMCs) can be separated from whole blood by density gradient centrifugation and cultured *in vitro* in the presence or absence of antigen. Antigen-specific cells will become activated and undergo a period of proliferation when cultured. During the final stages of culture, radiolabelled (^{3}H) thymidine is added, which will be incorporated into the DNA of newly dividing cells. At the end of the culture period the cells are harvested onto a filter mat and the amount of radioactivity is measured, which is provides an index of the stimulation that has occurred.

A positive control for this assay typically involves the use of substances called **mitogens**, which bind in a non-specific manner, stimulating a polyclonal proliferative response. The test generally entails using an antigen to which the animal has been previously exposed and thus carries immunological memory (e.g. a vaccine antigen). In practical terms, blood is collected into anticoagulant tubes (such as heparin or citrate) from the test animal. Although the assay can be performed with whole unfractionated blood, it is preferable to isolate the mononuclear cell fraction. This is readily achieved by **density gradient centrifugation** in which a diluted sample of whole blood is carefully layered over a separation medium (such as Ficoll-Hypaque 1.077) and, following centrifugation, the less dense **peripheral blood mononuclear cells** (**PBMCs**) can be obtained at the interface between the separation medium and the plasma.

The PBMCs can be seeded into the wells of microtitre plates in culture medium, the mitogen or antigen is added and plates are incubated, typically for 72–96 hours. There are different approaches for determining whether lymphocyte stimulation has occurred. The most commonly used method involves the addition of **radiolabelled (^{3}H) thymidine** into the culture for a period of time. The thymidine is incorporated into the DNA of dividing cells, such that at the end of the culture period, the amount of radioactive label within the cell pellet (as determined by a beta counter) is proportional to the amount of cell division. There are also non-radioactive, colourimetric alternatives, although these are not as widely utilized and are much less sensitive. Another means of determining a response to antigen is to **assay the culture supernatant** for cytokines that may have been secreted by the stimulated cells. This can be achieved either by ELISA, using bead-based multiplex assays using the Luminex technology, or by the enzyme-linked immunospot (**ELISPOT**) assay (**Figure 10.21**).

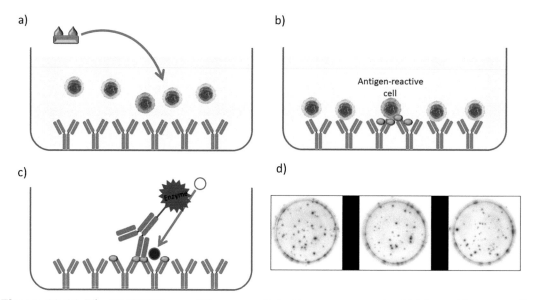

Figure 10.21 The ELISPOT assay. The enzyme-linked immunospot test is a highly sensitive means of detecting production of cytokines by single cells in a culture. (a) A capture antibody specific for the cytokine of interest is coated onto a membrane in the base of a microtitre well. (b) Lymphocytes are added and stimulated (e.g. by antigen or mitogen) so that some cells in the population produce cytokines. (c) Cells are washed from the plate and the bound molecule is identified by a second detecting antibody conjugated to an enzyme. Addition of substrate produces a colour change with each spot correlating to local cellular production of the cytokine. (d) Spots can be evaluated microscopically or with purpose-designed ELISPOT readers.

The cytokine ELISA method is the one adopted by the Animal and Plant Health Agency (APHA) for testing blood samples for reactivity to *Mycobacterium bovis* (bovine TB). Heparinized blood samples are cultured with *M. bovis* antigen (PPD-B) and if lymphocytes react (indicating prior exposure to this pathogen) they produce interferon-gamma, which can be detected in the supernatant using the BOVIGAM™ (bovine IFN-γ) ELISA test.

10.5 TESTS OF CYTOTOXIC FUNCTION

Much less commonly performed in veterinary medicine are tests for assaying **cytotoxicity**. These generally involve the use of **'target cells'** that will readily be killed by cytotoxic T lymphocytes or NK cells. Target cells are radio-labelled, often with ^{51}Cr, then incubated with the **'effector cells'** derived from the patient and after a period of culture, any free radioactive label within the culture supernatant can be measured as an index of destruction of the target cells (**Figure 10.22**).

10.6 TESTS OF PHAGOCYTIC CELL FUNCTION

A range of tests are available for assessing **phagocytic cell** activity/function. Neutrophils are readily isolated from blood samples by the use of density gradient centrifugation, as described previously, and monocytes may be separated from lymphocytes by virtue of their adhesion properties by short-term incubation on glass coverslips or in plastic wells/flasks. The most important functions of phagocytic cells

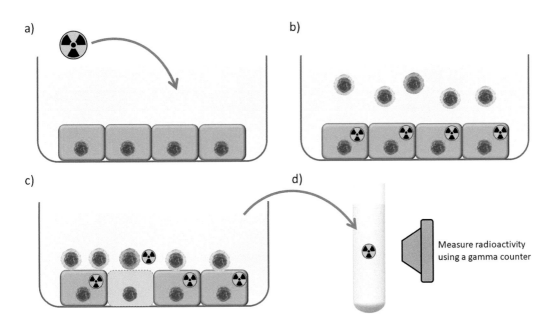

Figure 10.22 Cytotoxicity assay. (a) Target cells are cultured and radiolabelled (typically with ^{51}Chromium). (b) Effector cells are added, which may contain target-specific cytotoxic cells. (c) Effector cells exert their cytotoxic effect on the target cell, thereby releasing the radiolabel into the supernatant. (d) Supernatant is collected and the amount of radiation liberated is quantified using a gamma counter.

are their ability to **migrate towards a chemotactic stimulus**, to **phagocytose particles** and to **destroy** the particles that have been phagocytosed.

The chemotaxis of phagocytes can be detected in a variety of ways. The cells can be placed into the upper well of a '**Boyden chamber**', which is divided from the lower chamber by a permeable filter membrane. The chemotactic stimulus (e.g. molecules derived from bacteria or chemokine) is placed into the lower chamber and diffuses to establish a chemotactic gradient. Phagocytic cells that are attracted towards the gradient can become caught in the dividing membrane and may be identified by removing and staining the membrane after culture or by enumerating the cells that have migrated from the top to the bottom chamber (**Figure 10.23a**).

The phagocytic ability of these cells can also be readily assessed by incubating the purified cells with a target particle of optimum size for phagocytosis. Classically, **staphylococci** or **latex beads** have been used for this purpose and the phagocytosis is generally enhanced when these particles are appropriately **opsonized** by pre-incubation with fresh autologous serum as a source of antibody and complement. The relative phagocytic ability of cells from test and control animals can be determined by stopping the reaction and staining cytospin preparations of the culture in order to count the numbers of intracytoplasmic versus unphagocytosed particles. Alternatively, the process can be automated by the use of fluorescein-labelled particles that can be detected by flow cytometry (**Figure 10.23b**).

The efficiency of phagocyte **killing of bacteria** and the activation of the intracellular enzymatic pathways can also be measured. Viable bacteria (e.g. staphylococci) are used as targets and pre-incubated with the phagocytic

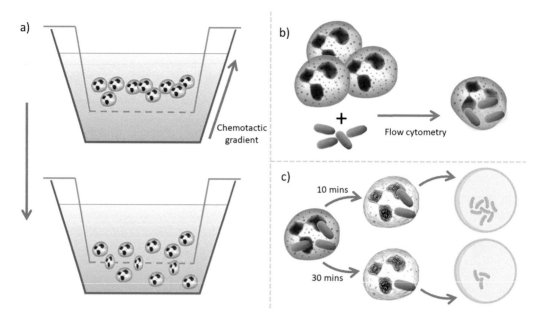

Figure 10.23 Testing of phagocyte function. (a) The Boyden chamber has two halves separated by a semi-permeable membrane. A chemotactic stimulus is added to the lower half of the chamber and purified phagocytic cells (e.g. neutrophils) to the upper well. The neutrophils migrate towards the source of the chemotactic stimulus. At the end of the culture period the membrane is removed and the number of adherent neutrophils counted or the number of cells that have migrated into the lower chamber can be enumerated. (b) A suspension of phagocytic cells is incubated with microparticles/bacteria labelled with a fluorochrome. After incubation and uptake of the particles, the neutrophil suspension can be analyzed by flow cytometry. (c) A suspension of phagocytic cells is incubated with bacteria. Over the period of incubation, increasing numbers of the phagocytosed bacteria will be killed. The progression of this killing can be monitored over time by lysing the neutrophils and culturing any viable intracellular bacteria on agar plates.

cells such that an optimum number of these are endocytosed. After a period of incubation, phagocytic cells can be removed from the culture and lysed to determine the number of viable organisms remaining at that time point relative to the total number of starting organisms (**Figure 10.23c**). Alternatively, the **respiratory burst** of the neutrophil can be measured after phagocytosis of particulate material by an enzymatic reaction producing a colour change (the nitroblue tetrazolium test) or by emission of light in a **chemiluminescence** reaction.

KEY POINTS

- Serology is the study of antigen–antibody interaction and is widely used in clinical veterinary practice.
- The antibody titre is a measure of the amount of antibody in a serum sample.
- Many serological tests require the use of a polyclonal antiserum or monoclonal antibodies.

- Antiserum is made by immunizing an animal with the antigen of interest and subsequently collecting serum (containing antibody) from that animal.
- Agglutination and precipitation assays involve direct interaction of antigen and antibody that may be observed visually.
- The CFT, HAI test, VN test, IFA test, ELISA, immunochromatography and western blotting all require the use of an indicator system to visualize the primary antigen–antibody interaction.
- Tests of lymphocyte or phagocytic cell function are clinically useful, but not commonly available in veterinary medicine.
- Lymphocytes may be stimulated to divide *in vitro* by incubating them with mitogens or specific antigens.
- Mononuclear cells can be separated from blood by density gradient centrifugation.
- Cellular division may be measured by quantifying the incorporation of radiolabelled thymidine into the nucleus of proliferating lymphocytes.
- Release of cytokines into the culture medium can also be evaluated by techniques such as ELISPOT.
- The ability of cytotoxic T cells or NK cells to kill a radiolabelled target cell can be measured *in vitro*.
- Assays can be used that determine whether phagocytic cells have migrated along a chemotactic gradient.
- The ability of phagocytic cells to take up particulate antigen and kill microbes can be assessed *in vitro*.

OBJECTIVES

At the end of this chapter, you should be able to:

- Discuss how the innate and adaptive immune systems respond to viral infection.
- Discuss how the innate and adaptive immune systems respond to a bacterial infection.
- Discuss how the adaptive immune system responds to helminth infection of the intestinal tract and how this effector response relates to type I hypersensitivity.
- Discuss how the adaptive immune system responds to an obligate intracellular protozoan infection.
- Discuss how the innate and adaptive immune systems respond to a systemic fungal infection.

11.1   INTRODUCTION

In the preceding chapters of this book, we described the separate components of a functioning innate and adaptive immune system. In this chapter, we take a more integrated approach to explain how these individual components interact with each other in a synergistic manner to generate a range of different **protective immune responses**. There has been selection pressure for evolutionary development of the immune system arising from emergence and co-evolution of an array of different pathogens. The immune system needs to keep pace with the strategies used by these organisms to invade, replicate and disseminate, in order to prevent persistent infection.

The concept that those interactions that occur between pathogens and the cells of innate immunity determine the nature of the subsequent adaptive immune response has already been highlighted. Additionally, many of the effector responses of the adaptive immune system serve to enhance innate defence systems. The aim of this chapter is to reinforce these concepts by discussing the fundamental role of the immune system in protecting animals from infectious disease. A series of examples will be used to illustrate these points and these will be taken from the major classes of infectious agent: viral, bacterial, fungal, protozoal and helminth parasite.

11.2   THE IMMUNE RESPONSE TO VIRAL INFECTION

Viruses are a highly successful class of pathogen, responsible for significant morbidity and mortality amongst both animal and human populations. As an example of a model immune

DOI: 10.1201/9781003310969-11

response to viral infection, we shall consider how the immune system might deal with a virus affecting the alimentary system, as might occur with, for example, **bovine rotavirus**. When infectious virus particles arrive at their target site they will encounter an array of **innate immune defences** relevant to that epithelial surface (**Figure 11.1**). In the case of the intestinal mucosa, these will include the **enterocyte barrier**, the **secretions** that coat the luminal surface of that barrier (including mucus and polyreactive immunoglobulins) and a range of **innate immune cells** that normally populate the epithelial compartment (e.g. the γδ T cells) and the underlying **lamina propria** (e.g. macrophages, dendritic cells and NK cells).

Viruses infect host cells by binding to one or more specific **cell surface receptors**. These are generally proteins or carbohydrates that the virus co-opts to gain access to the target cell. In this example, the rotavirus interacts with sialoglycans and transmembrane proteins on the enterocyte to enter the host cell (**Figure 11.2**). Once within the cell, the aim of the virus is to **replicate** in order to produce new virions that might then leave that cell (and in the process destroying the cell) to infect new targets. The means by which this is achieved depends on the

nature of the virus and its genomic material. Activation of cytoplasmic PRRs (such as RIG I-like receptors) activates the interferon regulatory factor (IRF) signalling pathway, which results in secretion of type I and type III IFNs. These antiviral interferons bind to interferon receptors on adjacent non-infected cells and stimulates them to produce an array of other proteins, which confer a measure of resistance to viral replication. The virus-infected cell may also display stress molecules on its surface and if that expression is concurrent with downregulation of MHC class I molecules, the infected cell becomes a **target for NK cells** in the vicinity, which are also sensitized by the presence of antiviral interferons. Alternatively, the infected cell may process and present virus antigen in the context of MHC molecules, which will be of significance later in the host response, when adaptive immunity reaches its full potential.

At this stage of the viral infection, **dendritic cells** in the lamina propria are able to sample virus antigen or even become infected themselves, allowing classical processing and presentation by these APCs. The interaction between virus and APC is mediated by viral **PAMPs** binding to dendritic cell **PRRs** that are typically cytoplasmic (e.g. RIG-1) or endosomal

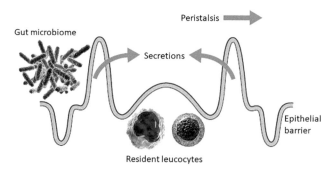

Figure 11.1 Innate immune defence of mucosal surfaces. Potential pathogens must initially evade the range of innate immune defences present at mucosal surfaces. In the case of the intestine, these include peristaltic movement, the luminal microflora, the mucosal secretions (mucus, enzymes, polyreactive immunoglobulins and complement molecules), the epithelial barrier, the intraepithelial γδ TCR T cells and innate immune cells of the lamina propria (dendritic cells, macrophages, mast cells and NK cells).

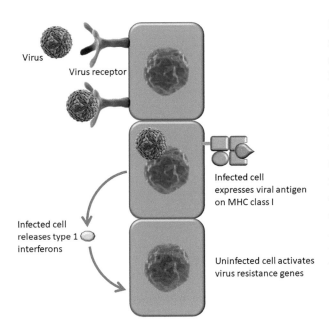

Virus

Virus receptor

Infected cell
expresses viral antigen
on MHC class I

Infected cell
releases type 1
interferons

Uninfected cell activates
virus resistance genes

Figure 11.2 Viral infection of a mucosal surface. Viruses enter host cells by targeting surface receptor molecules. Within the host cell the virus replicates by mechanisms determined by the nature of the genetic material of that virus. Newly produced virus particles seek to infect new host cells. Virus-infected cells secrete type I and III interferons (including IFN-α and IFN-β), which confer resistance to infection on nearby uninfected cells (paracrine effect). Virus-infected cells may express viral peptides associated with MHC class I molecules to seek a response from the adaptive immune system (via CD8+ T cells).

(e.g. TLR3 recognizes viral double-stranded RNA, TLR7 and TLR8 recognize single-stranded RNA and TLR9 recognizes DNA). These interactions and signalling pathways (via NF-κB and IRF) lead to selective gene activation in the APC. The dendritic cells migrate via the lymphatics to the regional **lymph nodes** so they can interact with cells of the adaptive immune system (specifically T lymphocytes). It is also possible for virus antigen to directly be taken up via M cells into Peyer's patches, where other dendritic cells are located.

Once the antigen-laden dendritic cells have entered the interfollicular areas of Peyer's patches or the paracortex of a lymph node, they interact with recirculating naïve or memory T cells, some of which bear receptors specific for the viral peptides. The interaction between the APC and the T cells will be guided by the range of co-stimulatory surface molecules and cytokines that have been induced in the dendritic cell following PRR–PAMP interaction. Cross-presentation of viral antigen permits presentation of viral peptides by both MHC class I and class II pathways (**Figure 11.3**), facilitating activation of CD4+ and CD8+ T cells. This

typically leads to generation of both Th1 cells that can help stimulate cell-mediated immunity in the form of CD8+ cytotoxic T cells as well as Th2 and Tfh cells that can help stimulate humoral immunity in the form of virus neutralizing antibody.

The Th1 and CTLs generated leave the lymph node in efferent lymph in order to enter the bloodstream and 'home' to the site of viral infection (in this case the intestinal mucosa). Once these adaptive immune cells arrive in the mucosa, the full **effector phase of adaptive immunity** will come into play (**Figure 11.4**). Th1-derived IFN-γ will amplify the effects of NK cells and CTLs, and such cell-mediated cytotoxicity is the major effector mechanism in the antiviral immune response.

There is a role for both IgG and IgA antibody in this immune response. Th2 and Tfh cells stimulate IgG production and can also help to ensure production of antiviral IgA. IgG+ plasma cells may largely be located within the medullary cords of lymph nodes and the bone marrow, producing circulating virus-neutralizing IgG that diffuses into the tissue fluid from the circulation. In contrast, IgA+ B cells, activated in the

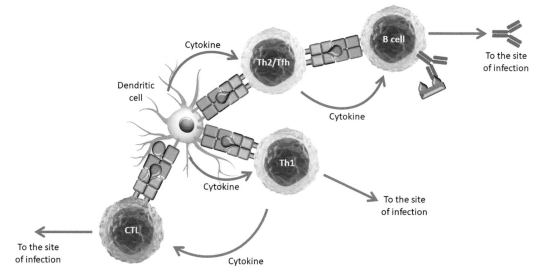

Figure 11.3 Induction of the antiviral adaptive immune response. Virus antigen presented by dendritic cells within the lymph node paracortex activates T helper cells, which in turn provide help for the activation of CD8+ cytotoxic T cells and B cell class switching to produce virus neutralizing antibody (typically IgG for systemic immunity or IgA for mucosal immunity).

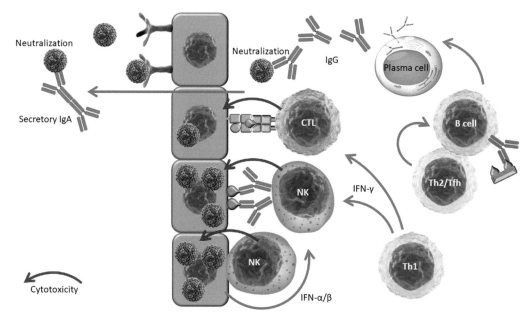

Figure 11.4 The effector immune response to virus infection. CD8+ T cells destroy virus-infected target cells following recognition of MHC class I presenting virus peptide. NK cells may contribute to the cytotoxic response, but they will only kill infected targets which have downregulated MHC class I expression. Both cells are functionally enhanced by Th1-derived IFN-γ and NK cells may also respond to IFN-α and IFN-β produced by infected cells. Th2 and Tfh cells enhance production of IgG and IgA antibodies. IgG antibody may contribute to NK cell killing via ADCC. IgA antibody may be actively secreted across the mucosal surface and inhibit virus–receptor interactions, while IgG antibody may diffuse from the plasma and neutralize as well as contribute to the destruction of virus particles (e.g. by complement-mediated lysis).

Peyer's patches and mesenteric lymph nodes, circulate back to the intestine and produce IgA locally within the mucosa after differentiating to plasma cells. IgA antibodies are transported across the epithelial barrier to the mucosal surface where they can neutralize virus binding to epithelial cells. IgG antibodies may also mediate NK cell ADCC or opsonize infected cells for macrophage phagocytosis. If this adaptive immune response is successful in eliminating the infection, then late-stage immunosuppression (mediated by **induced Treg cells**) and the development of **memory lymphocytes** (both T and B cells) will occur.

11.3 THE IMMUNE RESPONSE TO BACTERIAL INFECTION

To remain with the intestinal model, we will next consider the nature of immune responses that might be generated following exposure to an **enteric bacterial pathogen** such as *Escherichia coli* or *Salmonella* spp. On arrival in the intestinal tract, these organisms will immediately encounter the range of innate immune defences outlined in **Figure 11.1**. However, of particular importance in this context would be the presence of the **intestinal bacterial microflora/microbiome**, which will compete with the pathogen for space and nutrients, making colonization more challenging. Another element of innate immunity that may have greater relevance to this class of infection is the $\gamma\delta$ **T cell** within the enterocyte layer. These cells are well situated for early interaction with bacterial pathogens and are thought to be primarily activated in response to this type of organism.

Pathogenic bacteria often require an initial **receptor-mediated interaction** with host cells; for example, the K88 and K99 pili/adhesins of *E. coli* permit attachment to cell surface receptors on the luminal side of the enterocytes. Enteric pathogens such as *E. coli* or *S. enterica* serovar Typhimurium (S. *typhimurium*) may utilize a variety of different mechanisms to induce

disease, dependent on the strain of the microorganism. Some bacteria can produce locally active **enterotoxins** that bind host cell receptors/ion channels and lead to an osmotic imbalance and **secretory diarrhoea** (e.g. enterotoxigenic *E. coli*). Others may attach to and disrupt the epithelial surface (e.g. enteropathogenic *E. coli*, which cause 'attaching and effacing' lesions) or **invade** the intestinal mucosa and regional lymph nodes, leading to a local **pyogranulomatous inflammatory response** (e.g. enteroinvasive *E. coli* or *enterohaemorrhagic E. coli*). Some **Gram-negative bacteria** are also characterized by the ability to produce systemically absorbed **endotoxins** responsible for severe generalized disease (**endotoxaemia**) (**Figure 11.5**). Some **Gram-positive bacteria** (e.g. *Clostridium difficile* and *Clostridium botulinum*) produce **exotoxins** that can have a negative impact on the host either locally or systemically.

Once the mucosal surface has been colonized, the adaptive immune response will be required to help resolve the infection. Again, mucosal dendritic cells should sample bacterial antigen and this process involves the interaction of PRRs with a range of bacterial PAMPs. Dendritic cell surface PRRs involved in the recognition of bacterial PAMPs include TLRs 1, 2, 4, 5 and 6 (and in the case of gram-negative bacteria, particularly TLR4, which recognizes lipopolysaccharide and TLR5, which recognizes flagellin), while TLR9 recognizes bacterial CpG DNA within the endosome. Bacteria may also be transported across the epithelial barrier via **M cells**, within the dome epithelium overlying the **Peyer's patch**, as is the case for *S. typhimurium*. The activated dendritic cells will activate T cells and, in turn, follicular B cells in the Peyer's patch or **mesenteric lymph nodes**. The desired effector immune response includes the production of antigen-specific immunoglobulin and increased secretion of antimicrobial peptides, so naïve CD4+ T cell would typically differentiate into **Th2/Tfh (antibody-mediated immunity)** and **Th1/Th17 (cell-mediated immunity) effector cells** (**Figure 11.6**). The Th1 and

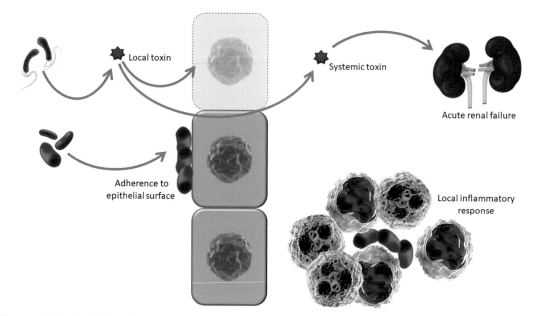

Figure 11.5 Pathological mechanisms utilized by enteropathic bacteria. Bacteria infecting the intestinal tract (e.g. *Escherichia coli*) may cause disease via several distinct mechanisms. Some strains may secrete locally active enterotoxins that mediate secretory diarrhoea. Others attach and efface the surface epithelium or invade into the lamina propria, inducing a pyogranulomatous inflammatory response. Some strains of bacteria elaborate toxins that have systemic effects, including causing acute renal failure (e.g. shiga-toxin producing *E. coli*; STEC/*E.coli* O157:H7).

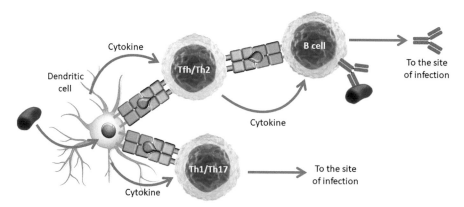

Figure 11.6 Induction of the antibacterial immune response. Dendritic cells acquire bacterial antigen at the site of infection and migrate to the paracortex of a lymph node. Here they can stimulate production of Th2 and Tfh lymphocytes, which act on those B cells that will class switch to the production of antigen-specific IgG or IgA. Generation of Th1 cells is also important for intracellular pathogens such as *Salmonella typhimurium* and *Mycobacterium paratuberculosis*, where they can assist macrophage activity, through release of IFN-γ. Generation of Th17 cells also aids mucosal epithelial defences.

Th17 cells, together with antigen-specific B cells, exit the Peyer's patches and mesenteric lymph node to home back to the mucosal surface. The Th17 cells release IL-17 and IL-22, which increases the secretion of antimicrobial peptides by enterocytes.

The effector humoral immune response involves the synthesis of specific **IgG (systemic protection) and IgA (mucosal immunity) antibodies**. For those organisms mediating pathology via toxin production (such as *Clostridium spp.*), antibody neutralization of the toxin will be important. IgG antibodies may also **opsonize** invasive organisms to enhance phagocytosis or facilitate **complement-mediated lysis** of the bacteria. IgA antibodies will be secreted into the intestinal lumen, where they can interfere with the function of bacterial adhesins (**Figure 11.7**).

In the case of an invasive and intracellular pathogen such as *S. typhimurium*, or *Mycobacterium avium paratuberculosis*, a protective host immune response would also require a robust cell-mediated immune response, involving activation of Th1 and CD8+ T cells.

11.4   THE IMMUNE RESPONSE TO HELMINTH INFECTION

The final example of an intestinal immune response is that related to protection from **helminth endoparasitism**. The physical size of such parasites is clearly a challenge for the immune system, but a series of strategies have been devised that contribute to a protective immune response. These strategies may sometimes be particularly successful, as in the

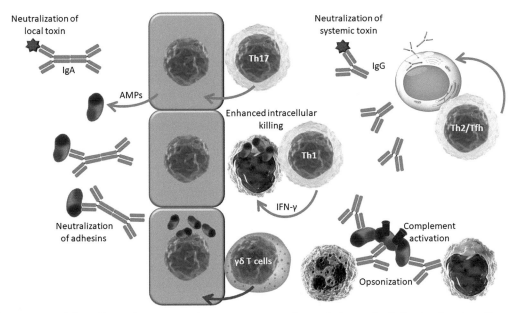

Figure 11.7 **The effector immune response to enteropathogenic bacteria.** Following homing of antigen-specific Th and B cells to the affected segment of intestinal mucosa, these cells mediate local production of IgG and IgA antibodies. In the lumen, secretory IgA antibodies neutralize toxins and interfere with the receptor-mediated attachment of bacteria to the epithelial surface. IgG antibodies may cause complement-mediated cytolysis or opsonize bacteria for phagocytosis. IgG antibodies may also bind to and neutralize enterotoxins or endotoxins. Th1 cells enhance the bactericidal activity of infected macrophages through IFN-γ signalling, and Th17 cells induce epithelial cells to release antimicrobial peptides (AMPs). γδ T cells may also destroy infected target cells that have upregulated stress proteins on their basal surface.

'self-cure phenomenon' in which there may be rapid elimination of adult *Haemonchus contortus* nematodes by sheep that mount a hypersensitivity reaction to antigens from developing larvae. We will now consider how the immune system responds to the presence of a large and biologically complex parasite that might be physically attached to the intestinal mucosa and feeding on host tissue or blood.

Helminth parasites release a range of antigenic molecules (**excretory–secretory [ES] proteins**), which may be sampled by APCs and processed and presented by these cells within the Peyer's patches or mesenteric lymph nodes. The ideal immune response in this circumstance is one in which the parasite PAMPs interact with the PRRs of the APC to permit activation **of CD4+ Th2 and Tfh effector** cells. **Class switching** in antigen-specific B cells will be directed by Th2-produced IL-4 predominantly towards **IgE** with some **IgG of restricted subclass**. Some may differentiate to become plasma cells in lymphoid tissue and produce circulating plasma antibody, but others may home back to the intestinal mucosa to generate local antibody.

The normal intestinal mucosa is rich in **tissue mast cells** and these are located immediately beneath the enterocyte layer in close association with small capillaries. In this effector immune response, parasite-specific IgE binds to Fcε receptors on the surface of these mast cells. When the host is exposed to further parasite ES antigen across the mucosal barrier it will **cross-link the surface bound IgE molecules**, leading to local mast cell **degranulation** and subsequent tissue oedema (**vasodilation**) and extravasation of serum proteins and leucocytes (**inflammation**). **Eosinophils** are prominent within these inflammatory infiltrates. One specific consequence of triggering an intense local tissue inflammatory response is that it makes the tissue a less attractive environment for the parasite, such that the organisms may even detach from the mucosal surface. In combination with this detachment, infiltrating T lymphocytes

secrete IL-13 that acts on **goblet cells**, causing them to increase in number, to **discharge mucus** into the intestinal lumen and to alter the chemical composition of the mucus. The mucus coats the outside of the detached parasite, making it much simpler for **peristaltic movement** to push the organisms along the intestinal tract for eventual **expulsion (Figure 11.8)**.

In concert with these effector mechanisms, the immune system also mediates a direct attack on helminth parasites within the tissues in a 'David versus Goliath' manoeuvre. IgE and IgG antibodies are able to coat the outer surface of the parasite and allow binding of cells (particularly eosinophils) expressing the appropriate **Fc receptor**. This brings these leucocytes into close apposition with the surface of the parasite and they **degranulate** locally, with the granule contents causing damage to the parasite. Eosinophils attracted and activated by locally secreted chemokines (eotaxins) release basic proteins that are toxic. In addition, macrophages differentiate into M2 macrophages that activate fibroblasts to produce more collagen resulting in tissue fibrosis. These anti-parasitic mechanisms occur when an invasive parasite is within or migrating through the tissues, rather than simply feeding from the surface (**Figure 11.9**).

11.5 THE IMMUNE RESPONSE TO PROTOZOAL INFECTION

Canine leishmaniasis results from infection with the protozoan parasite *Leishmania infantum*, which is also responsible for zoonotic disease of major significance to the human population in endemic areas such as Central Asia, Mediterranean countries and Central and South America. The domestic dog is the predominant reservoir species and transmission is via a range of sandfly vectors. The disease is endemic in Mediterranean countries and is of significant concern in traditionally non-endemic European countries and North America as a consequence of the movement of pet dogs.

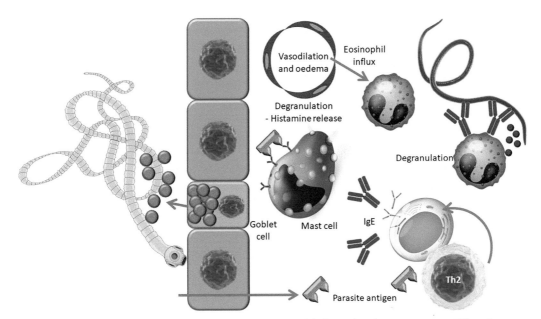

Figure 11.8 **The effector immune response to intestinal helminth infection.** Parasite ES antigens activate Th2 cells, which stimulate B cell class-switching to IgE. The IgE coats sub-epithelial mast cells, which degranulate when further ES antigen is absorbed. Mast cell degranulation leads to vasodilation, oedema and an inflammatory response characterized by recruitment of eosinophils. This inflammatory response may result in surface-feeding parasites detaching from the mucosa. T cell-derived cytokines (e.g. IL-13) concurrently stimulate goblet cells to produce mucus that coats the parasite, and these effects combine with peristaltic movement to help expel the parasite from the intestine. Parasite-specific IgE (and IgG) may directly coat the surface of tissue-migrating forms of helminths and allow targeting by leucocytes expressing appropriate Fc receptors (e.g. eosinophils, neutrophils, macrophages). This has the effect of bringing granulocytes into close apposition with the parasite surface and local degranulation of these cells may cause damage to the coat of the organism ('frustrated phagocytosis').

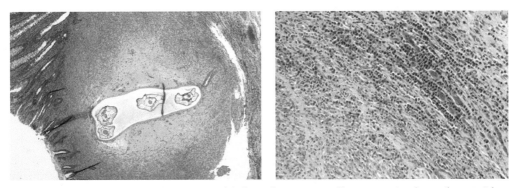

Figure 11.9 **Immune response to intestinal helminths.** Section of large intestine from a horse with migrating strongyle larvae and an intense localized inflammatory response. Eosinophils are prominent in the inflammatory infiltrate.

The bite from an infected sandfly transmits the promastigote stage of the organism into the cutaneous tissues of the dog. The organism is endocytosed by tissue macrophages, where it avoids intracellular killing mechanisms and transforms and multiplies as an amastigote, which is the obligate intracellular stage of its life cycle. Eventually, parasitized macrophages may rupture and release amastigotes that establish infection in new host phagocytic cells (**Figure 11.10**).

There are several outcomes to this infection in dogs. These outcomes are **genetically controlled** and reflect the nature of the host immune response to the parasite. Dogs of a genetically 'resistant' phenotype make an appropriate cell-mediated immune response that controls, but does not necessarily eliminate the infection. These animals do not usually develop clinical disease, but can remain subclinically infected carriers and a potential source of the infectious agent to sandflies. In contrast, dogs of the 'susceptible' genotype make an inappropriate, non-protective (humoral) immune response, allowing dissemination of the parasite and development of secondary immune-mediated sequelae. These dogs typically develop severe, multisystemic disease with high morbidity and mortality, if left untreated.

Development of a protective immune response in a resistant dog requires a biased cell-mediated immune response. In this situation, dermal dendritic cells capture antigen for processing and presentation within the regional cutaneous draining lymph node. Within this lymphoid tissue, the essential immune response that must be generated is that mediated by **CD4$^+$ Th1 lymphocytes (Figure 11.11)**. B cells switch to the appropriate IgG subclass associated with cell-mediated immunity. These effector populations then recirculate and home to cutaneous (or other) sites of infection. The effector phase of the protective immune response essentially involves the

Promastigotes

Amastigotes

Figure 11.10 The life cycle of *Leishmania infantum*. An infected sandfly bites the canine or human host in order to take in a blood meal and injects promastigotes into the dermis. The promatigotes are phagocytosed by macrophages and divide within these cells to form amastigotes. Parasitized macrophages may rupture and release amastigotes, which infect other host cells locally or systemically. Infected macrophages may also be taken up by naïve sandflies as they take a blood meal, and released amastigotes transform again to promastigotes and multiply in the gut of the sandfly. (Sandfly image courtesy of CDC/Frank Collins.)

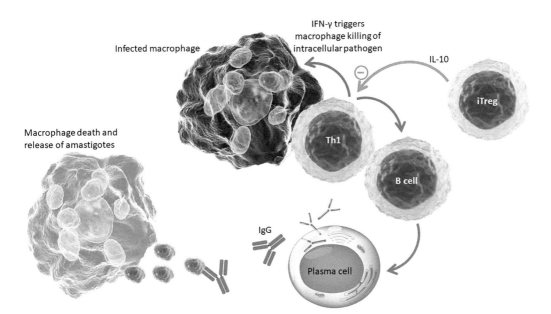

Figure 11.11 The effector immune response to *Leishmania* in resistant dogs. Resistant dogs mount a polarized immune response dominated by CD4⁺ Th1 effector cells. The IFN-γ secreted by these cells activates infected macrophages and permits them to destroy the amastigotes. Th1-related B cell production of IgG antibody also occurs. This specific IgG may be able to bind free amastigotes as they leave a ruptured macrophage leading to complement-mediated lysis. In the later stages of the immune response, IL-10-producing induced Treg cells seek to limit the potential for secondary immune-mediated pathology, but may permit persistence of a low-grade infection.

secretion of IFN-γ by the Th1 cells. This cytokine **activates infected macrophages**, which then permit those cells to destroy the amastigotes. **IgG antibody** produced systemically or locally probably has an adjunctive role in binding and destroying free amastigotes by complement-mediated lysis as they exit ruptured macrophages and attempt to infect new target cells.

In most resistant animals this effector immune response may not eradicate the infectious organisms and instead, there is a **low-grade persistent infection** with the potential for that animal to still act as a reservoir of infection or for recrudescence to occur later in life. Failure to eliminate the infection is thought to relate to generation of a **regulatory T cell response** during the later stages of the immune response. This may be mediated by **induced Treg cells**

secreting **IL-10**, although there is also evidence that Th1 cells may undergo a subsequent phenotypic shift to take on a more regulatory (IL-10 producing) role. The reason for such regulation is to prevent secondary immune-mediated pathology, as occurs in susceptible animals.

In dogs of a **susceptible** genotype, antigen presentation stimulates an inappropriate **Th2-mediated immune response**. This leads to the generation of high levels of **serum antibody** and inhibition of Th1 function, which permits uncontrolled infection within the macrophages. The range of **secondary immune-mediated sequelae** that characterize the clinical infection are caused by the development of circulating immune complexes (e.g. polyarthropathy, glomerulonephritis and uveitis) or to the induction of autoantibodies (e.g. anti-erythrocyte,

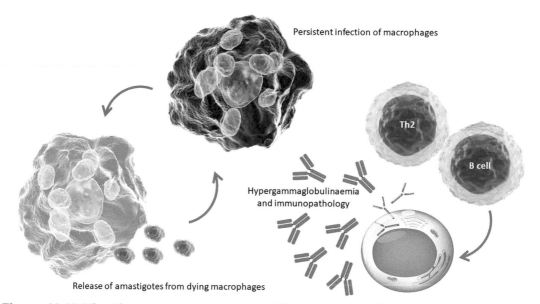

Persistent infection of macrophages

Th2

B cell

Hypergammaglobulinaemia
and immunopathology

Release of amastigotes from dying macrophages

Figure 11.12 **The effector immune response to *Leishmania* in susceptible dogs.** Susceptible dogs make an inappropriate polarized Th2 response to *Leishmania*, with the generation of high concentrations of serum antibody and suppression of Th1-mediated cellular immunity. The infection is allowed to disseminate widely and the antibody response underlies the secondary immune-mediated pathology that accounts for many of the clinical signs of active disease. These humoral effects include the induction of autoantibodies and the formation of circulating immune complexes.

anti-platelet and anti-nuclear antibodies) (**Figure 11.12**).

11.6   THE IMMUNE RESPONSE TO FUNGAL INFECTION

As with helminths, many fungal pathogens provide a challenging target for the immune system because of the relative colony size of the organisms. In this final example we shall consider the immune response of the dog to colonization of the nasal cavity and associated sinuses by the organism ***Aspergillus fumigatus*** (**Figure 11.13**). This fungus can form a large colony over the nasal mucosa, but the organisms tend not to infiltrate into the lamina propria. The colonies comprise a tangled mass of **fungal hyphae** with intermittent **conidia** that represent the sites of **spore** formation.

In order for such a colony to establish, the organism must broach the normal innate immune barriers of the upper respiratory tract, including the antimicrobial substances found within nasal secretions. Although innate phagocytic cells (neutrophils and macrophages) are capable of phagocytosing fungal spores, they appear unable effectively to kill these elements. The fungal hyphae are simply too large a target for effective phagocytosis.

The adaptive immune response to this infection is well characterized. Recent studies have shown upregulation of expression of genes encoding TLRs1–4 and 6–10 and NOD2 in the mucosa of dogs with sinonasal aspergillosis. It is likely that APCs carrying fungal antigen stimulate this response in regional lymphoid tissue such as the **nasopharyngeal tonsil** or **retropharyngeal lymph nodes**. The effector phase of the mucosal immune response involves infiltration by **CD4+ Th1** and probably **Th17 cells**, as determined by upregulation of gene expression for **IFN-γ and IL-23** in inflamed

Figure 11.13 Canine fungal rhinitis. This dog has a nasal discharge secondary to infection of the nasal sinuses by *Aspergillus fumigatus*. *A. fumigatus* fungi grow over the surface of the sinonasal mucosa as large plaques of hyphae with intermittent conidia formed of fungal spores. These structures are very large targets for the immune system.

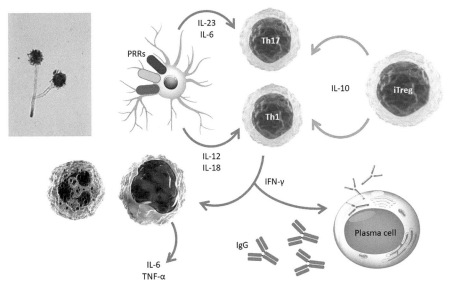

Figure 11.14 The effector immune response in canine sinonasal aspergillosis. The mucosa underlying the fungal plaque is heavily infiltrated by CD4+ and CD8+ T cells, macrophages and IgG-producing plasma cells. Analysis of cytokine gene expression in these inflammatory lesions is consistent with activation of Th1 and Th17 cells. The effector immune response may therefore involve IFN-γ enhancement of macrophage function and the production of local and systemic antigen-specific IgG. Antibody and complement molecules may coat fungal elements, forming a bridge to granulocytes that may degranulate to cause damage to the organisms. Antibody may also opsonize fungal spores or hyphal fragments for macrophage phagocytosis and destruction, once that phagocytic cell is signalled by IFN-γ. A regulatory process involving local IL-10 production is also recognized within these lesions. This may limit secondary immune-mediated tissue damage, but also permits persistence of the infection.

tissue. There is also marked **pyogranulomatous inflammation** along with infiltration of IgG-bearing plasma cells. Th1-derived IFN-γ most likely provides stimulation to macrophages to permit these cells to destroy any phagocytosed fungal spores. Antibody and complement molecules are likely to have an opsonizing effect for FcR- and CR1-bearing **granulocytes**. These cells may **degranulate** locally in a process of frustrated phagocytosis, and induce focal damage to the hyphae. Infected dogs generally mount a strong **serum IgG** antibody response to the organism, which is a useful diagnostic feature of this infection. The inflammatory response itself is likely to be responsible for the extensive tissue and bone destruction that can occur in this disease. Similar to observations in leishmaniasis, there is an additional **regulatory element** to the response, as there is concurrent upregulation of **IL-10** gene expression. Again, this is interpreted as an attempt by the adaptive immune system to prevent extensive tissue damage or the onset of systemic sequelae, but at the same time this facilitates **persistence of infection** and development of chronic sinonasal disease (**Figure 11.14**).

KEY POINTS

- Virus-infected cells produce type I interferons, such as IFN-α and IFN-β, to upregulate antiviral defences in neighbouring cells.
- NK cells and cytotoxic CD8+ T cells are important in the immune response to viral infection, detecting and destroying virus-infected cells.
- Virus-neutralizing antibody (IgG and/or IgA) is important for preventing attachment of virus to host cells, preventing transmission from host to host and from cell to cell (in the extracellular fluid).
- Intraepithelial γδ T cells may be important in the initial immune response to some bacterial pathogens of mucosal surfaces.
- Production of antibodies can neutralize bacterial virulence factors, including adhesins and toxins.
- Production of antibodies (IgM and IgG in particular) enhances antimicrobial innate defence systems, including phagocytosis and complement activation.
- Excretory–secretory antigens are important in initiating anti-parasite immune responses.
- Parasites induce Th2-regulated immunity with generation of IgE and sensitization of mast cells and eosinophils.
- Mast cell and eosinophil degranulation causes damage to the surface of parasites.
- The immune response to *Leishmania* is genetically determined and may be polarized to a Th1 (resistant) or Th2 (susceptible) phenotype.
- Late-stage regulation of the immune response to infectious agents is required in order to prevent secondary immune-mediated pathology, but this may also allow persistent infection. Regulation probably involves IL-10-producing induced regulatory T cells.
- Th17 cells may act with Th1 cells in the immune response to particular types of infectious agent.

IMMUNE EVASION BY PATHOGENS

OBJECTIVES

At the end of this chapter, you should be able to:

- Explain the various mechanisms that pathogens can use to avoid detection and/or destruction by the innate and adaptive immune systems.
- Describe the specific strategies used by retroviruses, pestiviruses and herpesviruses to evade or subvert the host immune response.
- Discuss the consequences of viral immune evasion for the host, in terms of persistent infection and clinical outcome.

12.1   INTRODUCTION

As we have seen in the previous chapter, the immune system has a number of defensive strategies and countermeasures designed to detect and eradicate a range of different pathogens. However, this is a biological arms race and the host: pathogen interaction is constantly evolving. Some pathogens have developed strategies to avoid, evade or subvert the immune system to gain an advantage, particularly in terms of their ability to replicate in the host and facilitate their transmissibility between hosts. In this chapter, we examine a number of biological mechanisms, whereby pathogens can infect and establish a more chronic or even persistent infection in animals despite an immune response by the host.

12.2   EVADING THE INNATE IMMUNE RESPONSE

12.2.1   Viral Inhibition of the Interferon Response

Viral double-stranded RNA is detected during replication in infected cells using specific host intracellular PRRs, such as RIG-1 and TLR-3. The cell signalling mechanisms that are activated lead to expression of type I and type III interferon genes and secretion of these antiviral cytokines. These act on IFN receptors on neighbouring cells (via a paracrine effect) to activate viral resistance genes and to increase synthesis of MHC class I proteins. Some viruses (e.g. bovine viral diarrhoea virus; BVDV, a pestivirus) produce non-structural (early) proteins that can interfere with cell signalling via this pathway

DOI: 10.1201/9781003310969-12

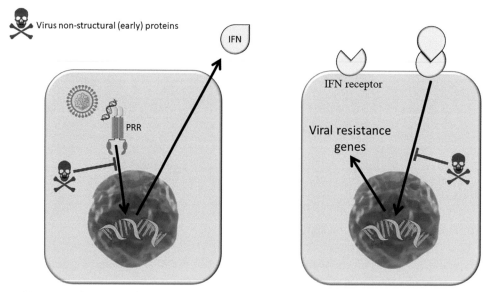

Figure 12.1 Interference with the interferon response. Under normal circumstances pattern-recognition receptors (PRRs) such as RIG-1 or TLR-3 will detect the presence of viral nucleic acid and signal to the host cell to synthesize and secrete type I and III interferons. Some viruses produce non-structural (early) proteins that block this system. Interferon acts on neighbouring cells and signals them to upregulate their viral resistance genes. Some viral non-structural (early) proteins block this signalling pathway.

to prevent production of IFN. Other viruses (e.g. poxviruses) can interfere with signalling through the IFN receptor, preventing the anti-viral response to these cytokines (**Figure 12.1**).

12.2.2 Viral Inhibition of NK Cells

Cell surface expression of MHC class I often declines in virus-infected cells, which makes them less susceptible to recognition and killing by CD8+ killer T cells (see more on this later). However, in doing this, it releases the inhibitory receptors on NK cells and simultaneously, virus infection can induce expression of ligands of their activating receptors. These changes make virus-infected cells more susceptible to killing by NK cells. Some viruses have developed strategies to overcome this, including actively preventing the expression of activating molecules on virus-infected cells and encoding MHC class I-like molecules (fakes) that can engage the inhibitory receptors on NK cells, which then protects the infected host cell from cytotoxicity.

12.2.3 Bacterial Subversion of Defensive Barriers

Bacteria have evolved strategies to colonize epithelial surfaces of the skin and mucous membranes by circumventing physical and/or biochemical defences. Teichoic acid molecules, embedded within the peptidoglycan layer of Gram-positive bacteria, such as *Staphylococcus spp.* may provide some degree of protection against the fatty acids present in sebum. thereby allowing these microorganisms to colonize the epidermis. Enteropathogenic bacteria may express adhesins (typically plasmid DNA encoded) to allow them to adhere to and colonize the mucosal epithelium of the alimentary tract. Other enteropathogenic bacteria can produce exotoxins (again, plasmid DNA encoded) that damage the mucosal epithelial barrier to facilitate their entry. Some Gram-positive bacteria alter the chemistry of their peptidoglycan layer (by N-deacetylation) to make them more resistant to the digestive effects of lysozyme, present in secretions such as saliva, sweat and tears. Bacteria can also defend themselves against the toxic effects of antimicrobial peptides (AMP) such as beta-defensins and cathelicidin, on their cell membrane by expressing a capsule. In addition, LPS molecules on the surface of some Gram-negative bacteria can bind to AMPs and prevent them from reaching the cell membrane.

12.2.4 Bacterial Evasion of Phagocytosis and the Complement System

Expression of an outer capsule, containing hyaluronic and/or sialic acids can protect bacteria from being endocytosed by phagocytic cells (**Figure 12.2A**). Expression of a capsule interferes with detection of bacterial PAMPs by PRRs expressed by phagocytic cells, including TLR-2 and TLR-4, as well as preventing the binding of complement activators (such as C3b and mannose-binding lectin). Growth of bacteria in a 'biofilm', whereby the microorganisms produce an extracellular polymer that facilitates their attachment and colonization, also, to some extent, makes them more resistant to phagocytosis and complement activation. *Mannheimia haemolytica*, an important bacterial cause of bovine respiratory disease, produces leukotoxin and *Staphylococci* produce leukocidins, which are toxic to neutrophils as they migrate into the site of infection, thereby destroying these phagocytic cells before they have had the opportunity to engulf the microorganisms. Following endocytosis, some bacteria are able to avoid or resist the killing and digestion processes, mediated by the respiratory burst and the action of lysosomal constituents. Some bacteria (including *Listeria monocytogenes*) can escape from the primary endosome into the cytoplasm before formation of the phagolysosome, thereby avoiding exposure to defensins and proteolytic enzymes (**Figure 12.2B**). Pathogenic mycobacteria (such as *M. bovis*) and *Salmonella spp.* inhibit the fusion of lysosomes to the primary phagosome, thereby allowing the microorganisms to survive and replicate in

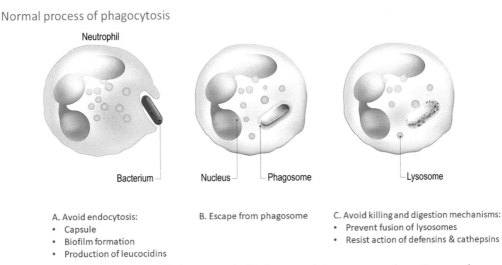

Normal process of phagocytosis

Neutrophil

Bacterium   Nucleus   Phagosome   Lysosome

A. Avoid endocytosis:
- Capsule
- Biofilm formation
- Production of leucocidins

B. Escape from phagosome

C. Avoid killing and digestion mechanisms:
- Prevent fusion of lysosomes
- Resist action of defensins & cathepsins

Figure 12.2 Avoidance/evasion of phagocytosis. Under normal circumstances, bacteria are endocytosed into a phagosome prior to microbial killing by reactive oxygen species and digestion via the action of lysosomal enzymes. Bacteria may avoid this by A. Avoiding endocytosis; B. Escaping from the endosome into the cytoplasm; or C. Preventing the formation of the phagolysosome and/or resisting the actions of lysosomal constituents, such as defensins and cathepsins.

the intracellular endosomal/vesicular compartment (**Figure 12.2C**). In addition, pathogenic mycobacteria express a waxy, hydrophobic cell wall containing mycolic acids (which makes them 'acid fast' upon staining with the Ziehl Neelsen method) that makes them resistant to the action of cathepsin enzymes present within the lysosomes.

12.3   EVADING THE ANTIBODY-MEDIATED ADAPTIVE IMMUNE RESPONSE

12.3.1   Avoidance of Extracellular Transmission

Following replication, release of new virus particles into the extracellular fluid allows viruses to spread from cell to cell within the tissues and also enables transmission from host to host. Virus-neutralizing antibody binds to virus surface antigens, preventing their attachment to target cells and effectively halting viral

infection within the host. Some viruses have adapted to spread, without entering the extracellular fluid and choose instead to undergo contiguous transmission from cell to cell, by disrupting the cell membranes between them (**Figure 12.3**). Syncytium formation is a cytopathic effect that can be seen on histopathology in some virus infections e.g. in the lungs of cattle infected with parainfluenza type 3 (PI-3) virus. Other viruses (e.g. retroviruses such as FeLV) avoid the extracellular fluid by integrating their DNA into the host genome to induce malignant transformation of the infected cell. Thus, the infected cell is forced to undergo uncontrolled proliferation, propagating the virus genome at the same time. The clinical outcome for these animals is that they are likely to succumb to neoplastic disease. Some retroviruses have integrated their DNA into germline cells and can therefore undergo vertical transmission. There are endogenous retrovirus sequences in the genomes of many mammalian

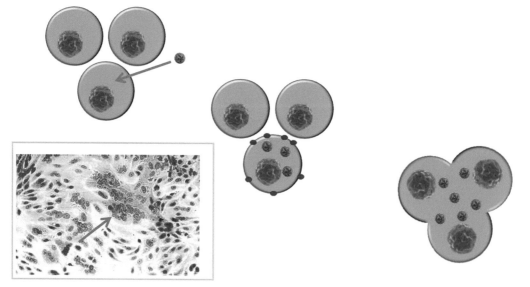

Figure 12.3 **Syncytia formation.** The virus enters a cell and starts to replicate. A fusion protein is expressed on the cell surface. Fusion of the plasma membrane of the infected cell with those adjacent to it leads to syncytium formation and the virus can undergo contiguous transmission. Inset: Syncytium formation in a cell line infected by respiratory syncytial virus (RSV). (Image by Dr Linda Stannard, courtesy of University of Cape Town, South Africa.)

species, including dog and pig, indicating an ancient viral integration event. This is one reason why there is concern about pig to human xenotransplantation, since this might lead to reactivation, adaptation and emergence of a porcine endogenous retrovirus in humans receiving such organs, which subsequently transmits to the rest of the population. However, advances in gene editing technologies has allowed scientists to 'excise' these endogenous retroviral sequences to create genetically modified pigs, such that this concern is ameliorated.

12.3.2 Antigen Decoys

Some viruses have evolved strategies to evade neutralization by the antibodies they elicit. Ebola virus produces a structural protein and a truncated soluble version. The soluble antigen is secreted in large quantities and saturates the binding sites of antibodies, such that when infectious virus particles are released, much of the antibody has already bound to the decoy

antigen and can no longer effectively neutralize the structural antigen on the surface of the virion. Other viruses express decoy antigens on their surface, which are much more immunogenic than the protein used for attachment (**Figure 12.4**). Thus, the antibody response in the host is distracted to focus on an antigen that is not important for virus infection. Some retroviruses use this strategy, which can impact on development of effective vaccines, whereby the subunit approach (i.e. only including those antigens that generate neutralizing antibody) is likely to give better levels of protective immunity.

12.3.3 Antigenic Variation

Antigenic drift (e.g. feline calicivirus) and antigenic shift (influenza viruses) are seen in some viruses. The basic principle is that genetic variation, either by random gene mutations, introduced during replication of the genome by low fidelity viral polymerase enzymes (antigenic

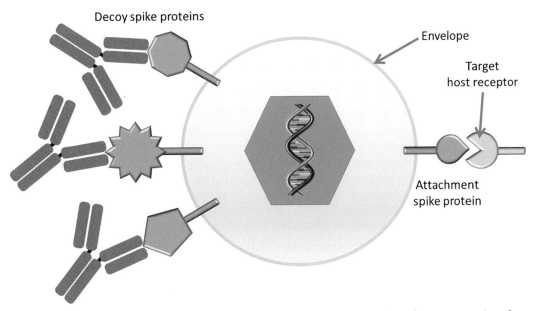

Figure 12.4 Antigen decoys. The virus expresses decoy antigens that are relatively immunogenic and stimulate production of antibodies that do not neutralize the virus. This 'distracts' the immune response away from the spike protein responsible for attachment to the host-cell receptor, which would normally induce virus-neutralizing antibodies.

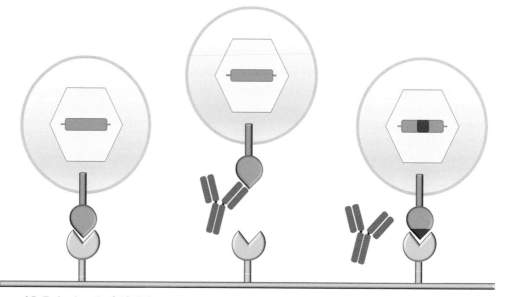

Figure 12.5 Antigenic drift. The virus expresses a spike protein that binds to the target host-cell receptor. Neutralizing antibody is subsequently produced that prevents attachment. Spontaneous mutations are acquired in the genome that alter the amino acid sequence of the spike protein, such that the original epitope(s) are altered and antibody no longer binds.

drift; **Figure 12.5**) or gene exchange between different viruses with a segmented genome (antigenic shift; **Figure 12.6**), can change the sequence of structural proteins and therefore the shape of antigenic epitopes. Thus, antibody produced against the original antigen, might no longer bind to the mutant/variant version and no longer neutralizes virus infectivity. Different strains of the virus can develop and evolve, which not only impacts on transmissibility and virulence (e.g. mutation of feline panleukopenia virus to become canine parvovirus), but also on protective immunity from vaccines (e.g. many different strains of feline calicivirus with variable protection from the immune response to the vaccine strain). In equine infectious anaemia (EIA), animals are persistently infected and sequential antigenic variants are produced, with each successive variant different enough to evade the immune response raised against the preceding one. Thus, in EIA, clinical signs occur in cycles, with each cycle being initiated by a new variant. In addition, generation of new vari-

ants during chronic infection can also lead to increasing virulence, affecting the severity and progression of the disease. Having said that, it is not uncommon for emerging viruses to mutate and generate variants (or 'escape mutants') that are less pathogenic to the host (albeit often more transmissible between hosts), as we have experienced during the SARS-CoV-2 pandemic.

Phase and antigenic variation in bacteria may also lead to some degree of antibody avoidance, particularly when this involves surface-expressed proteins (**Figure 12.7**). *Leptospira interrogans*, for example, may only express certain outer-membrane proteins (such as LipL41) when resident within the kidney tissue, whereas others (such as LipL32) are more constitutively expressed. Such outer membrane proteins may act as virulence factors, which not only facilitates bacterial colonization, but also delays recognition by the adaptive immune system. *Leptospira interrogans* also demonstrates antigenic variation in the outer lipopolysaccharide, which gives rise to various serovars/serogroups

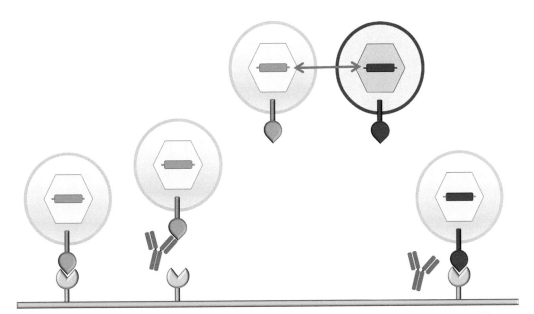

Figure 12.6 Antigenic shift. The virus expresses a spike protein that binds to the target host-cell receptor. Neutralizing antibody is subsequently produced that prevents attachment. Genetic reassortment between strains leads to emergence of chimaeric viruses that have acquired a novel spike protein, such that the original antibody no longer binds.

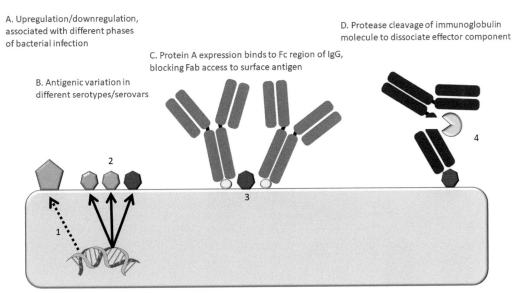

Figure 12.7 Microbial evasion of antibody. Various defence systems allow microbes to avoid and evade antibody binding and downstream antimicrobial effector mechanisms.

12.3.4 Microbial Antibody Defence Systems

Rather than avoiding antibody detection, some bacteria employ countermeasures to inhibit antibody binding or prevent downstream antimicrobial responses following antibody binding (**Figure 12.7**). Protein A, expressed on the surface of bacteria such as *Staphylococcus* *aureus*, binds to the Fc region of IgG, coating the organism with non-specific antibody and blocking the binding of opsonizing and complement fixing antibodies directed against its surface antigens. Some bacteria produce proteases that cleave immunoglobulin molecules after attachment, effectively dissociating the Fc region (responsible to binding to Fc-receptors and C1q in the case of IgG), such that antibody binding no longer leads to antimicrobial effector responses such as opsonization and complement activation.

12.4 EVADING THE CELL-MEDIATED ADAPTIVE IMMUNE RESPONSE

12.4.1 Viral Sequestration in Immune Privileged Sites

Some tissues (e.g. CNS and testes) are immune-privileged sites. The blood–brain barrier (BBB) in particular acts as a defensive structure to prevent transfer of pathogens from blood into the CNS, but also limits the ability of antibody and lymphocytes to gain access to the CSF/CNS. Furthermore, neurons express very little in the way of MHC class I. This is thought to be

a mechanism to protect the brain from inflammatory/destructive immunological processes that would likely lead to permanent damage or even death of the host during an inflammatory response to infection. Instead, the CNS tries to maintain sterility by virtue of the BBB. However, some viruses (e.g. rabies) have evolved to gain access to neurons or the CNS more generally, where they can evade immune detection. BVDV can persist in the testes of bulls for several years post-infection despite a systemic immune response.

12.4.2 Blockade of Antigen Presentation to CD8⁺ Killer T Cells by Immunoevasins

In infected cells, viral antigens are normally degraded by the proteasome complex in the cytoplasm to generate peptides, which enter the MHC class I processing pathway. This leads to expression of MHC class I:peptide complexes on the cell surface, allowing targeted detection and destruction by CD8⁺ killer T cells. Immunoevasins produced by some viruses can interfere with this pathway in a number of ways (**Figure 12.8**). Some viral immunoevasins block peptide entry into the ER by targeting the

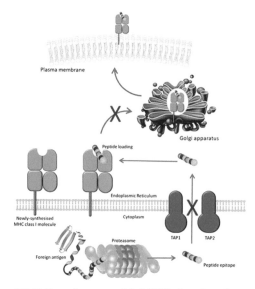

Retention of MHC-peptide complex in the ER and prevention of translocation to the cell surface

Blockade of TAP transporters, preventing peptide from entering ER and binding to MHC molecules

Figure 12.8 **Interference with MHC class I pathway.** Some viruses produce immunoevasins that block antigen presentation via the MHC class I pathway, thereby avoiding detection by CD8⁺ killer T cells.

TAP transporters. Other molecules can prevent MHC class I transport, leading to retention of MHC:peptide complexes in the ER. A herpesvirus protein misdirects newly synthesized MHC molecules into the cytoplasm for degradation.

12.4.3 Production of Superantigens

Superantigens act to cross-link MHC and TCR molecules non-specifically, as they bind outside the peptide-binding groove. This leads to non-specific activation and proliferation of antigen unreactive T cells, which acts to dilute the antigen-specific response against the virus. Thus, the immune system loses focus and the response is disorganized. There is evidence that some retroviruses use this strategy to evade the immune response.

12.4.4 Induced Resistance to Apoptosis

Upon recognition of a virus-infected cell, CD8⁺ killer T cells induce apoptosis using granzymes, FasL and cytotoxic cytokines. These all act via the caspase cascade to induce programmed cell death. Some viruses produce non-structural proteins that interfere with this killing response. Strategies include targeted inhibition of signalling responses through death receptors, inhibition of caspase enzymes through production of serpins or production of analogues of anti-apoptosis proteins (e.g. Bcl2).

12.4.5 Subversion of the Cytokine Response

The immune response is orchestrated by various cytokines. Viruses can use several strategies to subvert these cytokine responses (**Figure 12.9**). Production of cytokine-binding proteins and 'fake' cytokine receptors will antagonize cytokine responses in the host. Alternatively, production of 'fake' cytokine analogues can act as agonists. Orf virus and some herpesviruses produce an interleukin 10 homologue

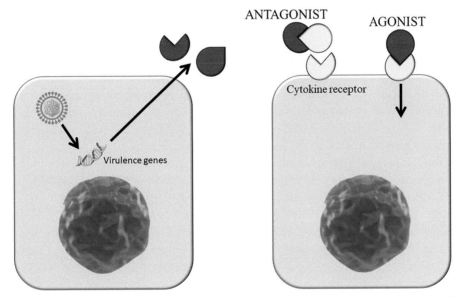

Figure 12.9 Subversion of cytokine responses. Some viruses produce non-structural proteins that are secreted and act as cytokine-binding proteins (antagonists) or 'fake' cytokines (agonists) to modulate/ downregulate the host immune response.

that mimics the immune system's natural immunosuppressive cytokine, normally produced by regulatory T cells. Production of this molecule dampens down the host immune response to infection and allows the virus to replicate more effectively.

12.5 SPECIFIC VIRAL STRATEGIES FOR IMMUNE EVASION

12.5.1 Direct Immunosuppression by Lentiviruses

Some viruses deliberately infect and kill host immune cells, to avoid adaptive immunity. Lentiviruses, such as Feline Immunodeficiency Virus (FIV) and HIV, infect and kill CD4+ T cells and cells of the monocyte/macrophage lineage. As the T helper cell numbers decline in the host, this has a major suppressive effect on both cell-mediated and antibody-mediated immunity. Cats infected with FIV become increasingly susceptible to opportunistic infections and cancer as immune surveillance of foreign antigen and altered self-antigen, associated with malignant transformation, diminishes.

12.5.2 Latency in Herpesviruses

Latency refers to the ability of a virus to become quiescent/dormant within the host, leading to persistent subclinical infection. After a period of active replication, the virus infects a host cell (often a different type to the primary target such as a nerve cell or lymphocyte), but does not replicate, thus avoiding production of viral PAMPs and antigens that can alert the immune system. Viral latency is also maintained by restricted expression of genes that have the capacity to kill the cell. This phenomenon is typically seen in herpesviruses. During periods of lowered immunity in the host, the virus can emerge to infect and replicate again in the target cells (recrudescence), causing a further phase of clinical disease. Once infected, the host will be prone to these reactivation events during periods of lowered immunity. In horses, infection with equine herpes virus typically leads to respiratory disease via infection/replication in mucosal epithelial cells. However, a lymphocyte-associated viraemia (particularly with EHV-1) can transport the virus to sites of secondary replication, whereby the virus infects and replicates in endothelial cells. Where this occurs in the pregnant uterus or central nervous system, this can lead to abortion or paralysis, respectively. Those animals that recover from respiratory disease (typically young animals) develop latency and are subsequently a potential source of infection to other horses, during periods of recrudescence. A similar syndrome occurs in cats infected with feline herpes virus 1 (feline viral rhinotracheitis virus), one of the agents responsible for 'cat flu'. Animals that have recovered from herpes flu often become latently infected and can demonstrate recrudescence, particularly during periods of lowered immunity caused by stress (e.g. a stay in a cattery) or treatment with corticosteroids. This can lead to the reappearance of clinical disease and signs of cat flu once again. Vaccination after infection and following development of latency will not usually protect against recrudescence and recurrence of clinical disease. Elephant endotheliotropic herpesvirus (EEHV) has probably host adapted in African elephants and infection is fairly benign, although when transmitted to Asian elephants, it can cause a highly fatal hemorrhagic disease.

12.5.3 Immune Tolerance in Pestivirus Infection

Some viruses exploit immune tolerance mechanisms, normally designed to prevent inappropriate responses against harmless and self-antigens. Bovine viral diarrhoea virus (BVDV) is a pestiviruses that infects cattle.

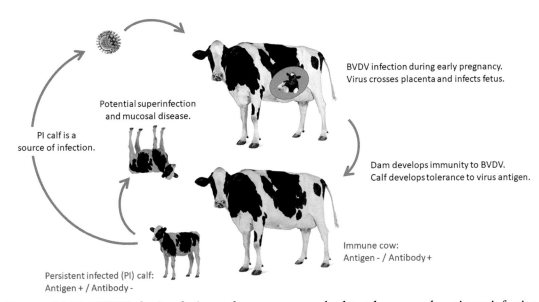

Potential superinfection and mucosal disease.

PI calf is a source of infection.

BVDV infection during early pregnancy. Virus crosses placenta and infects fetus.

Dam develops immunity to BVDV. Calf develops tolerance to virus antigen.

Immune cow: Antigen - / Antibody +

Persistent infected (PI) calf: Antigen + / Antibody -

Figure 12.10 **BVDV infection during early pregnancy can lead to tolerance and persistent infection of the calf.** If a pregnant cow is exposed to a non-cytopathic biotype of BVDV during pregnancy, the virus can cross the placenta and induce tolerance in the fetus. When the calf is born, it is persistently infected and is a source of infection to the rest of the herd. Mutation of the virus or exposure to a cytopathic biotype of BVDV can subsequently lead to high morbidity and mortality resulting from 'mucosal disease'.

There are different strains of BVDV, some of which are cytopathic and of high virulence/pathogenicity and others that are non-cytopathic and of low virulence/pathogenicity. Infection with non-cytopathic BVDV early in the course of pregnancy can lead to transplacental infection of the embryo/fetus. This may result in embryonic death, (resorption or abortion), but if infection occurs between 80 and 125 days of gestation (before development of immunocompetence) the fetus may survive, albeit with a persistent infection (**Figure 12.10**). Since viral antigen is expressed in the fetal primary lymphoid tissues, at a time when lymphocyte development is occurring, virus-reactive T and B cells are eliminated by clonal deletion as they would do to host self-protein. The developing calf becomes tolerant of the virus and is persistently infected (PI)

after birth, remaining antibody negative. Since the virus is of low virulence and causes little cellular damage, these PI calves appear relatively normal, but shed virus into the environment, acting as the major source of infection to other animals in the herd. A problem can subsequently develop in the PI calf, if the endogenous virus mutates to become more cytopathic or if the calf encounters another cytopathic biotype of BVDV in the environment, when extreme pathology (ulceration) occurs in the mucosal epithelium as the virus replicates and destroys cells undetected by the immune system, leading to the clinical syndrome known as 'mucosal disease'. A similar syndrome occurs with border disease virus in sheep where, in addition to persistent infected animals, lambs can be born with congenital defects known as 'hairy shakers'.

KEY POINTS

- Some viruses can subvert the host cytokine response, mediated by interferons or others, such as interleukin 10.
- Some bacteria can avoid the host defences associated with phagocytosis and complement activation.
- Some viruses avoid antibody detection by remaining in the intracellular fluid as much as possible.
- Some viruses avoid antibody binding to surface antigen by expressing decoy antigens or by regularly varying their antigenic structure.
- Some pathogens specifically infect immune-privileged sites.
- Viruses can produce immunoevasins that block antigen presentation by MHC molecules or by producing superantigens that subvert the interaction between MHC and the T cell receptor.
- Viruses can produce serpins that block the normal killing response to CD8+ T cells.
- Retrovirus infection can lead to malignant transformation or direct killing of lymphocytes resulting in immunosuppression.
- Herpesviruses can enter a state of latency to facilitate persistent infection in the host, with subsequent recrudescence leading to pathogen shedding and re-emergence of clinical signs.
- Pestiviruses may infect pregnant animals and induce a state of immunological tolerance during fetal development, leading to persistent infected animals being born and entering the population as a source of infection to other animals.

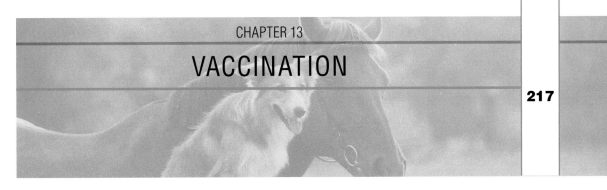

CHAPTER 13

VACCINATION

OBJECTIVES

At the end of this chapter, you should be able to:

- Discuss the difference between passive and active immunization.
- List the requirements of an ideal vaccine.
- Discuss the different vaccine technologies and their advantages/disadvantages.
- Define the term 'marker vaccine'.
- Understand why it is necessary to give multiple vaccines to young animals.
- Define 'core' and 'non-core' vaccines.
- Define the terms 'duration of immunity' and 'herd immunity'.
- List some possible vaccine side effects and understand why it is important to report these to an appropriate authority.

13.1 INTRODUCTION

Vaccination is the single most common application of immunology to clinical veterinary practice. It is the most effective medical intervention aimed at control and prevention of infectious diseases in animals. Vaccination of animals can also serve to protect the human population from zoonotic infectious diseases such as rabies. Greater understanding of basic and veterinary immunology along with molecular and technological advances continue to change the technology and practice of vaccination. It is crucial that practicing veterinarians keep up-to-date with current recommendations in vaccinology in order to provide the best standard of care for their patients. This chapter reviews the fundamentals of vaccination in animals.

13.2 PASSIVE IMMUNIZATION

Passive immunization involves the administration of **preformed antibodies** specific to a particular antigen in order to provide **immediate immunological protection**. The transfer of antibodies from the mother to her offspring (either transplacental or via colostrum) is the naturally occurring form of passive immunization. Passive immunization is rarely performed in veterinary practice. The best examples are the use of antibodies against tetanus and other clostridial toxins (antitoxins) and against snake venom (antivenoms). Such antisera are traditionally raised in a large animal species, such as a horse or sheep. Serum is collected from hyperimmunized animals and the immunoglobulin harvested from the serum (**Figure 13.1**). These

DOI: 10.1201/9781003310969-13

Figure 13.1 Passive immunization. In this process a large animal species (e.g. horse or sheep) is hyperimmunized with the antigen of interest (e.g. tetanus toxoid). The immune serum (anti-toxin) is harvested and the immunoglobulin fraction purified. This may be further treated to render it less immunogenic in the recipient animal to reduce the likelihood of inducing a hypersensitivity reaction on repeated administration.

immunoglobulins are immunogenic when injected into heterologous species and repeated administration of antitoxins can induce a **hypersensitivity reaction** in the recipient animal. The immunogenicity of the foreign immunoglobulin can be reduced by cleaving the Fc region from the protein and using just the F(ab')$_2$ component of the molecule. A further consideration with the use of antitoxins is that their administration will inhibit the ability of the recipient to mount their own endogenous antibody response to the target antigen. For these reasons, the use of passive immunization should be carefully considered and the minimal possible administration undertaken.

13.3   ACTIVE IMMUNIZATION

Vaccination usually refers to **active immunization** in which an antigen is administered to an individual to **induce an immune response** and, most importantly, **immunological memory** of that antigenic exposure. Active immunization of animals is mostly used to generate

immune responses that will protect against **infectious diseases**. The level of protection afforded by different vaccines varies. Some vaccines induce a very strong protective immunity, which prevents the vaccinated animal becoming infected by the organism (sterilizing immunity). In other cases, the vaccinated animal might become infected but remain asymptomatic or develop only a mild form of disease caused by the agent. These levels of protection form the basis of the **'claim'** for the vaccine and will be clearly defined on the **data sheet** (or in Europe the 'summary of product characteristics', SPC) that accompanies the product.

No vaccine currently marketed is perfect and consistently able to afford the highest level of protection to every single animal that receives it. The **properties of an ideal vaccine** include:

- The ability to induce the most **appropriate type of immune response** to afford protection from infection with the specific organism under consideration.

- The ability to induce a **long-lived immune response** and **immunological memory**.
- **Stable** without a requirement for specialized storage conditions (e.g. refrigeration).
- **Consistent in formulation** with minimal variation between batches.
- **Inexpensive** to produce in large quantity.
- **No systemic adverse effects and limited or no local reaction.**

Vaccination is a **medical procedure** and should only be performed when the ratio of risk and cost versus benefit is medically and economically favourable, and when an effective vaccine is available. For some vaccines, a single dose administration is sufficient to induce a protective immune response, whereas other vaccines require two or three doses, which further enhance or **boost** the immune response. The immunologic mechanisms that underlie the protection induced by vaccination can be different from those that are necessary to recover from natural infection. For example, while CD8+ T cells are critical to eliminate influenza virus from an infected animal, vaccine-induced antibodies play an important role in protection against influenza viral infection. For some diseases, such as canine distemper and rabies, specific thresholds of vaccine-induced antibody are correlated with protection. A range of formulations and technologies is available for veterinary vaccines, from relatively crude preparations, made using methods practiced by Louis Pasteur in the 19th century, to modern genetically engineered products. These will be reviewed in the next sections.

13.4 VACCINE TECHNOLOGIES

13.4.1 Live Attenuated Vaccines

This class of vaccine is the single **most commonly used** in veterinary medicine. These vaccines are based on the use of an **intact and viable organism** that has been 'attenuated' to reduce its virulence. Live attenuated organisms are capable of undergoing **low-level replication**

within the animal, but **do not induce significant tissue pathology or clinical disease**. Traditional methods of attenuation might involve passaging the organism multiple times through one or more types of cell culture or by selecting organisms for growth at lower temperatures than physiological body temperature to develop **temperature-sensitive** mutants. The mutations that underlie the attenuation often remain poorly defined. A more refined means of attenuation is to use molecular techniques to produce genetically modified organisms from which one or more **virulence genes have been deleted or modified**. An example of this approach is the development of vaccines for pseudorabies (Aujeszky's disease) in pigs in which the thymidine kinase gene has been deleted from the causative herpesvirus.

An alternative approach is to use an antigenically related organism that is able to induce immunity but not cause disease or an adverse reaction. An example of this approach is the use of an attenuated strain of canine adenovirus (CAV)-2 to protect against disease caused by both CAV-1 and CAV-2 without the risk of the dog developing 'blue eye' (corneal oedema secondary to immune complex uveitis) as an adverse effect of vaccination with CAV-1 (see Chapter 17). A heterologous vaccine incorporates an organism that is **antigenically related** to the target infectious agent, but is **adapted to another host** species. The use of cowpox (vaccinia) virus to protect against smallpox by Edward Jenner is an example of this strategy. An example in veterinary medicine is the use of human measles virus to protect against the related canine distemper virus (CDV). The measles virus vaccine can be used to immunize pups two to four weeks earlier than with a CDV vaccine, when the pups may still have maternally derived CDV antibody. However, this procedure is no longer widely practiced.

Live attenuated vaccines typically induce immune responses similar to that induced by natural infections including antibody responses and activation of CD4+ and CD8+ T cells. They are **highly immunogenic** and **relatively inexpensive**. The main drawbacks of live attenuated vaccines are **safety concerns**. The attenuated vaccine strain might undergo further mutations during replication in the vaccinated animal or 'recapture' virulence genes from field organisms by recombination, allowing them to **revert to virulence** and induce clinical disease rather than protection. Some live attenuated vaccines intended for injection (e.g. feline herpesvirus, FHV) may **induce disease** if they are accidentally aerosolized during administration or groomed by the animal from the cutaneous site of administration. The attenuated strains may also cause disease when inadvertently administered to immunosuppressed animals. The nature of large-scale manufacture of live attenuated vaccines also carries a theoretical risk that a particular vaccine might become **contaminated** with another unrelated organism. Finally, the organisms in live attenuated vaccines need to remain viable and this requires more specialized storage conditions (i.e. they are often formulated as freeze dried pellets that require refrigeration).

13.4.2 Inactivated Whole Organism Vaccines

Inactivated vaccines include an organism that is unable to replicate or induce pathology or clinical disease. There are various methods of inactivating infectious organisms, most often by treatment with **chemicals** such as formalin, alcohol or alkylating agents. It is critical that the chemical treatment does not destroy important antigenic epitopes of the organism. Killed bacteria included in vaccines may be referred to as 'bacterins'. The inactivated types of vaccine are relatively safe and may be used in pregnant animals. Because inactivated organisms are less immunogenic than live organisms, these vaccines often require adjuvants to stimulate an effective immune response. The inactivated organisms do not replicate intracellularly and are therefore less likely to induce a CD8+ T cell response, with protection more dependent on the induction of antibodies. Furthermore, they are less suitable for use in young animals, because they are more susceptible to interference by maternal antibodies.

13.4.3 Subunit Vaccines

A subunit vaccine does not contain an entire intact organism, but rather one or more specific **immunogenic structural proteins or secreted toxins** that are either isolated from cultures or produced by **recombinant DNA technology**. Bacterial toxins such as tetanus toxin are inactivated by treatment with formalin and are called **toxoids**. Expression of recombinant protein antigens is accomplished by inserting the gene encoding an immunogenic protein of interest (able to induce a protective immune response) into a plasmid for expression in bacteria, yeast, or insect cells. The **recombinant protein** can be harvested from cultures and incorporated into a vaccine (**Figure 13.2**). The recombinant protein can be further engineered to optimize its expression, stability and immunogenicity. Viral proteins sometimes self-assemble into virus-like particles, essentially virus particles that lack nucleic acid and nonstructural proteins. Subunit vaccines are extremely safe, but the proteins are often poorly immunogenic, and require an adjuvant to induce optimum immunity. Examples of such products are porcine circovirus-2 vaccines that contain recombinant capsid protein expressed in insect cells; a FeLV vaccine that incorporates the recombinant p45 antigen expressed in bacteria; and the *Borrelia* vaccine for dogs, which includes a recombinant version of the immunodominant outer surface protein A of the organism.

13.4.4 Recombinant Vector Vaccines

A more recent development in veterinary vaccinology is the use of **recombinant vector vaccines**. The vectors are viruses or bacteria that have been selected or genetically altered

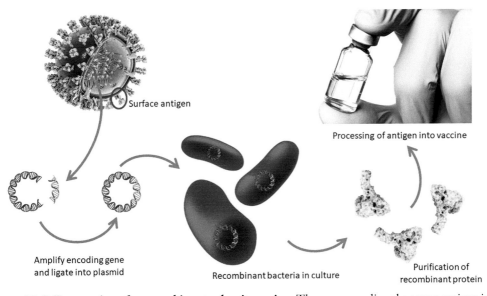

Surface antigen

Processing of antigen into vaccine

Amplify encoding gene and ligate into plasmid

Recombinant bacteria in culture

Purification of recombinant protein

Figure 13.2 Preparation of a recombinant subunit vaccine. The gene encoding the target antigen is identified and amplified. The coding sequence is then inserted into a bacterial plasmid and bacteria (typically *E. coli*) are transformed. The bacteria are cultured in bulk and the recombinant protein harvested, purified and processed into a vaccine. This type of recombinant subunit vaccine usually requires an adjuvant to be added to induce an effective immune response.

to prevent or limit their replication and not cause disease. The vector is modified to incorporate a gene from an unrelated pathogen that is expressed within infected host cells. The carrier organism is often readily taken up by dendritic cells and triggers both humoral and cell-mediated protective immune responses (**Figure 13.3**). Experimentally, a range of **carrier viruses** (e.g. pox viruses and adenoviruses) and bacteria (e.g. an attenuated strain of *Salmonella*) have been studied for systemic and mucosal delivery. Pre-existing or induced immunity to the vaccine vector can diminish the effectiveness of recombinant vector vaccines. In veterinary medicine, the **canarypox virus** has been successfully employed as a vector for genes derived from FeLV, canine distemper virus, rabies virus, West Nile virus and equine influenza virus. These vaccines induce potent protective immune responses and can induce **immunity in the face of levels of maternally derived immunoglobulin** that would normally block the effect of traditional live attenuated virus vaccines (see Chapter 14). The avian virus is unable to induce disease in the mammalian host and the vaccinated animal appears not to make a significant immune response to the canarypox, permitting repeated use of the vaccine. Recombinant vector vaccines are safe as they cannot **revert to virulence**.

13.4.5 Nucleic Acid Vaccines

Instead of viral or bacterial vectors, DNA and RNA that encode an antigen can themselves induce an immune response. They can be delivered directly as nucleic acid, plasmids or in lipid nanoparticles. The principle of this method involves the nucleic acid **transfecting host cells at the site of injection**, both dendritic cells and other cells. Proteins expressed within the dendritic cells can enter the MHC class I pathway of antigen processing, whereas apoptotic material released from dying cells can be taken up by dendritic cells for processing via the MHC

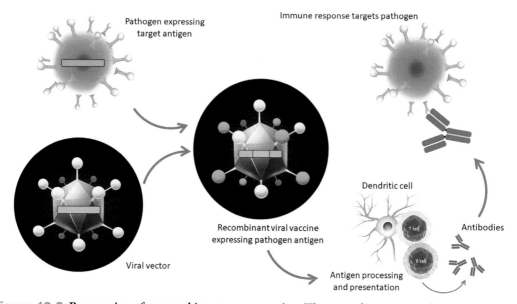

Figure 13.3 Preparation of a recombinant vector vaccine. The gene of interest encoding an antigen known to induce a protective immune response is taken from the pathogen and inserted into a viral vector (e.g. canarypox or adenovirus). On injection, the recombinant virus replicates, generating the antigen of interest. This antigen then stimulates an immune response, including production of neutralizing antibodies, that targets the original pathogen.

class II pathway or MHC class I pathway (cross-presentation) (**Figure 13.4**). Nucleic acid vaccination can trigger a **mixed cell-mediated and humoral immune response** that can provide effective protective immunity. Such vaccines may be used in young animals **in the face of maternally derived antibody**. The first licensed DNA vaccine for animals is used in commercial aquaculture to protect salmon from infectious hematopoietic necrosis virus. The vaccine consists of DNA plasmids that are injected intramuscularly into young fish. A DNA plasmid vaccine is also the basis of a therapeutic melanoma treatment for use in dogs (Chapter 20). mRNA vaccines have recently come to the fore in human medicine for SARS-CoV-2.

13.4.6 Marker Vaccines

Many infections are diagnosed by the detection of serum antibody as evidence of exposure (e.g. Lyme borreliosis, FIV), but as vaccination also induces serum antibody, it has traditionally been difficult to **discriminate between vaccine-induced antibodies and those stimulated by natural exposure**. Marker **vaccines** have been developed to make this distinction possible. A good example of such a product is the marker vaccine for infectious bovine rhinotracheitis. The gene encoding surface glycoprotein E has been deleted from the attenuated or inactivated herpesvirus contained in this product; therefore, if a cow demonstrates serum antibody to glycoprotein E, this must have been generated by exposure to field virus rather than by vaccination (**Figure 13.5**). Development of a marker vaccine requires parallel development of an **appropriate diagnostic test**. These vaccines are also known as **DIVA (differentiating infected from vaccinated animals)** vaccines.

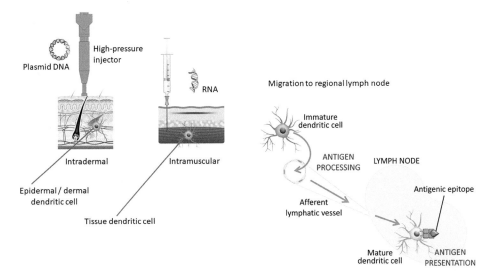

Figure 13.4 Administration of nucleic acid vaccine. The gene of interest from a pathogen, encoding an important antigen, is identified. The coding sequence can be cloned into a bacterial plasmid (pDNA), or alternatively processed to generate mRNA. The resulting nucleic acid (pDNA or mRNA) is administered, either via use of a high-pressure needleless injector or by intramuscular injection. The nucleic acid is taken up by tissue cells and dendritic cells and recombinant protein is generated *in vivo*. This protein antigen is subsequently processed and presented to the immune system to generate T cell-mediated and antibody-mediated immunity.

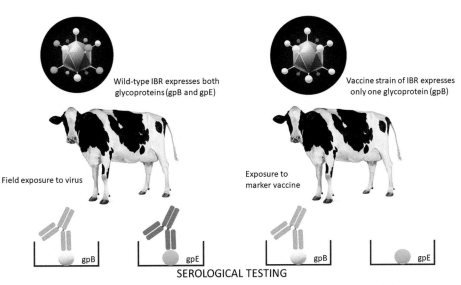

Figure 13.5 Preparation of a marker vaccine. A marker vaccine permits serological discrimination between vaccine-induced and natural exposure antibodies. The Infectious Bovine Rhinotracheitis (IBR) marker vaccine consists of a genetically modified herpesvirus that does not express glycoprotein E (gpE). A serological test (e.g. ELISA) can be performed to discriminate between field infection with the wild-type virus and vaccine immunity. Any cow that has serum antibody to gpE must have been exposed to the wild-type virus.

13.5   VACCINES FOR NON-INFECTIOUS DISEASES

Vaccines have been mainly designed to protect against the risk of infectious diseases and have been highly successful. Vaccination strategies are also being investigated for the treatment (**therapeutic vaccines**) and possibly prevention of **allergic**, **autoimmune** and **neoplastic diseases**. It should be noted that some vaccines against oncogenic viruses are aimed at preventing cancer caused by these viruses. A veterinary example is the Marek's disease vaccine, which protects against a herpesvirus that causes lymphoma in poultry and can arguably be considered the first cancer vaccine. In humans, the papillomavirus vaccine protects against cervical cancer. We will discuss the melanoma vaccine for dogs in Chapter 20. Another application of vaccine technology is to prevent 'boar taint' in pork production. A commercially available vaccine induces an immune response to gonadotropin-releasing hormone (GnRH), which removes the stimulus for production of luteinizing hormone and follicle-stimulating hormone and, in turn, the effects of these hormones on testosterone production. In effect, the vaccine produces an 'immunological castration' by inducing autoantibodies against GnRH, which has welfare benefits when compared with physical castration practised without anaesthesia or analgesia.

13.6   VACCINE ADJUVANTS

Vaccines that contain inactivated organisms or individual proteins such as toxoid or subunit vaccines generally do not induce an effective immune response by themselves. They are typically combined with substances known as **adjuvants** that enhance the magnitude and quality of the immune response. The term adjuvant was first coined by the French veterinarian Gaston Ramon (1886–1963), who was tasked with raising antibodies against diphtheria and tetanus toxins in horses for therapeutic use in human patients. He discovered that horses injected with proteins combined with materials that induce local inflammation developed higher antibody titers than injection of proteins alone. Adjuvants increase the immune response by one or more of the following mechanisms:

- Depot effect. A 'slow release' of antigen into the lymphatic system to provide more sustained antigen stimulation of lymphocytes in the drainage lymph node.
- Adsorb antigens and make them particulate. Particulate material is more easily taken up by antigen presenting cells through phagocytosis. In addition, adsorbed antigens are retained at the injection site from which they are gradually released.
- Induction of an inflammatory reaction at the injection site which enables the recruitment of antigen presenting cells.
- Activation of dendritic cells resulting in increased expression of MHC and costimulatory molecules CD80 and CD86 and secretion of cytokines.

The chemical and molecular nature of adjuvants used in human and animal vaccines is quite diverse. They include aluminium salts, oil-in-water and water-in-oil emulsions, synthetic polymers, saponins and liposomes. These adjuvants lack ligands of pattern recognition receptors (PRRs), but induce local cell injury at the site of injection resulting in the release of danger signals that can activate PRRs (**Figure 13.6**). Increasingly, adjuvants also contain molecules that are specifically designed to activate PRRs of innate immune cells such as monophosphoryl lipid A (a ligand of TLR4) and CpG DNA (ligand of TLR9).

13.7   ROUTES OF ADMINISTERING VACCINES

The majority of vaccines are administered by **intramuscular** or **subcutaneous** injection.

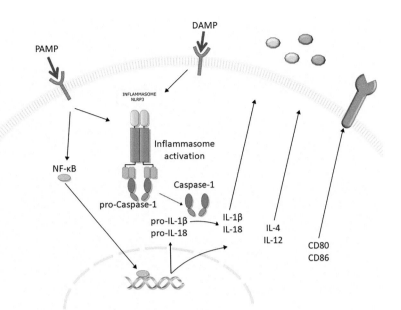

Figure 13.6 Adjuvant effect on the innate immune system. Inclusion of an adjuvant can enhance the adaptive immune response to the vaccine antigen through modulation of innate immunity. Stimulation of specific PRRs via PAMPs/DAMPs can elicit a pro-inflammatory cytokine response, as well as upregulation of other co-stimulatory signals such as CD80/CD86. Adjuvants can be used to stimulate production of other cytokines (such as IL-4 and IL-12) that play a role in T cell differentiation and thus influencing the balance between antibody-mediated (CD4+ Th2) and cell-mediated (CD4+ Th1) immunity.

This is quick and convenient and, when performed correctly, ensures that the entire dose is delivered. The injections are given in body sites that have low carcass value for food animals and are relatively safe for the person performing the injection. **Intradermal** vaccination is attractive because it can be performed with needle-free delivery systems avoiding risks of injury from needle sticks and disease transmission when needles are reused (**Figure 13.4**). It also places the vaccine in a tissue that is rich in dendritic cells and macrophages, which increases the immunogenicity of the vaccine. Needle-free injectors deliver a thin stream of the vaccine under high pressure through the epidermis into the dermis. Intradermal vaccines are currently available for use in pigs and in dogs.

Vaccination through injection induces a robust systemic immune response, but does not provide a strong response at mucosal sur-

faces, which is where many infectious organisms enter the body. Delivery of vaccines directly to mucosal surfaces stimulates a local immune response and can provide local as well as systemic immunity. However, it is challenging to overcome the natural tendency towards tolerance at mucosal surfaces. Inactivated and subunit vaccines are generally ineffective as there are currently limited options for adjuvants that can be applied safely and effectively to mucosal surfaces. However, there is an increasing number of live attenuated and recombinant vector vaccines that are specifically designed for mucosal delivery. **Intranasal vaccination** of individual animals is therefore practised in order to protect them from respiratory pathogens such as infectious bovine rhinotracheitis virus, *Streptococcus equi*, *Bordetella bronchiseptica*, canine parainfluenza virus, feline calicivirus and herpesvirus-1. As these are live vaccines, animals may show transient upper respiratory signs such as sneezing and increased nasal discharge. Mucosal delivery has also been used for delivery of vaccines to groups of animals; for example, by **aerosolization** of Newcastle disease vaccines for poultry or **administration in feed or water** for a range of poultry vaccines. This is less stressful to the animals and less labor-intensive as it does not require handling individual animals.

In ovo vaccination of poultry is accomplished by injecting live attenuated virus through the egg shell into the developing embryo two to three days before hatching. This process is fully automated and avoids the stress of handling individual chicks. Initially developed to vaccinate against Marek's disease, the technology is now also used to protect chickens against other diseases including infectious bursal disease, Newcastle disease and avian influenza. **Fish can be vaccinated** by **intramuscular or intracoelomic injection** or by **immersion**. Immersion vaccination is performed by bathing fish in a holding tank with a high concentration of live attenuated or inactivated vaccine. Antigens are taken up orally, via the gills or via the skin. The method is often less effective than injection, but it is much less labor intensive, when dealing with large numbers of fish.

13.8 BASIC PRINCIPLES OF VACCINATION

It is beyond the scope of this chapter to give details of all of the currently licensed veterinary vaccines and the recommended schedules by which these are administered. Instead, some general principles for vaccination are given here:

- The **efficacy of a vaccine** is the ability of the vaccine to prevent the occurrence of a specific disease outcome in vaccinated animals under controlled experimental conditions. Such outcomes may include death, clinical signs of disease, abortion or weight loss. The efficacy is determined in challenge studies in the animal species for which the vaccine is developed. Challenge studies are performed in a small number of animals under conditions that may differ from natural exposure in the field.

- The transmission and spread of an infectious disease is reduced when a high enough percentage of animals has developed protective immunity. This so-called **herd immunity** protects animals that are not immune because they have not been vaccinated or because they are unable to develop a protective immune response. The percentage at which herd immunity is achieved depends on the pathogen and the heterogeneity and density of the animal population, but is often assumed to be around 80%.

- The use of some vaccines is indicated **during pregnancy** in order to ensure good colostral antibody titres. However, some live attenuated vaccines may induce abortion or teratogenic effects. Therefore, vaccines should only be administered to pregnant animals if this procedure is supported by data sheet recommendations.

- Vaccines should only be used when the disease against which they protect is common and causes significant morbidity and/or mortality or when required to protect public health. For companion animals, a distinction is made between **core vaccines** that should be used in all animals (Table 13.1) and **non-core vaccines** that are only warranted when an animal has a significant likelihood of exposure to the pathogen because of lifestyle and geographic location. The decision of which non-core vaccines should be administered should arise from **consultation between veterinarian and client**.

- The **duration of immunity** (DOI) refers to the **length of time after vaccination that an animal maintains protective**

Table 13.1 Core Vaccines for Dogs, Cats and Horses

DOGS[1]	CATS[2]	HORSES[3]
Canine distemper virus	Feline parvovirus	Tetanus
Canine parvovirus	Feline herpesvirus	EEE/WEE[4]
Canine adenovirus	Feline calicivirus	West Nile virus
Rabies virus	Rabies virus	Rabies virus
	Feline Leukemia virus	

[1] American Animal Hospital Association (AAHA)—2017; [2] AAHA/American Association of Feline Practitioners—2020; [3] American Association of Equine Practitioners—2020; [4] Eastern and Western Equine Encephalomyelitis.

immunity. It varies greatly among vaccines and there is intense interest in understanding the factors that determine the longevity of the immune response. The DOI is demonstrated by challenging a group of vaccinated animals with the infectious agent (at a defined time point post vaccination) and determining if they are protected relative to unvaccinated controls. For some vaccines challenge experiments can be substituted by measuring serum antibody titres when a strong correlation between antibody titres and protection has been established. In general, adult animals should be vaccinated as infrequently as legal requirements, current best practice guidelines and quality of the vaccine permits.

13.9 VACCINATION FAILURE

Although licensed vaccines are generally effective, they can occasionally fail to induce a protective immune response. The reasons for this are diverse and can be grouped into three categories: related to **vaccine storage and administration**, related to factors associated with the recipient animal and related to factors associated with the **vaccine composition**.

The **single most common cause for failure of a vaccine** to protect an animal from infectious disease is not adhering to the manufacturer's recommendations for use and storage of the vaccine.

- Vaccines have an **optimum storage temperature** (described on the data sheet), which is typically between 2 and 8°C (domestic refrigerators should normally be at 4°C). These products should not be frozen nor positioned adjacent to the freezer compartment of the refrigerator, and the refrigerator temperature should be monitored regularly. Transportation of the vaccine should also be subject to continuation of the '**cold chain**'.

- **Freeze-dried vaccines** should be reconstituted immediately prior to its use, with appropriate diluent or liquid vaccine given concurrently (as per manufacturer's recommendations). It is not good practice to make up the vaccines in anticipation of their use, then delay their administration. Some vaccine components (e.g. CDV, FHV-1) are particularly labile in this regard.

- Animals can be vaccinated against different infectious diseases at the same time. The immune system is perfectly capable of responding to multiple antigens simultaneously. **Multivalent vaccines** are specifically formulated to induce protection against several organisms. They are effective as they are only licensed when protective efficacy can be demonstrated for all components. However, **different vaccines should not be mixed in the same syringe unless specified in the data sheet** as the different components may interfere with each other. An animal can be vaccinated with more than one vaccine at a time, as long as they are administered separately.

- **Vaccines should only be administered as specified in the data sheet.** Vaccines formulated for intranasal administration should not be injected intramuscularly or subcutaneously and vice versa.
- Injection sites should not be sterilized with alcohol as this may inactivate the vaccine.
- Vaccines should not be used beyond the expiration date and precise details of batch numbers and site of injection should be noted in the animal's medical record.
- Even when a vaccine is correctly stored and administered, it may not generate an effective immune response because of certain factors associated with the recipient.
- There is **normal variation in the immune response** between individual animals. These differences are in part **genetically determined**, but **environmental factors** such as **nutrition** and **overall health status** also play a role. Genetic differences may be related to MHC alleles and variations in the expression of genes that encode cytokines and other immune factors. In addition, variation in the immune response to vaccines in different dog breeds is correlated to their size. Small breed dogs tend to have a stronger response to vaccines than large breed dogs (**Figure 13.7**). Since all dogs should receive the same vaccine dose, smaller dogs are given a relatively greater amount of antigen in terms of their bodyweight, although since the response to vaccination occurs in the lymph node draining the site of injection, it is not clear whether there is any immunological difference in terms of that lymph node being present in a small dog versus a large dog. It is worth highlighting here that the Rottweiler is a breed known to make poor responses to vaccination (in particular for canine parvovirus and rabies).
- **Animals that are ill or medically immunosuppressed should not be** vaccinated as the immune response will be suboptimal.
- **Maternally derived antibody** will interfere with the ability of young animals to respond to the majority of currently available vaccines. Young animals therefore require a series of primary immunizations followed by a booster vaccine, generally given 12 months after the priming series or at 12 months of age. The reasons for this approach are discussed in Chapter 18.

If the vaccine strain of the organism is different from the strain that circulates in the environment, the vaccine-induced immune response may not protect against the disease. This is common with vaccines against RNA viruses such as influenza virus, because of the strong tendency of these viruses to mutate. To overcome this problem, veterinarians sometimes resort to **autogenous vaccines** especially in the swine, poultry and aquaculture industries. To prepare autogenous vaccines, material from diseased animals is sent to a commercial laboratory to isolate the infectious organism. The organism is inactivated and mixed with adjuvant for injection into animals in the same environment. Autogenous vaccines do not require efficacy testing and may not be sold commercially to other farms during the first 24 months after production.

13.10 ADVERSE CONSEQUENCES OF VACCINATION

Since the introduction of vaccination by Edward Jenner, there have been concerns about the risks of vaccination. As vaccines are typically administered to healthy animals, the acceptable risk of adverse reactions is very low. Furthermore, the success of vaccination has led to the nearly complete disappearance of serious and often fatal infectious diseases such as canine distemper to the point that few veterinarians and pet owners have encountered the disease. Over the

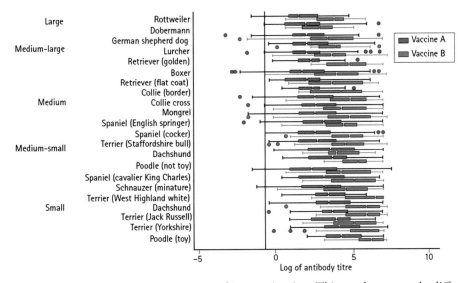

Figure 13.7 Breed differences in response to rabies vaccination. This graph presents the differences in serum antibody titres to rabies vaccination demonstrated by dogs of different breeds. These were all naïve animals receiving a primary course of vaccine for pet travel from the UK. Small breed dogs make the strongest responses and Rottweilers and Dobermanns achieve the lowest antibody titres. (Reprinted with permission from Kennedy LJ, Lunt M, Barnes A et al. (2007) Factors influencing the antibody response of dogs vaccinated against rabies. *Vaccine* **25**:8500–8507.)

past 25 years there has been mounting concern, amplified by media reporting and social media platforms, about the potential risks of vaccination in humans and companion animals. These concerns have sometimes led to reduced vaccination coverage and the unfortunate re-emergence of vaccine-preventable diseases.

At the outset it should be emphasized that **adverse effects following vaccination are rare** and that the **benefits of vaccination far outweigh any risks**. Hundreds of millions of animals are vaccinated every year, but very few adverse consequences are reported. It is recognized that the passive nature of reporting schemes is a barrier to gathering high-quality data on adverse effects, but where quantification of risk has been attempted, figures of 50 adverse events (no matter how minor) or less per 10,000 doses of vaccine administered have been calculated for dogs and cats. Just as there is individual variability in the immune response to vaccines, there is also **variability in the frequency and**

severity of adverse reactions, which are partially genetically determined. In dogs, small breeds generate a stronger immune response to vaccines, but also have a higher incidence of adverse reactions than large breed dogs.

Vaccination-associated adverse events refers to any localized or systemic reaction that occurs **following vaccination** and **does not necessarily imply causality**. Such events should be **reported to regulatory agencies and to the vaccine manufacturer**, and may prompt investigations to identify a possible cause. This may reveal an error in the production process, such as inadequate inactivation of a vaccine organism or contamination of a live attenuated vaccine with an unrelated pathogen. Such errors are extremely rare and can be corrected quickly when reported and recognized.

While uncommon, a wide spectrum of adverse effects can occur as a result of the innate and adaptive immune response to vaccines. The single most frequent event is the onset of **transient pyrexia and lethargy** in the few days post vaccination. This effect relates to initiation of the immune response and cytokine production and is therefore not unexpected. Another potential occurrence involves a **type I hypersensitivity response within minutes of administration** of a vaccine (**Figure 13.8**). These usually manifest (in cats and dogs) as an acute episode of pruritus or cutaneous oedema (often facial oedema) that resolves spontaneously. Rarely, vaccination induces a life threatening anaphylactic reaction. Type I hypersensitivity reactions may occur again with subsequent vaccination. Studies in dogs have revealed that this type of response is often related to the **presence of excipient proteins** such as bovine serum albumin (BSA) within vaccines. The induction of IgE antibodies against BSA may lead to an allergic reaction.

Vaccines have been suspected to be associated with **adverse events that are based on type II or type III hypersensitivity reactions**. Examples of type II hypersensitivity reactions are immune-mediated haemolytic anaemia (IMHA) and immune-mediated thrombocytopenia (ITP). However, there is no convincing evidence that vaccination causes either IMHA or ITP in animals. An example of a type III hypersensitivity reaction is blue eye (corneal oedema) after vaccination with live attenuated canine adenovirus-1. The replacement of canine adenovirus-1 by canine adenovirus-2 has subsequently eliminated this adverse vaccine reaction. Vasculitis resulting in ischemic dermatopathy with alopecia may occur following rabies vaccination, predominantly in small breed dogs. Although the pathogenesis is poorly understood, the presence of complement in the wall of affected blood vessels is suggestive of a type III hypersensitivity reaction.

The most serious vaccine-associated adverse event in animals is the **feline injection site**

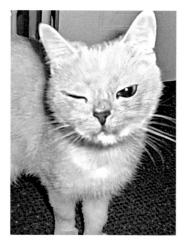

Figure 13.8 Hypersensitivity reactions following vaccination. Occasionally vaccinated animals can develop IgE-mediated type I hypersensitivity reactions post-vaccination and develop sudden onset urticaria or facial oedema (as shown in this cat and dog). Such reactions are thought to represent hypersensitivity reactions to excipients/other proteins, such as bovine serum albumin, contained within the vaccine, either through deliberate addition or residual from the cell culture system used to generate the vaccine antigen.

sarcoma (**FISS**). This malignant neoplasm is an **invasive sarcoma that can metastasize** and that develops at **sites related to previous injection**. The pathogenesis of these tumours is thought to involve a **chronic inflammatory response**, which in some animals leads to local transformation of mesenchymal cells and development of malignancy. Histologically, these tumours are **most often fibrosarcomas**, but differentiation of mesenchyme towards other matrix elements may occur (**Figure 13.9**). The development of FISS is not related to any particular type of vaccine and, indeed, has also been reported following injection of non-biological drugs and in apparent response to suture material and microchips. **The risk of FISS is low with an estimated frequency of less than 3 per 10,000 vaccinated cats.** By comparison, the risk of anesthetic death in cats is estimated at 11 per 10,000 cats. **Surgical excision and management of FISS is challenging and the tumours carry a poor prognosis.** Complete surgical resection is particularly challenging when vaccines are injected in the interscapular region, which was the traditional site of vaccine administration. Guidelines from the American Association of Feline Practitioners recommend that vaccines are injected in the distal limbs to allow for amputation if a malignant tumour develops. Cats may also be vaccinated safely into the skin of the tail and they show seroconversion to rabies virus and feline parvovirus after vaccination at these alternative sites. Injection site sarcomas have also been reported in other animal species, including dogs, ferrets and a horse, but their occurrence is exceedingly rare.

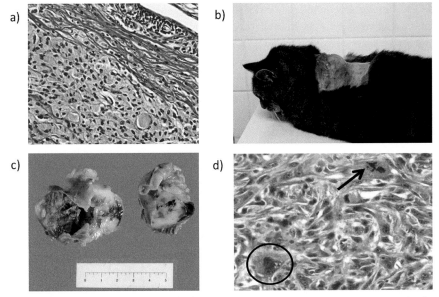

Figure 13.9 **Adjuvant-associated inflammation and feline injection site sarcoma.** (a) This histological section reveals the chronic granulomatous inflammatory response that can persist for a long period after administration of an adjuvanted vaccine. (b) Cat affected with feline injection-site sarcoma. (Photograph courtesy of Prof. Johannes Hirschberger, Ludwig Maximilians University, Munich, Germany) (c) A surgically excised mass from the interscapular region of a cat. The mass consists of firm white tissue with a necrotic cavitated centre. The tumour tissue infiltrates muscle at the margins of the excision. (d) Histological section from the mass. The tumour comprises a highly pleomorphic population of spindle cells including multinucleate cells (circled, bottom left of image). A bizarre mitotic figure is present in one cell (arrow).

There is no evidence of development of tumours following vaccination in people.

Although the immune response induced by vaccines is usually protective, in some instances the immune response aggravates the disease following naturally occurring infection. Antibody-dependent enhancement of disease has been reported for feline infectious peritonitis (FIP). The coronavirus that causes FIP replicates in macrophages. Non-neutralizing antibodies induced by vaccination can form immune complexes with the virus that facilitate the uptake by macrophages. There is an intranasal FIP vaccine comprised of a live temperature-sensitive variant of the virus. The induction of IgA would prevent infection of vaccinated animals with FIP. However, the vaccine is currently not recommended by the AAHA/AAFP guidelines. Vaccine-associated enhancement of respiratory disease has been observed following vaccination against influenza and mycoplasma infections in swine. An incomplete match between the vaccine strains and the circulating field strains of the organisms may result in a suboptimal immune response and formation of immune complexes that contribute to inflammation in the lungs.

KEY POINTS

- Passive immunization involves administration of pre-formed antibodies to provide immediate protection, typically from the effects of a toxin or venom.
- Active immunization involves administration of an antigen (typically an infectious agent or component of that agent) in order to induce an immune response and immunological memory.
- The ideal vaccine is cheap to produce, consistent in formulation, stable with a long shelf life, and able to induce an appropriate and long-lived immune response with good immunological memory, in the absence of side effects.
- Live attenuated vaccines are common. Attenuation is achieved by growing an organism under suboptimal conditions or by genetically modifying it to eliminate virulence gene(s).
- A heterologous vaccine is antigenically related to the organism of interest, but adapted to another host species.
- Inactivated vaccines contain antigenically intact whole organisms that cannot replicate.
- Subunit vaccines contain a relevant structural antigen or secreted molecule from a pathogen and require an adjuvant.
- Recombinant vector vaccines use a carrier virus or bacterium to transport a gene encoding an antigen from a pathogen able to induce a protective immune response.
- Nucleic acid vaccines directly carry the genetic information encoding a protein of interest from the pathogen. The gene transfects host cells and antigen is expressed within them.
- Adjuvants are commonly used in inactivated and subunit vaccines.
- Adjuvants are heterogeneous materials that enhance the immune response by adsorbing antigens, inducing inflammation at the injection site and by activating dendritic cells.
- Marker vaccines allow discrimination of serological responses due to vaccine from those due to field exposure to the organism.

- A vaccine that induces an immune response against GnRH is used to prevent boar-taint of pork.
- A core vaccine is one that every animal should receive as the disease that it protects from is widespread and causes significant morbidity or mortality or is an important threat to public health.
- A non-core vaccine protects from a disease that not every animal may be exposed to.
- Herd immunity ensures that an entire population has the greatest chance of resisting infection from endemic pathogens.
- Neonatal vaccination requires multiple immunizations in order to avoid the effects of maternally derived antibodies.
- Vaccines with the longest DOI should be used. DOI is the time after vaccination that the animal remains protected from disease.
- Vaccines should only be used in pregnant animals if licensed for such use.
- Sick or immunosuppressed animals should not be vaccinated.
- Genetic background may determine how individual animals respond to vaccines.
- Vaccines should be stored and reconstituted according to manufacturer's instructions.
- Adverse effects following vaccination are rare.
- Adverse reactions are generally mild and transient, but can occasionally be severe and life threatening (e.g. anaphylaxis, FISS).
- Although rare, vaccines sometimes enhance the severity of disease caused by certain viruses and bacteria.
- Detailed records of vaccination should be kept and adverse reactions should be reported.
- Reporting of a suspected adverse reaction following vaccination does not imply causality, but allows an investigation and, if necessary, correction of a problem.

IMMUNE SYSTEM ONTOGENY AND NEONATAL IMMUNOLOGY

OBJECTIVES

At the end of this chapter, you should be able to:

- Briefly describe the progressive development of the immune system *in utero*.
- Describe the general composition of colostrum and milk.
- Discuss the absorption of antibodies from colostrum.
- Discuss how passive transfer of maternal immunity interferes with vaccination of young animals.
- Describe the causes and effects of failure of passive transfer.
- Describe the pathogenesis of neonatal isoerythrolysis and indicate the species in which this disease occurs.
- Understand that development of the immune system continues during the early life of animals.

14.1   INTRODUCTION

This chapter considers various aspects of immunology related to the *in utero* development of the immune system and to the range of specific immunological conditions that are seen in newborn animals.

14.2   IMMUNE SYSTEM ONTOGENY

During embryological development *in utero* there is progressive expansion of the different elements of the immune system. This is variably well characterized for different domestic animals, but follows a general pattern. For example, in species with a relatively long gestation period, such as cattle, development of **primary lymphoid tissue** (thymus and bone marrow) occurs within the first trimester and **T and B**

cells appear within the fetal circulation at that time. During the second and third trimesters there is development of **secondary lymphoid tissues** (e.g. lymph nodes and MALT) and from this time onwards the fetus is capable of mounting a **humoral immune response** to a range of potential pathogens that might be encountered *in utero*. As will be discussed later, newborn animals are often born without appreciable levels of circulating immunoglobulin, but *in utero* antigenic challenge may lead to the newborn having detectable antigen-specific immunoglobulin.

14.3   PASSIVE TRANSFER OF MATERNAL IMMUNE PROTECTION

A basic facet of animal husbandry in veterinary species is the requirement for newborn animals to

DOI: 10.1201/9781003310969-14

ingest **colostrum** from the mother, during a narrow window immediately after parturition. This transfer of maternal immunoglobulin (together with other proteins, lymphocytes and cytokines) represents **passive transfer of immunity** and confers temporary immune protection upon the newborn animal until such time as it is capable of generating their own immune response to antigenic challenge. This fundamental feature of the early life of domestic animals is directly related to the type of placentation that occurs in these species, which largely impedes direct transfer of immunoglobulins *in utero* (**Figure 14.1**).

The **epitheliochorial placentation** of the pig and the **synepitheliochorial placentation** of the cow represent a substantial barrier to passage of immunoglobulins *in utero*, such that there is no possibility for passive transfer of immunity in these species during pregnancy. In contrast, the **endotheliochorial placentation** in dogs and cats permits transfer of a small quantity of IgG, such that pups and kittens may be born with a reasonable level of serum immunoglobulin from birth. Colostrum-derived immunoglobulin is not essential for human and other primates because their **haemochorial placentation** permits transfer of maternally derived antibodies across the chorionic epithelium *in utero*, which means that neonates are born with serum IgG concentrations approaching those of the adult.

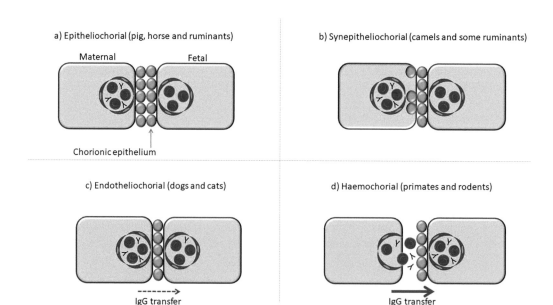

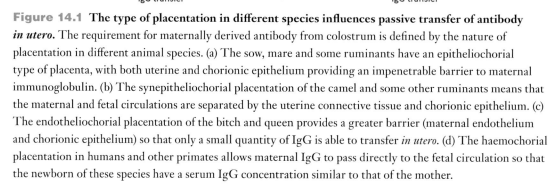

Figure 14.1 **The type of placentation in different species influences passive transfer of antibody** *in utero.* The requirement for maternally derived antibody from colostrum is defined by the nature of placentation in different animal species. (a) The sow, mare and some ruminants have an epitheliochorial type of placenta, with both uterine and chorionic epithelium providing an impenetrable barrier to maternal immunoglobulin. (b) The synepitheliochorial placentation of the camel and some other ruminants means that the maternal and fetal circulations are separated by the uterine connective tissue and chorionic epithelium. (c) The endotheliochorial placentation of the bitch and queen provides a greater barrier (maternal endothelium and chorionic epithelium) so that only a small quantity of IgG is able to transfer *in utero*. (d) The haemochorial placentation in humans and other primates allows maternal IgG to pass directly to the fetal circulation so that the newborn of these species have a serum IgG concentration similar to that of the mother.

The most important proteins within colostrum are **immunoglobulins**. These broadly originate from the **maternal blood** (particularly IgG) or from local production by plasma cells within the **mammary gland lymphoid tissue** (notably IgM and IgA). Colostrum is particularly **enriched in IgG**, with relatively lower concentrations of IgM and IgA (**Figure 14.2**). In contrast, milk has a considerably reduced total concentration of immunoglobulin and this largely derives from local plasma cells rather than the maternal circulation. The relative proportion of immunoglobulin classes may also change; for example, bovine colostrum and milk are both dominated by IgG, but in non-ruminant species the concentration of IgA in milk is much greater than that of IgG or IgM.

Recent studies have characterized the non-immunoglobulin components of colostrum. Bovine colostrum contains **CD4⁺, CD8⁺ and γδTCR⁺ T cells** and the CD8⁺ subset is the source of the significant concentration of **IFN-γ** within this secretion. Studies comparing calves that received natural colostrum with those that received cell-free colostrum have shown that the transfer of these maternal lymphocytes has a beneficial effect on the development of the neonatal immune system. Calves receiving natural colostrum had a greater number of circulating monocytes, with more effective antigen-presenting capacity and increased MHC class I expression on circulating lymphocytes. A study in piglets from sows vaccinated with keyhole limpet haemocyanin (KLH) demonstrated colostral transfer of antigen-specific lymphocytes found in the blood and mesenteric lymph nodes of the piglets. **Complement components** are also contained within colostrum and are passively absorbed, but are not thought to be biologically active within the neonate.

The process of absorption of immunoglobulin (IgG in particular) from colostrum is made possible by a series of modifications that are in place from approximately **6–24 hours after birth**. The colostral proteins would normally be rapidly degraded within the intestinal lumen, but they are protected by the relatively **low proteolytic activity** in this environment during the perinatal period together with the action of **colostral enzyme inhibitors**. Maternally

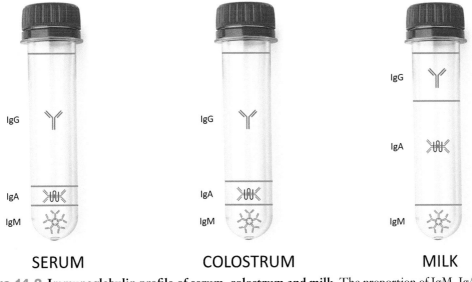

SERUM   COLOSTRUM   MILK

Figure 14.2 Immunoglobulin profile of serum, colostrum and milk. The proportion of IgM, IgA and IgG in colostrum is similar to that seen in serum. In non-ruminant species there is a greater proportion of IgA in the milk.

derived IgA has the additional protection conferred by the secretory piece. Immunoglobulin transfer across the neonatal intestinal epithelium is mediated by active transport via transient expression of specific Fc receptors (**FcRn**) by **enterocytes**. Absorbed immunoglobulin may pass directly into the vasculature, but a proportion of this immunoglobulin may first enter lacteals and subsequently transfer from the lymphatic system to the blood (**Figure 14.3**). Absorption of maternal immunoglobulin results in elevation of neonatal serum immunoglobulin concentration, with a peak level between 12 and 24 hours. A proportion of this absorbed immunoglobulin is lost into the urine as the glomeruli of the neonate have increased permeability during the first 24 hours of life. This results in a **transient proteinuria** in the neonate. There are species differences in the nature of this passive transfer. For example, in ruminants, IgG, IgM and IgA are all absorbed, but a proportion of IgA is re-secreted

into the intestinal lumen to afford protection to this mucosal surface. In contrast, in horses and pigs the colostral IgA largely remains within the intestinal tract and the principle absorption is of IgG and IgM. In all of these species the 'window of opportunity' for absorption of colostral immunoglobulin is up to 24 hours after birth, but for pups and kittens this period may extend for up to 48–72 hours. Once acquired, passive immunity from maternally derived antibody, will normally persist for two to three months and will confer protection against infection even in those animals that have a genetic immunodeficiency. After this time, the young animal will start to generate its own immune response to antigenic challenge or, in the case of genetic immunodeficiency, clinical signs of infection start to become apparent.

Chicks also require maternally derived immunity for protection in neonatal life. **Maternal IgY** is transferred from the blood to the **yolk** of developing eggs in the ovary and is

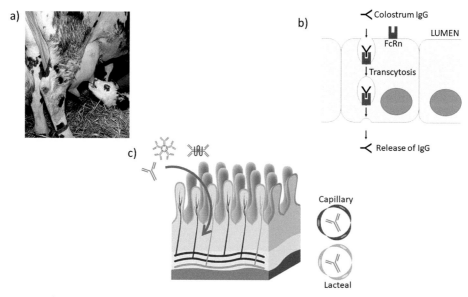

Figure 14.3 **Absorption of colostral immunoglobulin.** (a) Ingestion of colostrum by the neonate shortly after birth. (b) Uptake of colostral IgG is mediated by the FcRn, which is transiently expressed by intestinal enterocytes. (c) IgG may be directly absorbed to the circulation or taken up initially into the lacteals and passed from the lymphatics to the blood. Serum immunoglobulin concentration in the neonate peaks between 12 and 24 hours after birth, after which time FcRn expression diminishes.

highly concentrated in this fluid. The embryonic chick absorbs this IgY into its circulation. As the egg passes through the oviduct the albumin acquires IgM and IgA from local secretions and these immunoglobulins enter the amniotic fluid to be swallowed by the embryo to provide local mucosal immunity.

14.4 NEONATAL VACCINATION

Although vaccination was considered in detail in Chapter 13, it is appropriate to consider the principles of vaccination of young animals at this point. While passive transfer of colostral immunoglobulin is crucial for protection of the neonate from infectious diseases during this early stage of life, it is a 'double-edged sword' and poses a specific problem to these vulnerable animals in that maternally derived antibody **inhibits an active immune response**. The consequences for the neonate are that these animals are unable to make antibody responses to vaccines until such time as the maternally derived

antibody has waned. There is then a period of 'cross-over' while the young animal begins to synthesize its own immunoglobulin as the last of the maternal antibody is lost. This period is often termed the **window of susceptibility** or the **immunity gap** and is defined as the period of time when there is no longer sufficient passive immunity to afford protection from infection, but when there is still enough of this maternal antibody to prevent the young animal from mounting its own protective immune response to a vaccine (**Figure 14.4**).

The timing and duration of the window of susceptibility may **vary widely between individual animals** and even between **animals within the same litter**. This relates to factors including the amount of antibody in maternal colostrum and the amount of colostrum ingested and absorbed by individuals. For example, one animal within a litter may have a window of susceptibility between 10 and 12 weeks of age, but another, which ingested less colostrum or colostrum of lesser quality, may lose

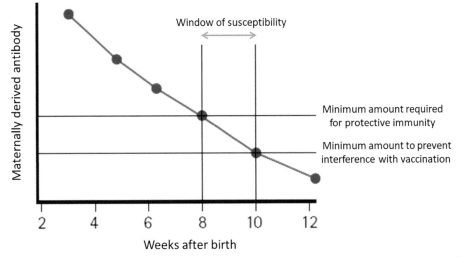

Figure 14.4 The 'window of susceptibility'. Maternally derived antibody degrades over time and once this neutralizing antibody has waned the young animal is able to make its own active immune response. The 'window of susceptibility' (also known as the 'immunity gap') is the time during which there is insufficient passive immunity to provide full protection, but still a sufficient level of antibodies to prevent the animal responding to vaccination. In the example shown, the 'window of susceptibility' is between eight and ten weeks of age.

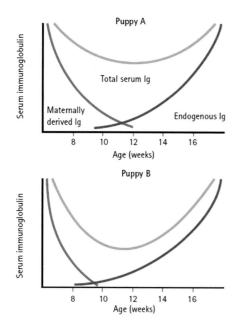

Figure 14.5 **Individual variation in the 'window of susceptibility'.** The two puppies in this example are from the same litter, but each ingested a different amount of colostrum. Pup A received more colostrum so has a 'window of susceptibility' between 10 and 12 weeks of age. In contrast, pup B received less colostrum and maternal immunoglobulin is lost sooner, with a 'window of susceptibility' between eight and ten weeks of age. If these pups received only a single primary vaccination at ten weeks of age, pup A would not respond. For this reason, administration of a primary course of vaccines at 8, 12 and 16 weeks of age is designed to ensure a primary response in all pups in the litter. This primary response must be 'boosted' with a further injection when they reach 12 months of age.

protective immunity sooner and have a window of susceptibility between six and eight weeks of age (**Figure 14.5**). The consequences of this are that within a litter or between different litters, individuals may be able to **respond to vaccination at different ages**. This creates some degree of difficulty in identifying the precise 'window' for individual animals and has led to the development of vaccination protocols whereby **young animals receive multiple vaccinations** in order that **at least one of these may stimulate an effective immune response**. In fact, for pups and kittens, current recommendations are for three such immunizations, typically at 8, 12 and again at 14–16 weeks of age, as it has been recognized that a proportion of these animals will still have blocking levels of maternally derived antibody at 12 weeks of age. Although it is stated for some vaccines that two doses may be given at eight weeks and ten weeks, this 'early finish' protocol potentially increases the chances of vaccine failure, where the levels of maternally derived antibodies are still relatively high at ten weeks old for some individuals and they fail to respond appropriately to both vaccine doses.

The **booster vaccine given one year after the primary course** (or often at one year of age)

is a crucial element in a vaccine protocol, because the primary response to a single vaccine dose is often insufficient to induce long term protection. Such vaccination protocols give all young animals the best chance of responding to vaccination and developing protective immunity. **Serological testing** of young animals post-vaccination can be used in order to confirm that there is a **'protective antibody titre'** against core vaccine components such as CDV and CPV. Such testing regimes are readily justified as there is excellent correlation between the serum antibody titre and protection against these viral infections.

14.5 FAILURE OF PASSIVE TRANSFER

Despite the importance of passively acquired maternal immunity, some neonatal animals fail to receive adequate colostrum and are thus highly **susceptible to infectious disease**. Reasons for failure of passive transfer include:

- **Premature birth**, such that the dam has not produced sufficient colostrum.
- **Premature lactation** with loss of colostrum.

- **Production of poor-quality colostrum** (determined by maternal nutritional and vaccinal status).
- **Failure of the neonate to suckle** within the first 24 hours of life.
- **Failure of the neonate to absorb** colostrum due to loss of the adaptive mechanisms.

For valuable animals (e.g. Thoroughbred racehorses) it is common to test for the serum immunoglobulin concentration after 24 hours to establish whether there has been adequate colostral antibody uptake. This testing might involve a precise measurement by methods such as the ELISA or single radial immunodiffusion (**SRID) test** (**Figure 14.6**) or the use of rapid semi-quantitative **commercially available immunoassays** that may be used 'patient side' (e.g. SNAP® Foal IgG test, IDEXX Laboratories). Simple turbidometric precipitation

methods have also been widely employed in studies of circulating maternally derived antibody in newborn calves.

14.6 NEONATAL ISOERYTHROLYSIS

Neonatal isoerythrolysis (NI) (also known as isoimmune haemolytic anaemia or haemolytic disease of the newborn) is an immunological disorder affecting newborn animals and an uncommon deleterious effect of maternal transfer of passive immunity. In this syndrome, the colostrum contains antibodies specific for antigens expressed on the surface of the erythrocytes of the newborn animal. Absorption of these antibodies therefore leads to immune-mediated destruction and haemolytic anaemia. The dam may have such **isoantibodies**, which arise spontaneously, perhaps due to exposure to cross-reactive environmental antigens. Alternatively, the dam may be actively sensitized to

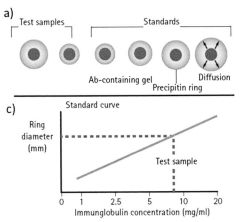

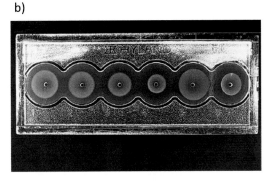

Figure 14.6 Single radial immunodiffusion. (a) SRID is a precipitation test used to quantify the amount of protein within a sample, in this context the amount of IgG within the serum of a newborn foal after receiving colostrum. Antiserum specific to equine IgG is mixed with agarose and the gel allowed to set on a plastic plate. A series of wells is cut into the gel. A set of IgG standards of known concentration are loaded into four of the wells and test samples from two foals are loaded to the remaining wells. The gel is incubated for 24 hours during which time IgG diffuses from the well to precipitate with the antiserum impregnated into the gel. (b) The precipitation is seen as a visible ring surrounding each well and the diameter of the ring is proportional to the concentration of Ig in each sample. (c) The standards are used to plot a standard curve on a semi-logarithmic scale and the concentration of IgG in the foal serum can be determined by interpolation from the curve.

foreign erythrocyte antigens. This most often occurs when the **blood groups of the dam and sire differ**, such that the fetus expresses a blood group antigen that is foreign to the mother. At the time of parturition, the dam is exposed to fetal blood during placental separation and this can lead to sensitization. If the same blood group mismatch occurs in subsequent pregnancies, NI can result.

Haemolytic disease of the newborn occurs in humans as a result of mismatch of rhesus factor, expressed on human erythrocytes. In animals, NI is most common in **foals and kittens** and much less common in other species. In the horse the disease occurs when a mare that lacks expression of the Qa and Aa blood group antigens (Qa⁻Aa⁻) is mated with a stallion that is Qa⁺Aa⁺ (**Figure 14.7**). The clinical presentation is of anaemia with jaundiced mucous membranes (**Figure 14.8**). It is possible to determine the likelihood of NI before parturition by performing a simple agglutination test with serum taken from the mare, mixed with washed eryth-rocytes from the stallion. Alternatively, incubation of mare serum with washed newborn foal erythrocytes will provide the same information, but this would need to be undertaken before the foal is allowed to suckle. NI can be prevented by **restricting access of the newborn foal to the dam's colostrum**, but an **alternative source of colostrum** must be provided for that animal to avoid failure of passive transfer.

NI in kittens also relates to the nature of the feline blood group antigens. Cats may be of blood groups A, B or AB. Type A cats may occasionally have low-titre anti-B isoantibodies, but type B cats invariably have high-titre anti-A antibodies, mostly of the IgM class (and type AB cats have neither). NI therefore typically occurs when a type B queen with anti-A antibody gives birth to kittens of blood type A or AB. As the type B blood group is more prevalent in certain breeds (e.g. Birman, Rex, British shorthair, Abyssinian, Persian and Somali), the disease **more frequently arises within these particular breeds**. Feline NI presents

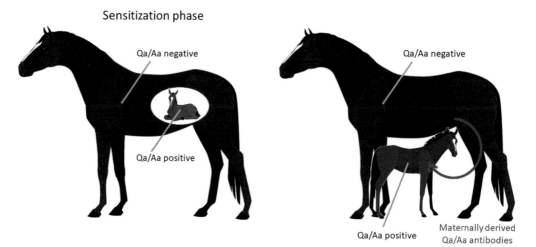

Sensitization phase

Qa/Aa negative

Qa/Aa positive

Qa/Aa negative

Qa/Aa positive

Maternally derived Qa/Aa antibodies

Figure 14.7 Susceptibility to neonatal isoerythrolysis in foals. Mares that are negative for the Qa/Aa red blood cell antigens are susceptible when mated with a sire that is Qa/Aa positive. If the fetus inherits and expresses the Qa/Aa antigens, the mare may become sensitized, particularly at parturition, when exposed to the foal's blood during placental separation. If subsequent pregnancies also have a mismatch in these antigens, anti-Qa or anti-Aa antibodies will likely be present in the colostrum and cause NI when ingested and absorbed by the foal.

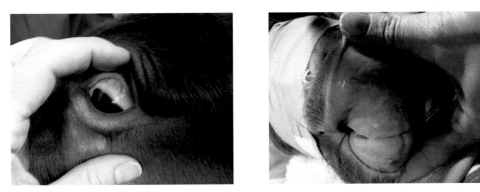

Figure 14.8 **Foal affected with neonatal isoerythrolysis, showing jaundice and anaemia.** (Images courtesy of Imogen Johns, B and W Equine Hospital.)

clinically as a severe haemolytic anaemia with icterus, haemoglobinuria, weakness and lethargy, failure to suckle and death within the first week of life. A milder, subclinical form is also recognized, where agglutination of erythrocytes by IgM antibody (cold agglutinin) leads to occlusion of peripheral capillaries and ischaemic necrosis (e.g. of the tail tip) within the first few weeks of life.

Breeders of susceptible cat breeds are often aware of this condition and will routinely **blood type their breeding animals** to avoid mating mismatched pairs. Testing serum from the queen with washed erythrocytes from the tom cat will provide an antenatal diagnosis. Prevention is by restricting access to colostrum from the dam and providing an **alternative source of colostrum**, which is challenging for kittens. Allowing kittens to suckle a type A foster mother may also be performed, although feline milk is not as immunoglobulin rich as colostrum. Alternatively, artificial colostrum may be prepared by mixing cat serum (from an appropriate vaccinated and blood-typed donor) with commercial milk replacer.

A recently emerged disease in Europe, related to colostral antibodies is **bovine neonatal pancytopenia** (BNP) or 'bleeding calf syndrome'. Affected newborn calves develop bone marrow depletion and pancytopenia, which is expressed clinically as haemorrhage secondary to thrombocytopenia (**Figure 14.9**). The disease is thought to occur following ingestion of colostrum containing alloantibodies that bind to neonatal circulating leucocytes and, presumably, to bone marrow precursor cells. These antibodies may have specificity for MHC class I molecules. The dams of affected calves have often been vaccinated against BVDV and it has been suggested that antigens derived from the bovine cell line used in manufacturing this inactivated vaccine (the Madin–Darby bovine kidney cell line) may be responsible for triggering antibody production in vaccinated dams.

14.7 EARLY LIFE IMMUNE DEVELOPMENT

The immune system of young animals undergoes significant development from birth, during the first months of life and in most species a functional immune system is not fully developed until around 12 months of age. As noted previously, this development might be influenced by proteins and cells derived from colostrum. The number of lymphocytes in the blood increases over the first few months of life and the relative proportions of lymphocyte subsets changes. For example, in pups and kittens the CD4:CD8 ratio gradually decreases as the CD8+ T cell

a)

b)

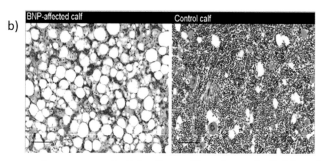

Figure 14.9 **Calf affected with bovine neonatal pancytopenia ('bleeding calf syndrome').** (a) Clinical signs of haemorrhage in an affected calf. (b) Bone marrow depletion of stem cells (left panel) compared with a highly cellular normal/control calf bone marrow (right panel). (Images courtesy of Charlotte R Bell, Roslin Institute, University of Edinburgh, originally published in, Bell CR, et al. (December 2010) Idiopathic bovine neonatal pancytopenia in a Scottish beef herd *Veterinary Record: Journal of the British Veterinary Association* 167(24): 938–940.

population expands. The responsiveness of these cells to mitogen stimulation increases over this time. In one study, blood lymphocyte proliferation and cytokine production were monitored in pups between four and ten weeks of life. Initial cytokine profiles were dominated by IL-10, which progressively declined with replacement by IFN-γ. The same dominance of IL-10 was seen in cultures taken from the dams of these pups during pregnancy, suggesting a relatively immunosuppressive environment *in utero* with a gradual switch to Th1 immunoregulation with increasing exposure to microbes and other environmental antigens during early life. It is thought that environmental (saprophytic) mycobacteria and other similar microorganisms might provide important 'cues' for this transition from a Th2/Treg dominated phase to a more balanced Th1/Th2 profile. Once maternal immunoglobulin has waned, it may take up to 12 months for normal levels of endogenous serum IgG, IgM and IgA to be achieved. The process of thymic involution typically begins around six months of age and there is a progressive decline over the following one to two years. However, there is evidence for residual thymic output throughout adult life in dogs, although maintenance of naïve T cell production seems to vary comparing different dog breeds and comparing individuals within a breed.

KEY POINTS

- There is progressive development of the immune system *in utero* and the fetus is capable of an immune response if challenged.
- Newborn domestic animals should receive colostrum within the first 24 hours of life to protect them from infectious disease.
- The nature of placentation of domestic animals often prevents transfer of maternal immunoglobulin *in utero*.
- Colostrum is enriched in immunoglobulin; milk contains less immunoglobulin, but the proportions of IgG, IgM and IgA in these secretions vary between different species.
- Colostrum also contains lymphocytes and cytokines, which may confer benefit to the neonate.
- Colostral immunoglobulin is protected from digestion by relatively low proteolytic activity in the neonatal gut and the presence of colostral enzyme inhibitors.
- Neonatal enterocytes express FcRn that mediates active transport of colostral immunoglobulin, particularly IgG, into the blood.
- Some absorbed colostral immunoglobulin is lost through the glomerulus in the first 24 hours of life.
- Maternally derived antibody inhibits the ability of the neonate to mount its own immune response.
- The 'window of susceptibility' or 'immunity gap' is the time between the point when there is insufficient maternal antibody remaining to provide passive immunity from infectious disease and the point at which the neonate can mount an effective immune response to antigenic challenge (e.g. vaccination).
- The 'window of susceptibility' varies between individual animals, even within a litter.
- Neonates must receive multiple vaccinations to ensure that the immune response is adequately primed.
- The first booster vaccine, administered around one year of age, is an integral part of the vaccination schedule, as it 'boosts' the primary immune response.
- Failure of passive transfer of colostral immunoglobulin renders the neonate susceptible to infection and serological testing after 24 hours of life is justified in valuable animals.
- Primary congenital immunodeficiency becomes clinically manifest after loss of maternal immunity.
- Neonatal isoerythrolysis occurs when the colostrum contains antibody to erythrocyte antigens expressed by the newborn; these antibodies cause immune-mediated haemolytic anaemia.
- Isoantibodies may arise spontaneously or be induced following first parturition or incompatible blood transfusion.
- NI is most common in foals and kittens.
- NI can be predicted and prevented by denying access to the dam's colostrum, although an alternative source of colostrum must be provided to prevent failure of passive transfer.
- The immune system of young animals continues to develop until adulthood.

CHAPTER 15

IMMUNOLOGICAL TOLERANCE AND IMMUNE SUPPRESSION

OBJECTIVES

At the end of this chapter, you should be able to:

- Define what is meant by immunological tolerance.
- Explain the mechanisms involved in central tolerance.
- Explain the mechanisms involved in peripheral tolerance.
- Give examples of immune-privileged sites.
- Define what is meant by oral tolerance.
- Describe how the immune system of the pregnant female tolerates the presence of fetal alloantigens.
- Outline how antibody plays a role in downregulation of an immune response.
- Explain the concept of the neuroendocrine–immunological loop.

15.1 INTRODUCTION

It is important that the immune system reacts to the presence of foreign antigen associated with a pathogen, during an infection. However, it is equally important that the immune system does not react to self-antigens or foreign antigens associated with harmless environmental components, particularly those that are encountered at mucosal surfaces (e.g. dietary protein). Pregnancy represents a special situation, whereby a 'foreign' organism is permitted to grow within the host. Additionally, the immune response to an infection can potentially lead to some degree of tissue damage (innocent bystander) and it is important that this response is appropriate and measured. Thus, there is a complex system of immune regulation, with tolerance and immune suppression mechanisms playing an important role in maintaining health, which is equally as important as those effector mechanisms designed to react to and eradicate infectious agents. In this chapter, we explore these mechanisms and in later chapters, we consider the consequences for the animal when immunological tolerance and immune suppression goes wrong.

15.2 TOLERANCE AND TOLERANCE MECHANISMS

Immunological tolerance represents the lack of an adaptive immune response to a specific antigen. Lymphocyte antigen receptors are designed to detect epitopes on protein antigens. However, since the strategy for generation of such receptors involves random assortment of V(D)J gene segments encoding their

DOI: 10.1201/9781003310969-15

variable region, some lymphocytes will express receptors that are specific for self-antigens or harmless environmental antigens (including dietary proteins). Immunological tolerance is designed to prevent those lymphocytes from becoming activated, while at the same time allowing lymphocytes that recognize foreign antigen from pathogens to go about their business unhindered (**Figure 15.1**). Whether or not an antigen induces an immune response or tolerance depends on the conditions under which the immune system is exposed to that antigen. By eliminating or inactivating B and T lymphocytes that react inappropriately to harmless foreign antigen or self-antigen, the immune system avoids the development of allergic and autoimmune diseases, respectively. In addition, when the host has recovered from an infection, there is a requirement to 'stand down' the effector cell response to minimize any collateral damage to tissues, which could potentially occur if pro-inflammatory processes were to persist for longer than is necessary. This chapter discusses the

mechanisms that underlie immunological tolerance and immune suppression, as well as some specific conditions in which tolerance occurs.

The immune system has developed a multimodal approach to tolerance that occurs in both primary and secondary lymphoid tissues (**Figure 15.2**). Developing lymphocytes in the primary lymphoid tissues may express high affinity receptors for self-antigens and are typically eliminated (**negative selection/clonal deletion**) during **central tolerance**. However, a substantial proportion of autoreactive lymphocytes escape negative selection, because not all organ-specific antigens are expressed in the primary lymphoid tissues and because alternative processing of antigens by inflammatory cells, tissue proteases and microbial proteases, may generate or expose **cryptic epitopes**. Additionally, harmless environmental antigens are only encountered after birth and when the lymphocytes have left the primary lymphoid organs. Sequestration of antigens in certain tissues (**immune-privileged sites**) may mean

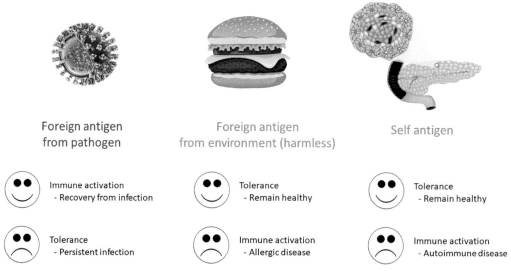

Figure 15.1 Antigen exposure to the immune system. The immune system is designed to react to antigen during the course of an infection leading to elimination of the pathogen and development of long-term immunity. However, the immune system is also exposed to a range of other antigens (including dietary and self-antigens), against which it should be tolerant, to avoid inappropriate immune responses.

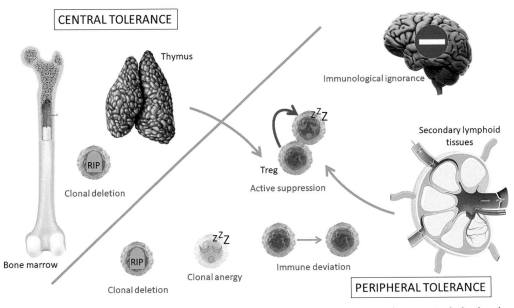

Figure 15.2 Mechanisms of immunological tolerance. Central tolerance mechanisms include clonal deletion of autoreactive T cells as part of negative selection and production of natural regulatory T cells (nTregs). Peripheral tolerance mechanisms include immunological ignorance, clonal deletion, clonal anergy, immune deviation and active suppression, the latter mediated by different phenotypes of regulatory T cells.

that the lymphocytes do not come into contact with particular antigens and there is a state of **immunological ignorance**. Alternatively, lymphocyte responses must be controlled in the secondary lymphoid organs and other tissues (**peripheral tolerance**) in order to prevent allergic and autoimmune reactions and a range of mechanisms come into play that contribute to this process, including **clonal deletion, clonal anergy, immune deviation** and, probably most importantly, **active suppression**. The role of these mechanisms likely depends on the specific tissue environment and conditions under which the lymphocytes interact with antigens.

15.3 CENTRAL TOLERANCE

In the bone marrow, developing B lymphocytes that recognize self-antigen within this tissue undergo apoptosis. However, autoreactive B cells can undergo further rearrangement of the light chain gene segments to create a different antigen-specificity (**receptor editing**). If this process is not successful in eliminating self-reactivity, the B cell will undergo induced cell death. This is a form of **clonal deletion,** but is not particularly efficient in eradicating autoreactive B cells. B cells that react primarily with soluble antigens in the periphery are permanently inactivated (**anergy**). Since class-switching and generation of memory cells in the secondary lymphoid tissues is dependent upon interaction of B cells with T helper cells, the immune system can accept the presence of autoreactive B cells, provided there is robust T cell tolerance in place. In fact, transient production of autoantibodies (mainly IgM) can be seen during some infections and following some vaccinations, but these are inconsequential and do not normally lead to any adverse effects in terms of pathology or clinical signs.

In the thymus, T cells that express a T cell receptor with high affinity for self-peptide presented by MHC molecules expressed on medullary epithelial cells and dendritic cells in the thymic medulla undergo apoptosis (**negative selection/clonal deletion**). The promiscuous transcription factor AIRE (autoimmune regulator) induces expression of many different tissue-specific antigens such as those found in the pancreatic islets, retina and central nervous system. This allows identification and elimination of T cells that are reactive towards many different self-antigens from various tissues. If negative selection in the thymus were too exuberant, this might impact on the T cell repertoire available for the immune response to infection, therefore there is a balance to be made between elimination of potentially harmful T cells and maintaining a diverse population of circulating naïve T cells for protection against infection. In fact, not all self-reactive CD4+ T cells undergo apoptosis. Those with moderate affinity to self-peptide may be rescued, but forced to differentiate into natural regulatory T cells (nTregs) (**Figure 15.3**). These cells show

markers associated with antigenic stimulation in the thymus, including CD25 (the α chain of the IL-2 receptor) and CTLA4 (cytotoxic T lymphocyte antigen 4). They also express the transcription factor Foxp3 (forkhead box p3), which effectively prevents the cell from generating a pro-inflammatory response if it were to become activated in the periphery and, instead, activation of nTregs leads to production of more anti-inflammatory/immunosuppressive cytokines such as interleukin 10. While nTregs demonstrate a paracrine effect in downregulating effector T cell functions, direct cell:cell interaction also seems to be required for them to demonstrate their full immunoregulatory capacity, although the nature of these direct interactions remain to be fully elucidated.

15.4   PERIPHERAL TOLERANCE

The negative selection process that underlies central tolerance does not eliminate all

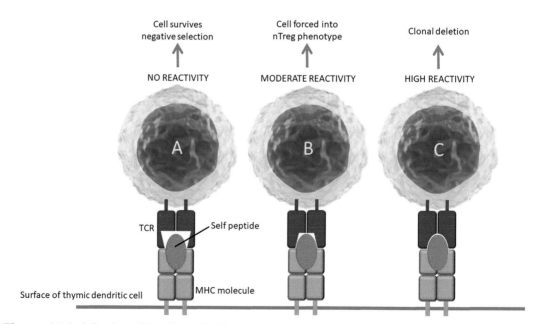

Figure 15.3 Selection of T cells in the thymus. TCRs are selected for reactivity to self-peptides. T cell A binds to MHC but does not react to self-peptide and is selected to enter the circulating lymphocyte pool. Cell C is autoreactive and is deleted by induced cell death (apoptosis). Cell B is somewhat self-reactive but is rescued from apoptosis and forced into becoming a natural T regulatory cell (nTreg).

self-reactive B and T cells, and these 'escapees' are subsequently kept in check by peripheral tolerance mechanisms. Some organs are less accessible to the immune system and have a locally immunosuppressive environment that limits immune responses. These **immune-privileged sites** include the **central nervous system, eye and testis**. Access of lymphocytes into these organs is restricted by vascular barriers. The **blood–brain barrier** is formed by endothelial cells connected by tight junctions, pericytes and astrocytes. Many neurons, glia cells and endothelial cells express **Fas-ligand (FasL)**. Binding of FasL to Fas expressed on activated T cells triggers their apoptosis. The immune-privileged nature of the testis may serve to avoid immune recognition of **antigens expressed on spermatids and spermatozoa** as these are not expressed until puberty. Tolerance is achieved by a structural barrier formed by **tight junctions between Sertoli cells** and the creation of an **immunosuppressive interstitial environment** by the secretion of cytokines such as **TGF-β** by Sertoli cells, Leydig cells, macrophages and Tregs. Several mechanisms contribute to the immunologically tolerant environment of the eye. This includes a **blood–retinal barrier** similar to the blood–brain barrier, and secretion of **immunosuppressive neuropeptides, TGF-β** and **indoleaminedioxygenase (IDO)**. The corneal endothelium, anterior uveal cells and retinal pigment epithelial cells express **PD-L1**, which suppresses activation of T cells by binding to PD-1, and many cells in the eye express **FasL**.

The immunosuppressive and anti-inflammatory environment helps to protect these tissues from damage associated with the inflammatory response, which could permanently impact on tissue function, in light of their relative inability to repair and regenerate. However, a breakdown of these anatomical and functional barriers as a result of trauma or infections can initiate an immune-mediated response against self-antigens that contributes to further tissue damage. An example of this is **Equine Recurrent Uveitis** (ERU), a chronic disease characterized by recurring episodes of inflammation of the eye that is thought to be the most common cause of blindness in horses. The disease is probably triggered by a prior infection and is often associated with subclinical or clinical **leptospirosis**. Anti-leptospiral antibodies in ocular fluids cross-react with eye proteins. The inflamed eye becomes infiltrated by CD4+ T cells that recognize retinal antigens. The progressive relapsing nature of the disease may be due to **epitope spreading** in which the immune system reacts with an increasing number of epitopes and proteins (see more on this later). The immunological nature of ERU is supported by the clinical response to immunosuppressive drug treatment.

Self-reactive T and B cells in the periphery can be eliminated through apoptosis/clonal deletion. The presentation of self-peptides to T cells in the absence of costimulatory signals alters the balance between pro-apoptotic (BIM) and anti-apoptotic (Bcl-2) protein signals resulting in activation of the intrinsic pathway of apoptosis. Reactivation of T cells by repeated exposure to self-antigens induces the expression of FasL, which can bind to Fas on the same cell, triggering the extrinsic pathway of apoptosis. The interaction between FasL on follicular helper T cells (Tfh) and Fas on autoreactive B cells removes these B cells in germinal centres.

As described in Chapter 7, the activation of T cells requires two signals; **signal 1** consists of the MHC/peptide complex, recognized by the T cell receptor and **signal 2** consists of costimulatory molecules **CD80 and CD86** which interact with **CD28** on T cells (**Figure 15.4a**). The recognition of MHC/peptide in the absence of signal 2 results in anergy or unresponsiveness (**Figure 15.4b**). In the absence of infection or tissue injury, dendritic cells in non-lymphoid and lymphoid tissues express MHC molecules with self-peptides and low levels of CD80 and CD86 which prevents the activation of autoreactive T cells. In addition to costimulatory

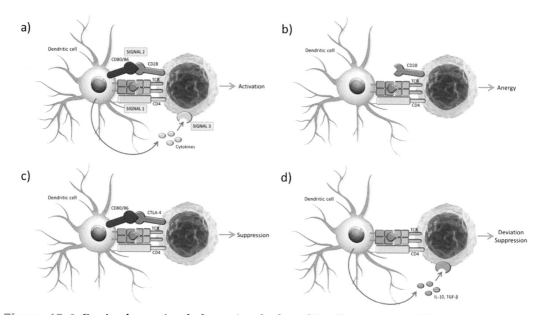

Figure 15.4 Costimulatory signals determine the fate of T cell responses. (a) Three signals are required for activation of naïve T cells. (b) In the absence of the co-stimulatory signal via CD80/86 interacting with CD28 (signal 2), the naïve T cell will become anergic/unresponsive. (c) Upregulation of CTLA-4 on the surface of activated T cells leads to a downregulatory signal upon ligation with CD80/CD86 and suppression of T cell activity. (d) Antigen presentation within an immunosuppressive cytokine environment (signal 3) leads to differentiation of the T cells to a more regulatory phenotype.

molecules, the presence of inhibitory receptors may determine whether T cells are activated and differentiate into effector T cells. The molecule **CTLA-4** (CD152) is **upregulated on activated T cells** and is constitutively expressed by Tregs. Activated effector T cells lose their expression of CD28 and this is replaced by CTLA-4, which binds to CD80 and CD86 in a similar manner to CD28, but with a much higher affinity. The binding of CTLA-4 results in generation of a downregulatory signal and suppression of T cell activity (**Figure 15.4c**). The molecule **PD-1** is also **expressed on activated T cells** and engages with either **PD-L1**, which can be found on a wide variety of cells including tumour cells or with **PD-L2**, expressed on antigen presenting cells. The binding of PD-1 to its ligands results in inhibition of the effector functions of T cells.

In Chapter 7, we discussed the generation of T helper cells of different phenotypes, which influence the balance between cell mediated (Th1 and Th17) and antibody-mediated (Th2) immunity. These cells are cross-regulatory by virtue of the specific cytokines they produce. Th1-derived IFN-γ is inhibitory towards Th2 function and Th2-derived IL-4 and IL-13 inhibit Th1 function. This effect may sometimes (but not always) lead to polarization of the immune response, with a dominant cellular or humoral effector phase, and this effect is known as **immune deviation**. This is important for certain types of infection, for example Th1-biased response for intracellular (vesicular) pathogens such as *Leishmania infantum* and Th2-biased response for production of IgE against helminth parasites. However, the former cell-mediated immune response is relatively pro-inflammatory and once a robust Th1 response has contained and controlled the infection, there appears to be a phenotypic change whereby the IFN-γ-producing Th1 cell alters its cytokine expression profile and begins to secrete immunosuppressive IL-10. This form of 'self-regulation' may be important in preventing the Th1 cells from mediating secondary immunopathology, but has the downside that this form of suppression may also prevent total elimination of the infectious agent, allowing a low-grade persistent infection to occur.

One of the most important mechanism of peripheral tolerance is **active suppression**. The best studied suppressor cells are the **CD4+ regulatory T cells (Tregs)**, which recognize self-antigens via their TCRαβ. **Natural Tregs** develop in the thymus, whereas **induced Tregs (iTreg: Foxp3+) and type 1 regulatory T cells (Tr1: Foxp3-)** differentiate from naïve CD4+ T cells during an immune response in the periphery. In this instance, it is 'signal 3' that plays a major role, whereby antigen presentation within an immunosuppressive cytokine environment (particularly in the presence of IL-10 and TGF-β) leads to T cell differentiation towards a more regulatory phenotype (**Figure 15.4d**) Several mechanisms have been identified by which these regulatory T cells suppress the activity of effector T cells. Following engagement of the TCR, Tregs can engulf MHC/self-peptide complexes from dendritic cells (a process called trogocytosis), which reduces their availability for self-reactive naïve T cells and results in **antigen-specific suppression**. Tregs can also suppress the activation of CD4+ T cells in a paracrine manner ('**bystander suppression**'). Tregs constitutively express high levels of CTLA-4 and binding of their CTLA-4 to CD80 and CD86 on dendritic cells decreases the expression of these costimulatory molecules and competitively reduces their availability for binding of CD28 by naïve T cells. As Tregs stably express CD25 (the α-chain of the IL-2 receptor) and do not secrete IL-2 themselves, they can act as an '**IL-2 sink**' which limits the availability of this critical growth factor for self-reactive CD4+ and CD8+ T cells. Furthermore,

Tregs can actively suppress immune responses by secreting **immunosuppressive cytokines, including IL-10, IL-35 and TGF-β**, and by the production of **indoleaminedioxygenase (IDO)**, which catalyzes the metabolism of tryptophan, depleting this essential amino acid and generating the tolerogenic compound kynurenine.

In addition to Tregs, several other immune cells can fulfil immunosuppressive roles. These include IL-10-secreting Foxp3- CD4+ T cells, subsets of CD8+ T cells, IL-10-secreting regulatory B cells and subsets of TCRγδ T cells. The removal of apoptotic cells during normal cell turnover by M2 macrophages and immature dendritic cells induces tolerogenic signals including the secretion of IL-10 and TGF-β. Myeloid-derived suppressor cells (MDSCs) are monocyte-like or granulocyte-like cells that contribute to the immunosuppressive environment of certain tumours through the secretion of IL-10 and TGF-β.

15.5 MUCOSAL TOLERANCE

Mucosal tolerance has enormous biological significance in preventing the intestinal immune system from responding to harmless environmental components, such as dietary proteins and antigens associated with the gut microbiome. In fact, failure of this tolerance mechanism is likely to underlie disorders such as **food allergy** (dietary hypersensitivity) and various forms of **inflammatory bowel disease**. It is remarkable that the intestinal immune system is able to selectively induce tolerance against such antigens, while at the same time retaining the ability to mount an active protective immune response against mucosal pathogens.

The mechanisms underlying mucosal tolerance have been widely investigated. To some extent, the nature of the mucosal immune response helps to prevent deleterious proinflammatory reactions, with MALT designed for antibody (particularly IgA and IgE)

production, with a predominance of B lymphocytes supported by a microenvironment that favours differentiation of naïve CD4+ T cells to Tfh, Th2 and Th3 phenotypes. Th3 cells, in particular, are important for class-switching to IgA, via their production of TGF-β, a cytokine that also demonstrates immunoregulatory properties. In addition, the mucosal immune system seems to be amenable to development of induced Tregs, when exposed to foreign antigen over a period of time, during which there are few markers of 'danger' (e.g. PAMPs or DAMPs). The ability to discriminate between harmless and harmful antigen may also relate to the route by which the immune system is exposed. **Particulate antigens** to which an active immune response is induced are more likely to be taken up by **microfold (M) cells** overlying the Peyer's patches. In contrast, soluble antigens are more likely to be absorbed across the epithelial surface and are more likely to generate a tolerogenic response (**Figure 15.5**). This tolerance may not be absolute, as most normal individuals have detectable serum IgG or IgA antibody specific for dietary antigens. The tolerizing antigen

must be processed and presented by dendritic cells, but the consequence of such presentation may be variable. Some T cells that recognize processed antigen may undergo **apoptosis** (clonal deletion) and others might recognize antigen but fail to become activated, as not all three signals required for T cell activation are received (**anergy**). More importantly, the presentation of tolerogenic antigen may induce specific **regulatory T cells**. Clonal deletion and anergy might be relevant at the level of the intestinal mucosa, but the systemic aspects of oral tolerance (i.e. failure to respond to antigen subsequently injected) are likely to be mediated by these regulatory cells, which may migrate to systemic lymphoid tissues (e.g. the spleen).

The phenomenon of oral tolerance is readily induced in domestic animals. In one study, dogs were fed a diet containing ovalbumin (OVA) over the first weeks of life and immunized systemically with OVA and Der p1, an antigen derived from the house dust mite *Dermatophagoides pteronyssinus*. Dogs previously fed OVA failed to demonstrate serum antibodies to that antigen, while they mounted a conventional serological

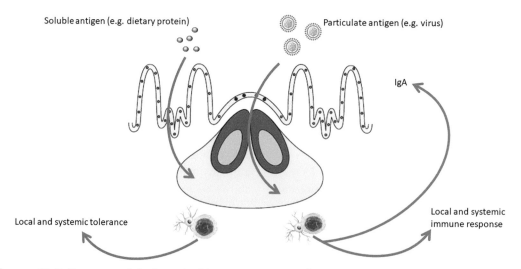

Figure 15.5 Exposure of the intestinal immune system to foreign antigen. Particulate antigen is most likely to be taken up by M cells overlying the Peyer's patch, leading to induction of an active immune response. Soluble antigen may be absorbed directly across the epithelial surface and may trigger a more regulatory immune response.

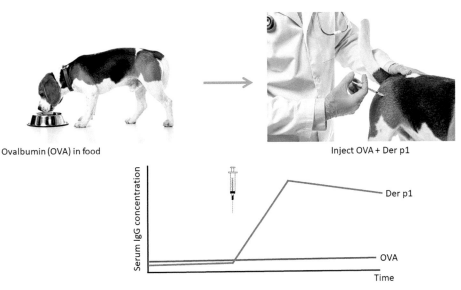

Ovalbumin (OVA) in food

Inject OVA + Der p1

Serum IgG concentration

Der p1

OVA

Time

Figure 15.6 **Canine oral tolerance.** In this experiment young dogs were fed with ovalbumin (OVA) for a ten-day period. Two weeks after this feeding period the dogs were injected with a mixture of OVA and the antigen Der p1 in adjuvant and serum antibodies measured over time. The dogs failed to mount an effective serum IgG response to OVA, but made a conventional antibody response to Der p1.

response to the Der p1 component of the immunization (**Figure 15.6**). The potential for inducing tolerance by feeding antigen was not lost on clinical immunologists, who saw this as a possible means of treating an array of allergic or autoimmune diseases. Several clinical trials have now been performed whereby autoantigens have been fed to patients affected with autoimmune diseases in an attempt to influence the progression of their disease. For example, myelin basic protein has been fed to patients with multiple sclerosis, type II collagen to rheumatoid arthritis patients, retinal S antigen to patients with uveitis and insulin to those with type I insulin-dependent diabetes mellitus. The overall outcomes from these trials have been equivocal and somewhat disappointing, although some individual patients do appear to benefit. It seems that induction of mucosal tolerance prior to systemic exposure to antigen is quite effective, but this is not the case when it is attempted after a systemic immune response has already become established, as is the case in patients who already have an autoimmune condition.

15.6 FETOMATERNAL AND NEONATAL TOLERANCE

In 1953, the immunologist Sir Peter Medawar recognized that the state of **pregnancy is an immunological paradox**, as the fetus expresses paternally derived antigens, equivalent to an incompatible allograft. This should mean that the maternal immune system would act to target and destroy the maternal–fetal interface, leading to abortion. It is now well-recognized that the decidua of women who undergo spontaneous abortion does contain increased numbers of NK cells and CD8+ cytotoxic T cells that might be expected to mediate fetal loss. The immune system also plays an intrinsic role in damaging the placenta, leading to abortion in domestic livestock induced by pathogens such as *Toxoplasma gondii*, *Neospora caninum*, *Chlamydophila abortus*, *Listeria monocytogenes* and BVDV. For example, in ovine enzootic abortion caused by *Chlamydophila abortus*, replication of the organism in placental tissue leads to an inflammatory response and activation of both fetal and maternal inflammatory cells. The effects of fetal macrophages (producing TNF-α) and maternal macrophages and T cells (producing IFN-γ) contribute to disruption of the placenta and fetal loss.

In order to prevent maternal immune destruction of the placental interface, a number of adaptations must be in place to enable the mother to 'tolerate' the fetal allograft. During the early stages of pregnancy in ruminants, cells of the embryonal trophectoderm secrete **IFN-τ, a type I interferon**. IFN-τ is important for **maternal recognition of pregnancy**. It acts on the corpus luteum to persist and secrete progesterone, which is essential for the **maintenance of pregnancy**. As a type I IFN, IFN-τ also has immunomodulatory activities and may contribute to local immunosuppression. Placental tissue is enriched for **immunosuppressive cytokines** such as IL-10 and TGF-β with parallel inhibition of pro-inflammatory cytokines such as TNF-α and IFN-γ. The expression of polymorphic **MHC I and II molecules** is reduced or absent from embryonal and fetal cells. Instead, these cells express nonpolymorphic, 'non-classical' MHC class I molecules which act to inhibit activation of NK cells. Additionally, **Fas ligand** may be expressed within placental tissue, such that infiltrating T cells (expressing Fas) receive signals for apoptotic death. Human studies show significantly more induced Treg cells within the placenta of women undergoing induced abortion than in the placenta of women undergoing spontaneous abortion, suggesting that **active maternal immune regulation** is a key event in survival of the fetus. This effect has been modelled in mice, in that there are significantly more induced Treg cells in the decidua of female mice mated with allogeneic (genetically dissimilar) males than is seen in female mice mated with males of the same strain (i.e. syngeneic).

This state of maternal immunomodulation has systemic as well as local effects. In order to protect the fetus, the maternal immune system undergoes a shift towards Treg cells and Th2 activity. The most convincing manifestation of this immune deviation occurs in women with the Th1-mediated autoimmune disease rheumatoid arthritis. During pregnancy, the inflammatory polyarthritis often goes into clinical remission as a result of the altered T cell balance, but disease relapses after parturition. The systemic Th2 bias in the mother also extends to the fetus, and it has been clearly shown that newborn mice, human infants and canine pups have an immune system skewed towards Th2 reactivity (referred to as the 'neonatal Th2 bias'). In contrast, foals appear to have the reverse pattern with a Th1 bias in neonatal life.

Dizygotic twin cattle may be of the same or opposite sex and are likely to be genetically dissimilar. However, since **vascular anastomoses** between their placentas are common, the blood of the calves mixes and the calves are **genetically chimeras**. Exposure to each other's **alloantigens** *in utero* leads to tolerance, which can be demonstrated by the absence of skin allograft rejection between the calves later in life. This phenomenon was first described in 1945 and provides a good example of **neonatal tolerance**, whereby exposure of the developing immune system to foreign antigen either *in utero* or during early neonatal life leads to the induction of tolerance, such that antigenic challenge later in life fails to induce an immune response. This effect has been replicated experimentally by **immunizing neonatal laboratory rodents** with antigen and demonstrating tolerance in later life or by producing transgenic mice in which a foreign gene is expressed *in utero* and again leads to tolerance to the protein product of the gene. The mechanisms underlying such tolerance likely include clonal deletion and the induction of natural Treg cells.

An important veterinary example of neonatal tolerance is that which develops to infection of pregnant cows with **bovine viral diarrhoea virus** (BVDV), which was discussed briefly in Chapter 12. The virus occurs in two biotypes, non-cytopathic and cytopathic, based on their effects in cell culture. The virus readily transfers across the placenta of pregnant cows and may cause abortion or congenital

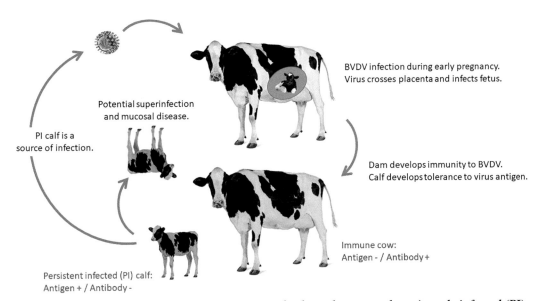

BVDV infection during early pregnancy.
Virus crosses placenta and infects fetus.

Potential superinfection
and mucosal disease.

PI calf is a
source of infection.

Dam develops immunity to BVDV.
Calf develops tolerance to virus antigen.

Immune cow:
Antigen - / Antibody +

Persistent infected (PI) calf:
Antigen + / Antibody -

Figure 15.7 **BVDV infection of pregnant cows can lead to tolerance and persistently infected (PI) calves.** Transplacental spread of a non-cytopathic strain of BVDV can lead to tolerance in the fetus and a persistent infected (PI) calf is born. This is a major source of infection in the herd and is also susceptible to developing mucosal disease upon mutation of the virus or superinfection with a cytopathic biotype of BVDV.

malformations such as cerebellar hypoplasia, hydranencephaly, retinal dysplasia and skeletal malformations. Infection with a non-cytopathic biotype of BVDV before the immune system of the calf has developed (i.e. before 125 days of gestation) can result in immune tolerance to the virus and the calf becomes persistently infected. These **persistently infected (PI) animals** are viraemic and remain seronegative to the virus. They continually shed virus, acting as a source of infection to other animals within the herd. Exposure of these PI animals to a cytopathic strain of BVDV or mutation of the endogenous virus to become more cytopathic causes extensive cell death in mucosal epithelial and lymphoid tissues resulting in fatal **mucosal disease**, characterized by erosions and ulcerations of the oral and nasal mucosa and throughout the gastrointestinal tract.

15.7 ANTIBODY-MEDIATED SUPPRESSION

Antibody can have downregulatory effects at several different levels of the immune response. At the simplest level, antibody may be involved in the neutralization/elimination of antigen, thereby masking the epitopes available for interaction with lymphocyte antigen receptors (**Figure 15.8a**). Maternally derived antibody in particular can have this effect and suppress the immune response to vaccine antigen in neonatal animals. Once formed, **immune complexes** of antigen and antibody are generally removed by phagocytic cells such as **macrophages**. There is evidence that the uptake of immune complexes may induce macrophages to release **immunosuppressive mediators** (e.g. prostaglandin E_2, IL-10 or TGF-β) that, in turn, inhibit the function of lymphoid cells (**Figure 15.8b**). Another means of antibody-mediated suppression acts at the level of a naïve antigen-specific B cell. Since naïve B cells carry inhibitory Fc receptors, it is possible for antibody derived from a different B cell with distinct epitope-binding specificity to occupy receptors on the surface of a B cell (**Figure 15.8c**). The presence of intact antigen may **cross-link** the B cell's own **surface membrane immunoglobulin with the**

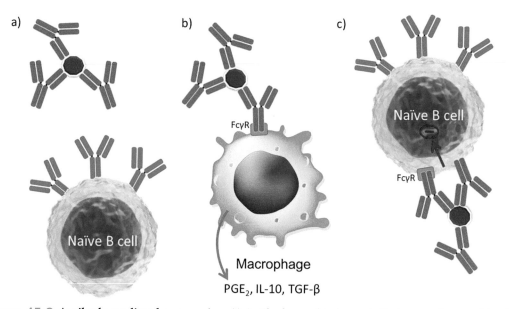

Figure 15.8 Antibody-mediated suppression. (a) Antibody may bind to and eliminate antigen, making it unavailable for detection by lymphocyte antigen receptors. (b) Alternatively, immune complexes of antigen and antibody that are opsonophagocytosed by macrophages may induce those cells to release immunosuppressive mediators. (c) Finally, antibody may occupy B cell surface Fc receptors such that antigen may cross-link the BCR with the FcR-bound antibody and inhibit that B cell.

Fc receptor-bound antibody. This induces a negative signalling event that prevents that B cell from becoming activated. An example of this phenomenon with relevance to human medicine is the management of **haemolytic disease of the newborn**. This arises when a mother who is Rhesus blood group antigen negative (Rh⁻) carries a child who has inherited the Rhesus blood group antigen from the father. At the time of parturition at the end of the first pregnancy there is an opportunity for Rh⁺ fetal red cells to enter the maternal circulation and for the mother to mount an immune response to the Rh antigen. During second and subsequent pregnancies, these maternal anti-Rh antibodies may cross the placenta and cause haemolytic destruction of fetal red cells, with severe consequences for the developing child. The anti-Rh immune response can be effectively prevented at the time of first parturition by injecting pre-formed anti-Rh antibody into the mother. These antibodies in part bind and destroy Rh⁺ red cells, but they may also work through occupying Fc receptors on those B cells in the maternal repertoire able to recognize the Rh antigen and, through cross-linking them, prevent activation of those B cells.

15.8   THE NEUROENDOCRINE-IMMUNOLOGICAL LOOP

This is the name given to describe the concept that these three key body systems are inter-linked, such that soluble mediators from one system are able to **cross-regulate** aspects of function of the other systems (**Figure 15.9**). There are numerous examples of such cross-regulation and some involve suppression of immune responses:

- The range of macrophage-derived pro-inflammatory cytokines (IL-1, IL-6 and TNF-α) is responsible for the induction of pyrexia by influencing the hypothalamic body temperature regulatory centre.
- Chronic neurological stress leads to elevated release of endogenous glucocorticoid from the adrenal cortex, which has an

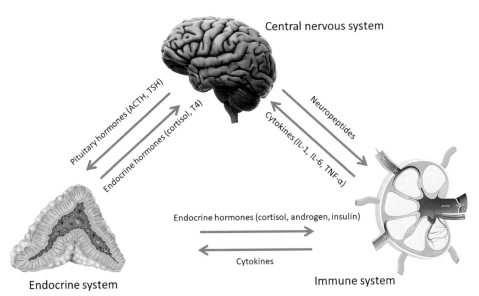

Figure 15.9 **The neuroendocrine–immunological loop.** Soluble mediators produced within each of these three body systems may cross-regulate elements of the other systems.

Figure 15.10 **Hyperadrenocorticism.** This dog displays the classical changes of hyperadrenocorticism, including skin atrophy and muscle weakness. Elevated endogenous glucocorticoid in this disease causes atrophy of lymphoid tissue and circulating lymphopenia.

immunosuppressive effect. This explains the common occurrence of coughs and colds in students around the time of examinations!

- Similarly, in the endocrine disease hyperadrenocorticism, inappropriate elevation of endogenous glucocorticoid or excessive iatrogenic administration of medical immunosuppression leads to atrophy of lymphoid tissue, blood leucopenia and susceptibility to secondary infection (**Figure 15.10**).

- As discussed previously, the hormonal changes during pregnancy imbalance the immune system.

KEY POINTS

- Immunological tolerance is mechanism to prevent inappropriate immune responses to specific antigens such as self-proteins and harmless environmental components.
- Tolerance can be divided into central tolerance and peripheral tolerance.
- Central tolerance involves the deletion of autoreactive B and T lymphocytes in the primary lymphoid organs.
- Peripheral tolerance comprises immunological ignorance, clonal deletion, clonal anergy, immune deviation and suppression by Tregs and other cells.
- Immunological ignorance occurs whereby antigens are sequestered in immunoprivileged sites include the CNS, eye and testis and are not visible to the immune system.
- T cell signalling via signal 2 (CD80/86 interaction with CD28) and signal 3 (cytokine) can lead to anergy, immune deviation
- Tregs act by decreasing MHC/peptide and costimulatory molecules on antigen presenting cells; consuming IL-2 and secreting IL-10, TGF-β and indoleaminedioxygenase.
- Mucosal tolerance is an active event involving the induction of apoptosis, anergy or regulatory T cells against environmental (particularly dietary) antigens.
- Breakdown of mucosal tolerance may lead to the development of food allergy or inflammatory bowel disease.
- Oral tolerance is failure to respond to systemic immunization with an antigen that has been previously delivered orally.
- Fetomaternal tolerance involves suppression of the maternal immune response to paternal alloantigens expressed by the fetus by diverse mechanisms including the action of induced Treg cells, reduced expression of MHC, and presence of immunosuppressive cytokines.
- Neonatal tolerance refers to exposure to antigen *in utero* or during early neonatal life such that challenge with that antigen later in life fails to trigger an immune response.
- Antibody may be regulatory by eliminating antigen or by binding a B cell's inhibitory FcRs to cross-link with the cell's BCR.
- The concept of the neuroendocrine–immunological loop indicates that these three body systems may cross-regulate each other.

IMMUNODEFICIENCY

OBJECTIVES

At the end of this chapter, you should be able to:

- Define primary and secondary immunodeficiency.
- List the clinical features that might indicate the presence of an underlying primary immunodeficiency disease.
- Review the spectrum of primary immunodeficiency diseases in animals.
- Discuss the specific gene mutations that give rise to SCID, cyclic haematopoiesis, canine leucocyte adhesion deficiency (CLAD) and the trapped neutrophil syndrome of border collies.
- Describe novel experimental therapies that have been tested in canine models of primary immunodeficiency for the benefit of human patients.
- List the characteristics of immunosenescence.
- Describe how secondary immunodeficiency may be induced by chronic infectious, inflammatory or neoplastic diseases.
- Discuss the clinical and immunological changes that occur in feline immunodeficiency virus infection.

16.1   INTRODUCTION

Immunodeficiency may be defined as **impairment** in function of one or more parts of the immune system that renders an animal more **susceptible to infectious disease**. Two broad types of immunodeficiency are recognized. **Primary immunodeficiencies** are caused by a genetic defect typically in an immune response gene. Such diseases are **inherited** and clinical signs often become apparent in **young animals** after maternally derived antibody has waned. There is a spectrum of such disorders, with some mutations consistently leading to increased morbidity and mortality and others related to only mild and chronic clinical presentation. In contrast, **secondary immunodeficiencies** occur in animals that have previously had normal immune function and may be caused by a variety of conditions including old age, malnutrition, infection, medical therapy or the presence of chronic disease.

16.2   PRIMARY IMMUNODEFICIENCY

A range of primary immunodeficiency syndromes are recognized in animals and homologues for many of these exist in humans. These

DOI: 10.1201/9781003310969-16

diseases are generally **breed-associated**, clearly **inherited** and have clinical and immunological abnormalities consistent with **immune dysfunction**. However, for several of these animal immunodeficiencies the precise genetic mutation has not been established and many of these disorders remain **putative immunodeficiencies** until such evidence is provided. Primary immunodeficiency disorders are **relatively rare** and, with some exceptions, lack of available cases and lack of research funding has made these disorders difficult to investigate. There are, however, some very well-studied animal immunodeficiencies, for which the precise genetic basis has been established. These diseases sometimes serve as animal models for the equivalent human disorders and colonies of affected animals may be kept for the purpose of developing therapies. It has proven possible to reverse immunodeficient states by the use of **gene therapy** in such models. Gene therapy will likely not be used to treat individual animal patients, but the molecular knowledge can be used to develop **diagnostic tests** to identify homozygous affected and heterozygous carrier animals (since many immunodeficiencies are inherited in an **autosomal recessive** manner), so as not to breed these animals and gradually eliminate the trait from the population. In fact, this screening and selective breeding approach has already proven successful with regard to bovine and canine leukocyte adhesion deficiency.

Primary immunodeficiency diseases can affect immune system development at different levels (**Figure 16.1**). A mutation that inhibits the development of both T and B lymphocytes (e.g. SCID) will have much more severe consequences for the animal than a mutation that selectively impairs the production of complement factor C3 or IgA, for example. Some of the genetic mutations that give rise to immunodeficiency are linked with **other congenital abnormalities**, one of the best recognized being the *FOXN1* mutation, associated with thymic aplasia and hypotrichosis (the absence of a full hair coat) in the so-called nude phenotype in humans, cats and experimental rodents. Immunodeficiency disease might also be associated with concurrent autoimmunity, allergy or immune system neoplasia.

There are certain clinical and historical features of disease that might be indicative of the presence of an underlying primary immunodeficiency in an animal. These include:

- Disease presenting in a particular **breed**.
- **Disease occurring in a young animal** with onset shortly after the expected time of loss of maternally derived immunity. Littermates may develop similar signs of disease.
- Disease occurring in male littermates only. Some immune function genes are located on the X chromosome and mutations may lead to X-linked immunodeficiency in males.
- **Chronic recurrent infection** with failure of infection to respond to standard antimicrobial treatment or relapse soon after completion of a course of antimicrobial therapy.
- Infection with **opportunistic pathogens** (e.g. *Aspergillus*, *Pneumocystis*).
- **Hypoplasia of lymphoid tissues (e.g. lymph nodes, tonsils), persistent lymphopenia, or hypogammaglobulinaemia.**
- **Failure to respond appropriately to vaccination.**

The diagnosis of primary immunodeficiency disease brings with it particular challenges for the veterinarian. The various 'red flags' described earlier might be recognized, but it is often difficult to progress the diagnosis in terms of defining a specific immunological defect. This often relates to the lack of availability of appropriate diagnostic tests. Although in animals it is entirely possible to perform a range of immune function testing, such tests are rarely

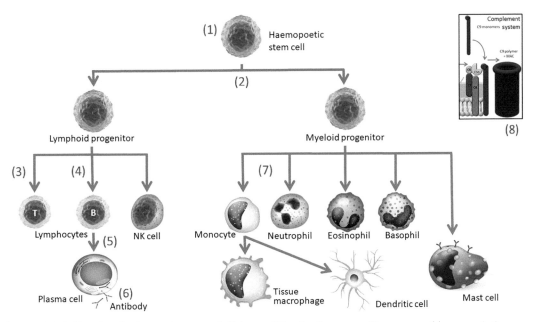

Figure 16.1 Levels of primary immunodeficiency. The development of immune and haemopoietic cells involves progressive maturation from bone marrow stem cell precursors. Genetic mutations underlying primary immunodeficiency can involve molecules involved in different stages of this maturation process. The range of immunodeficiency disorders encompasses (1) failure of the pluripotent stem cell, (2) failure of committed stem cells, (3) failure of T cell development, (4) failure of B cell development, (5) failure of B cell maturation to plasma cells, (6) failure of production of selected immunoglobulin class/es, (7) failure to produce functional phagocytic cells or (8) failure of production of one or more complement proteins.

commercially available. For a small number of conditions, where the specific genetic mutation has been determined, **PCR-based diagnostic tests** are available that may provide a definitive diagnosis.

Immunodeficiency diseases are recognized in a wide range of domestic animals and have been exploited in several **inbred strains of mice** for research purposes. Spontaneously arising primary immunodeficiencies are most common and best described in **dogs and horses**. There are about 30 canine immunodeficiency diseases, but only a few of these have been defined at the genetic/molecular level. Identification of the specific gene and mutation associated with the disease syndrome does not always easily explain the pathogenesis of the immunodeficiency, but it allows the development of genetic tests that may be used to eliminate the disease

(or at least reduce its prevalence) from the population. It is not intended that we cover the full spectrum of primary immunodeficiency disorders in this chapter, but those disorders that are better understood are discussed, starting with those that affect primarily the innate immune system, followed by examples of defects of the adaptive immune system.

16.2.1 Cyclic Haematopoiesis

Canine **cyclic haematopoiesis** (cyclic neutropenia or the 'grey collie syndrome') occurs in **collie pups** that grow poorly and have diluted ('grey') coat colour. They have a cyclic (every 12 days for approximately three days) neutropenia, monocytopenia, thrombocytopenia and reticulocytopenia (**Figure 16.2**) that is associated with episodes of infection (pyrexia, diarrhoea, conjunctivitis, gingivitis or arthritis),

Figure 16.2 Cyclic haematopoiesis in the grey collie. Affected collie dogs typically show diluted ('grey') coat colour. The graph shows the typical 12-day cycle in blood neutrophils (blue line) and monocytes (red line) that characterizes this immunodeficiency syndrome. (Redrawn after Benson KF, Li F-Q, Person RE et al. (2003) Mutations associated with neutropenia in dogs and humans disrupt intracellular transport of neutrophil elastase. *Nature Genetics* **35**:90–96.)

epistaxis and gingival haemorrhage. The disease shows **autosomal recessive** inheritance and involves a **mutation in the *AP3B1* gene, encoding a protein involved in the formation of melanosomes, platelet granules and lysosomes.** Experimental studies of affected dogs have shown that administration of **granulocyte colony-stimulating factor** (G-CSF) or **stem cell factor** can eliminate the cyclic neutropenia, while bone marrow transplantation can cure the disease.

16.2.2   Neutrophil Disorders

A range of disorders specifically affecting neutrophils is recognized in animals. The **Pelger Huët anomaly** presents as a clinical pathology finding with **reduced segmentation of neutrophil nuclei as well as megakaryocytes** on cytology, but this does not seem to have any functional consequences for the animals, who have normal blood neutrophil counts and are generally **clinically normal**. The anomaly is recorded in horses, dogs, cats and rabbits and appears to be usually inherited in autosomal dominant manner.

Chediak Higashi syndrome is characterized by **abnormal giant granules in the cytoplasm of neutrophils, NK cells, platelets,** melanocytes and other granule-containing cells. This autosomal recessive disorder is caused by a mutation in the *LYST* gene, which encodes a molecule involved in the fusion of lysosomal membranes. Affected animals have defective neutrophil and NK cell function and increased susceptibility to infection, as well as ocular and cutaneous hypopigmentation and increased tendency to bleeding. The disease has been reported in cats, cattle, Aleutian mink, white tigers and killer whales.

The '**trapped neutrophil syndrome**' (TNS) of **border collie** dogs is an **autosomal recessive** disease that appears widespread in this breed internationally. TNS presents clinically in dogs from 6 to 12 weeks of age that show stunted growth, fever, diarrhoea, lameness and effusive polyarthritis. Affected dogs demonstrate mild to moderate neutropenia, combined with myeloid hyperplasia in the bone marrow suggesting a **failure to release neutrophils from the bone marrow** (myelokathexis). The disease is caused by a four base pair deletion in exon 19 of the VPS13B gene (vacuolar protein sorting 13 homologue B), which encodes a transmembrane protein that is involved in vesicle-mediated transport and sorting of proteins in cells. A commercial diagnostic PCR test is available for dogs.

16.2.3 Leucocyte Adhesion Deficiency

Leucocyte adhesion deficiency (LAD) is best described in cattle (bovine leucocyte adhesion deficiency; BLAD) and in dogs (canine leucocyte adhesion deficiency; CLAD). **BLAD in Holstein-Friesian cattle is an autosomal recessive** disorder caused by a **mutation in the gene encoding the integrin β2 chain, CD18**. CD18 combines with CD11a, CD11b, CD11c and CD11d to form various cell adhesion molecules. Failure to express CD18 leads to a lack of expression of CD11a/CD18 (LFA-1) and CD11b/CD18 (Complement Receptor 3, CR3) and these defects mean that circulating neutrophils cannot adhere to vascular endothelium and are therefore unable to migrate into tissues (**Figure 16.3**). In addition, the absence of CD11b/C18 further impairs complement-mediated phagocytosis. Consequently, affected animals develop a marked **neutrophilia** and become susceptible to severe **multisystemic infections** in the **absence of recruitment of these cells into tissues**. The clinical presentation is of stunted growth, recurring fever, chronic pneumonia and diarrhoea, and ulcerative gingivostomatitis. Affected animals have poor wound healing and dermatitis. **CLAD is an autosomal recessive** disease that primarily affects the **red Irish setter** (and less commonly the red and white Irish setter). Similar to BLAD, affected dogs fail to express CD18 and develop persistent neutrophilia. The clinical signs include recurrent fever, omphalophlebitis, gingivostomatitis and osteomyelitis. A **PCR diagnostic test** is available to detect affected and carrier animals, which has resulted in virtual elimination of this genetic disorder.

16.2.4 Complement Deficiency

Complement **C3 deficiency** is recognized in the **Brittany spaniel** breed. Affected dogs were part of an experimental colony kept for research into hereditary spinal muscular atrophy. These

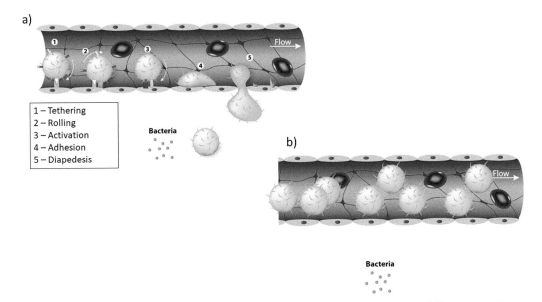

Figure 16.3 Leukocyte adhesion deficiency. (a) Under normal circumstances neutrophils express cell-adhesion molecules containing the CD18 molecule. This allows them to be targeted for entry into infected tissues as part of the inflammatory response. (b) In LAD, there is a mutation in the CD18 gene that means that the processes of adhesion and diapedesis do not occur and the neutrophils remain in the blood, despite the presence of infection in the tissues.

dogs have a single base deletion in the C3 gene, which is inherited in an autosomal recessive manner. The absence of C3 means that the complement system fails and these animals are **susceptible to infection** as well as being prone to developing immune complex glomerulonephritis. **Yorkshire pigs** have been identified with an autosomal recessive **factor H deficiency**. Factor H controls the activation of the alternative C3 convertase. In the absence of factor H, excessive cleavage of C3 results in secondary C3 deficiency and deposition of C3 fragments in the mesangium and glomerular basement membrane (**C3 glomerulopathy**). Affected pigs eventually die from **renal failure**.

16.2.5 Severe Combined Immunodeficiency

Severe combined immunodeficiency (SCID) is a defect in the adaptive immune system that affects both the humoral and cell-mediated immune responses. It has been recognized in both dogs and horses and because children may be afflicted by the same disorder, the animal diseases have received much research attention. SCID in the **Jack Russell terrier** is caused by a **mutation in the *PRKDC* gene** that encodes a **DNA-dependent protein kinase** involved in the recombination of VDJ gene segments in the generation of lymphocyte antigen receptors. Failure to develop these receptors prevents the development of B and T lymphocytes from lymphoid precursors and affected animals are agammaglobulinemic. The DNA-dependent protein kinase is also involved in general DNA repair and cells from SCID dogs have increased sensitivity to ionizing radiation. **X-linked SCID** occurs in the Bassett hound and Cardigan Welsh corgi and is caused by mutations in the *IL2RG* gene that encodes the **γ chain** of the **IL-2 receptor (Figure 16.4)** and is also a component of the receptor **for the cytokines**

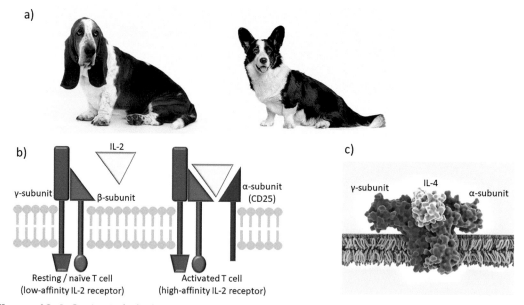

Figure 16.4 Canine X-linked SCID. (a) This syndrome is seen in Basset hounds and Cardigan Welsh corgis. (b) Resting T cells express a low-affinity form of the IL-2 receptor composed of a β and a γ chain. On activation of the T cell, expression of the α chain (CD25) leads to formation of a high-affinity receptor that permits IL-2 binding and signal transduction. The γ chain is shared with receptors for other cytokines including (c) IL-4, IL-7, IL-9, IL-15 and IL-21. In the form of SCID affecting these two canine breeds, a (different) mutation in the γ chain means that these receptors are non-functional and there is failure of the immune response.

IL-4, IL-7, IL-9, IL-15 and IL-21. Inhibition of the action of these key cytokines means that the T cell response to antigen is inhibited and consequently no immune response is generated. Affected animals also lack functional NK cells, but do have normal B cells that, however, do not receive signals from T cells for isotype switching resulting in normal or increased serum IgM and greatly reduced IgG and IgA. The clinical consequences of SCID involve bacterial and viral infection after loss of maternally derived antibody. Affected animals are lymphopenic and have hypoplasia of lymphoid tissue. Experimentally, these diseases can be treated by approaches such as **bone marrow transplantation, transplantation of heterologous stem cells** following partial myeloablation by irradiation, or **gene therapy** in which affected dogs are given a retroviral vector containing the canine *IL2RG* gene. In one study, three of four dogs receiving gene therapy developed relatively normal numbers of blood T and B cells, normal serum IgG concentration and were able to seroconvert following vaccination.

SCID is also well described in **Arab horses**, where the disease is caused by a five base-pair deletion in the *PRKDC* gene resulting in lymphopenia and agammaglobulinemia. Equine

SCID will be described further in a case study (see Chapter 22). Spontaneous mutations in the *Artemis* gene caused SCID in a line of Yorkshire pigs. Like the DNA dependent protein kinase, Artemis is involved in V(D)J recombination and in DNA repair, with affected animals lacking functional B and T cells and showing an increased sensitivity to ionizing radiation.

16.2.6 Thymic Aplasia

A deletion mutation in the transcription factor *FOXN1* causes **thymic aplasia**, with consequent **deficiency of T lymphocytes** and a failure of development of the hair coat. '**Nude' rats and mice (Figure 16.5)** have proven useful immunological tools, as the absence of T cell immunity means that they can be used to investigate the role of these cells in immune responses *in vivo*. A four base-pair deletion in the *FOXN1* gene has been reported in **Birman kittens** that were born hairless and athymic.

16.2.7 Hereditary Zinc Deficiency

Metalloproteins require zinc for their proper folding and enzymatic activity. A mutation in the *SCL39A4* gene that encodes an **intestinal zinc transport protein** underlies **acrodermatitis**

Figure 16.5 Athymic nude mouse. These inbred mice lack a thymus and must be kept under strict biosecurity conditions as they are immunodeficient, lacking function T cells.

enteropathica in human patients and has been identified in **Black Pied Danish and Friesian cattle** and **Turkish Van cats**. Affected animals develop cutaneous disease with alopecia and parakeratosis and diarrhoea. Defects in T cell function have been documented in affected cattle. Oral supplementation with zinc results in improvement of the skin condition and recovery of the T cell function.

16.2.8   Lethal Acrodermatitis

Lethal acrodermatitis of bull terrier dogs is an autosomal recessive disease in which affected dogs have stunted growth, severe cutaneous parakeratosis and skin, intestinal and respiratory infections (**Figure 16.6**). There is reduced T cell mitogen responsiveness and low serum IgA concentration. Although the clinical signs and pathology resemble those associated with zinc deficiency, oral supplementation with zinc does not improve the clinical condition of these

animals. The syndrome is associated with a mutation in the *MLKN1* gene which results in a truncated variant of the intracellular protein, muskelin 1. How the defect in muskelin 1 causes the clinical signs associated with lethal acrodermatitis remains to be established.

16.2.9   Selective Immunoglobulin Deficiency

A deficiency in production of specific immunoglobulin isotypes is recognized in a number of species including man. **Selective IgA deficiency** is the most common human primary immunodeficiency disorder affecting approximately one in every 700 people in Western countries. IgA deficiency, characterized by subnormal serum IgA concentrations, is also regarded as relatively common in dogs. It is most frequently found in the **Shar-Pei** breed, while other commonly affected breeds include **German shepherd dogs** (**Figure 16.7**), **Golden retrievers** and

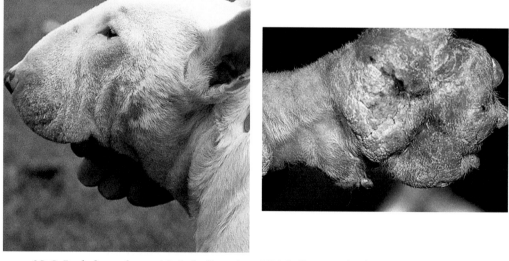

Figure 16.6 Lethal acrodermatitis in bull terriers. This bull terrier dog has stunted growth, submandibular lymphadenopathy and hyperkeratosis of the footpads, a combination of signs typical of the disorder termed 'lethal acrodermatitis'. These dogs are immunodeficient and prone to diarrhoea and respiratory infections.

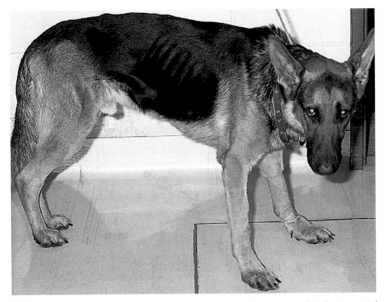

Figure 16.7 IgA deficiency. This German shepherd dog has a disseminated infection with the fungus *Aspergillus terreus*. The susceptibility of dogs of this breed to infectious, inflammatory and immune-mediated diseases has long been thought to reflect an underlying immunodeficiency. The most studied candidate for such a defect is IgA deficiency, but the evidence for this abnormality is inconsistent across different studies.

Labrador retrievers. In the German shepherd dog breed, the prevalence of selective IgA deficiency appears to be <3% and a genetic association has been suggested with a marker (FH2361) on chromosome 33. IgA deficiency is associated with increased risk of mucosal infections, gastroenteritis and possibly allergic (atopic dermatitis) and autoimmune (exocrine pancreatic insufficiency) disease.

Persistent subnormal concentrations of **serum IgG** are thought to play a role in the pathogenesis of a complex disorder of young **Weimaraner dogs** (**Figure 16.8**) and have been recognized in young **Cavalier King Charles spaniels** affected with *Pneumocystis carinii* pneumonia. **Pneumocystosis** is a hallmark infection of immunodeficient people and animals and was a major cause of mortality in human acquired immune deficiency syndrome (AIDS) patients. Selective **IgG2 deficiency** is also reported in **red Danish cattle** and is associated with pneumonia and gangrenous mastitis.

Foal immunodeficiency syndrome occurs in foals of Fell pony and Dales breeds and is characterized by severe nonregenerative anaemia, B cell deficiency and lack of immunoglobulin production. Affected foals are susceptible to respiratory and enteric infections and sepsis. This autosomal recessive disorder is associated with a mutation in the *SLC5A3* gene that encodes an inositol transporter, but the exact pathogenesis remains to be established. However, the identification of the specific gene mutation allows for genetic testing to identify carriers and eventually eliminate the mutation from the susceptible population.

Common variable immunodeficiency in horses is an unusual immunodeficiency as it occurs in adult animals. The syndrome affects male and female horses of different breeds. They develop fever and recurrent infections and have low to undetectable concentrations of serum IgG and IgM and undetectable B cells (but normal numbers of T cells). The genetic defect has not been identified.

Figure 16.8 Weimaraner immunodeficiency. This Weimaraner dog has a history of chronic recurrent infections with profound neutrophilia. The dog has consistently subnormal serum IgG concentration and similarly affected siblings are found within the pedigree. This poorly characterized disease appears often to be triggered by vaccination and may develop following the initial of the puppy series of vaccines or after the 12-month booster.

16.3   SECONDARY IMMUNODEFICIENCY

In contrast to the rarity of primary immunodeficiency, secondary immunodeficiency can affect animals of any breed and is relatively common. Secondary immunodeficiencies can affect adult animals that have had normal immune function until they undergo some form of physiological or pathological change. Causes of secondary immunodeficiency are reviewed next.

16.3.1   Immunosenescence
Immunosenescence is the term used to describe an **age-related decline** in immune function and is a normal physiological change in older animals and people. Due to advances in veterinary healthcare, companion animals are enjoying an increasingly longer life span and an entire discipline of **geriatric medicine** has recently developed. The age-related decline in immune function has been reasonably well characterized in dogs, cats and horses and similar trends are apparent. In general, there is a relative **decrease in circulating CD4+ T cells** and a relative **increase in CD8+ T cells**, with an overall reduced CD4:CD8 ratio. These circulating T cells are **predominantly memory cells** with relatively **few naïve cells** remaining in older animals. T cell function *in vitro* (e.g. mitogen-induced proliferation) and the ability to mount a cutaneous DTH response both decline in the elderly animal. Older animals are perfectly capable of mounting recall immune responses to antigens, but it may be more difficult to induce primary responses (e.g. to vaccines) in this population. It is suggested that immunosenescence is one factor underlying the increased susceptibility of older animals to infection, autoimmune disease and neoplasia.

16.3.2 Medical Immunosuppression

Secondary immunodeficiency might be deliberately induced by the veterinarian when **immunosuppressive therapy** (see Chapter 21) is used to control autoimmune disease or when **chemotherapy** is used in the management of cancer. Although this medical intervention may counteract the disease process in the patient, the major side effect of such treatments is secondary immunosuppression and increased susceptibility to infection (**Figure 16.9**).

16.3.3 Infections

The single best example of infection-associated secondary immunodeficiency is caused by **Feline Immunodeficiency Virus (FIV)** infection. FIV is a **T lymphotropic retrovirus** that infects lymphocytes and cells of the monocyte/macrophage lineage, and has been extensively investigated as an **animal model for HIV infection in humans (Figure 16.10)**. Infected cats have an acute phase of mild illness during which time there is a progressive **decline in blood CD4⁺ T cells**. Cats will then become asymptomatic, but during this second phase of disease there is a

continued decline in circulating $CD4^+$ T cells, which may occur over several years. During the third stage of disease there is a recurrence of mild illness, which progresses to more severe terminal stage 4–5 disease. The terminal illness is similar to human AIDS and is a **chronic, multisystemic disease** that may include gingivostomatitis, respiratory tract infection, enteritis, dermatitis, weight loss, pyrexia and lymphadenomegaly. Neurological disease and lymphoma may also develop and a range of secondary infections have been identified. FIV infection is typically diagnosed by demonstration of FIV serum antibody, with commercially available in-house test kits available for this purpose and may be confirmed by demonstration of virus in circulating lymphocytes or PCR-based amplification of viral nucleic acid. **Concurrent FeLV infection** should also be considered and FeLV may be immunosuppressive in its own right, due to depletion of infected T cells.

Some bacteria secrete toxins that specifically target leucocytes. An example is the leukotoxin of *Mannheimia haemolytica*, the cause of a bacterial bronchopneumonia in cattle, which binds to

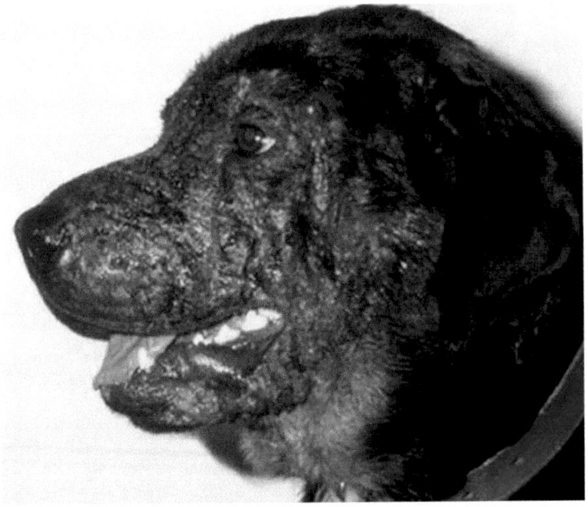

Figure 16.9 Medical immunosuppression. This dog has been treated with immunosuppressive doses of glucocorticoid for immune-mediated disease. The immunosuppression has permitted emergence of a severe secondary *Demodex canis* infection. (Photo courtesy Susan Shaw.)

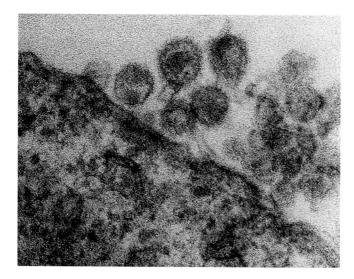

Figure 16.10 **Feline immunodeficiency virus.** This transmission electron microscopy image shows FIV particles budding from the cell membrane of an infected lymphocyte. FIV infection represents a good example of secondary retrovirus-induced immunodeficiency in domestic animals.

the CD11b/CD18 integrin receptor. Following internalization, the leukotoxin triggers death of neutrophils and monocytes.

16.3.4   Chronic Disease

Any animal affected by chronic infectious, inflammatory or neoplastic disease is likely to have a degree of secondary immune suppression and increased susceptibility to infection. Some **infectious agents** (e.g. CDV, canine and feline parvovirus, FIV and FeLV, porcine circovirus-2, equine herpesvirus-1, bovine viral diarrhoea virus) may cause **direct depletion of lymphoid tissue**. Other infections are associated with the production of circulating **immunosuppressive factors** that appear to inhibit lymphocyte blastogenic responses. Such inhibition of lymphocyte function has been demonstrated in diseases such as demodicosis, deep pyoderma, pyometra and disseminated aspergillosis in the dog.

16.3.5   Stress

Chronic stress is also immunosuppressive and follows an elevation in endogenous glucocorticoid production from the adrenal gland. A similar effect is seen is **hyperadrenocorticism**, in which there is circulating lymphopenia and increased susceptibility to opportunistic infection. Stress-induced immune suppression

is likely to play a major role in susceptibility to infectious disease in intensively reared livestock. Animals housed indoors at high stocking density or animals transported over long distances in close confines are considered at risk for such immune suppression. **High-intensity exercise** is also immunosuppressive, although milder exercise can actually enhance a range of immune functions. Young untrained horses show a range of changes in their immune parameters after intensive exercise, including reduced proliferative responses of blood lymphocytes, reduced neutrophil function, reduced CD4:CD8 ratio and reduced NK cell function. These effects are associated with elevated plasma cortisol. Interestingly, these effects are not as marked in older horses that may have lower stress responses.

16.3.6   Malnutrition

Severe malnutrition leads to increased susceptibility to infection due to **impairment of T cell function**, but usually sparing of B cell activity and immunoglobulin production. These effects are thought to be related in part to **leptin**, an adipokine (cytokine produced by adipocytes) related to body fat mass. An animal suffering malnutrition will have loss of body adipose tissue reserve and reduced concentrations of leptin. As leptin is also immunostimulatory (macrophage and Th1 function) and pro-inflammatory, starvation is associated with immune suppression. In addition, starvation may cause immune deficiency by changing the composition of the intestinal microbiome.

Deficiencies in trace elements such as zinc and selenium can cause immunodeficiency. In addition, vitamins A, C, D and E are important for proper function of the immune system. The role of vitamin D in maintaining a healthy immune system is increasingly recognized.

KEY POINTS

- Primary immunodeficiency usually involves a mutation in a gene encoding a key immunological molecule; such diseases are inherited and usually observed in young animals.
- Secondary immunodeficiency occurs in an animal with previously normal immune function that is now affected by advancing age, medical therapy, malnutrition or chronic disease.
- Primary immunodeficiencies are rare, but illustrate the relative importance of particular cells or molecules in immune responses and protection against infectious diseases.
- Primary immunodeficiency may affect the immune system at different levels of development and consequently there is a range of clinical presentation from asymptomatic to severe life-threatening disease.
- Primary immunodeficiency may be suspected clinically when dealing with chronic recurrent multisystemic infection in related littermate animals of a particular breed.
- Diagnosis of primary immunodeficiency is challenging in the absence of a specific genetic/molecular diagnostic test for the precise gene mutation.
- Primary immunodeficiencies are most commonly recognized in dogs and horses.
- Cyclic haematopoiesis of the grey collie is a mutation in a gene encoding a molecule related to neutrophil function.
- The 'trapped neutrophil syndrome' of the border collie involves neutropenia due to failure of bone marrow release of neutrophils into the circulation.
- Leucocyte adhesion deficiency (LAD) occurs in cattle (BLAD) and in dogs (CLAD). Affected animals show a marked neutrophilia, as these cells cannot migrate into tissues, because of a mutation in the gene encoding an integrin (CD18) that mediates the leucocyte–endothelial interaction.
- SCID occurs in dogs and horses. The disease involves a mutation in the DNA-dependent protein kinase gene (Jack Russell terrier, Arab horse) or an X-linked cytokine receptor gene (Bassett hound, Cardigan Welsh corgi).
- Selective deficiencies of immunoglobulin classes are recognized in dogs, horses and cattle.
- Immunosenescence represents an age-related decline in immune function, which is consistently seen in mammalian species.
- Key features of immunosenescence are reduced $CD4^+$ T cells, increased $CD8^+$ T cells, reduced CD4:CD8 ratio and a depletion of circulating naïve T cells with expansion of the memory T cell pool.

- Medical immunosuppression or chemotherapy can lead to secondary immunodeficiency.
- FIV is a T lymphotropic retrovirus that gradually depletes CD4$^+$ T cells, leading to terminal chronic multisystemic disease akin to human AIDS.
- Any chronic infectious, inflammatory or neoplastic disease can lead to secondary suppression of immune function.
- Malnutrition is immunosuppressive.

HYPERSENSITIVITY REACTIONS

OBJECTIVES

At the end of this chapter, you should be able to:

- Define hypersensitivity reactions.
- Discuss the Gell and Coombs classification of hypersensitivity and the mechanisms underlying types I–IV hypersensitivity.
- Give examples whereby hypersensitivity mechanisms are used in protective immune responses.
- Give examples of where hypersensitivity mechanisms lead to pathological immune responses and tissue damage.
- Understand that hypersensitivity mechanisms may occur sequentially or concurrently in some instances.

17.1 INTRODUCTION

The components of the adaptive immune system interact to produce distinct immune responses that protect against various types of infectious agent. The elimination of pathogens is often associated with some degree of tissue damage ('collateral damage'/'friendly fire'), but the degree of injury is usually proportionate to the damage inflicted by the infection. However, when these immune responses are directed against innocuous antigens, the immune response itself is responsible for the tissue injury and disease, and these responses are referred to as **hypersensitivity reactions**. This occurs when the immune response is directed against environmental antigens, components of the normal microbiome, or against self-antigens. A hypersensitivity reaction requires

sensitization to an antigen by prior exposure. Once an individual is sensitized, subsequent **re-exposure** to that antigen can lead to an **inappropriately excessive immune response**. Over time and with repeated exposure to the antigen, the hypersensitivity reaction often results in structural and functional changes in the tissue (**tissue remodelling**) which underlies the clinical manifestations of the disease. Hypersensitivity reactions have a strong heritable component with genetically susceptible individuals predisposed to mounting an inappropriate response. Repeated exposure to some environmental antigens (termed '**allergens**') may lead to immunological sensitization and an inappropriate hypersensitivity reaction (**allergic response**) on re-exposure. Many allergens are ubiquitous environmental substances such as grass pollens, house dust mites or dietary

DOI: 10.1201/9781003310969-17

proteins. The extrapolation of hypersensitivity reactions to allergic disease is somewhat unfortunate, because the same fundamental mechanisms often underlie productive and protective immune responses. In fact, these immune mechanisms have evolved to deal with particular types of pathogen, but have become subverted into the occasional generation of pathological immune responses in some individuals.

17.2 THE GELL AND COOMBS CLASSIFICATION OF HYPERSENSITIVITY

In 1963, P.G.H. Gell and R.A.A. Coombs published the influential text *Clinical Aspects of Immunology*, which laid the foundation for the development of the discipline of clinical immunology. It is worth noting that Robin Coombs (who also gave his name to the Coombs test) was in fact a veterinary graduate from the University of Edinburgh. Within their publication, Gell and Coombs proposed the classification of hypersensitivity reactions that is still used to this day. Their **classification scheme** groups immunopathological mechanisms into four main categories:

- Type I or immediate hypersensitivity mediated by IgE antibodies
- Type II or antibody-mediated hypersensitivity mediated by IgG or IgM
- Type III or immune complex hypersensitivity mediated by IgG or IgM
- Type IV or delayed-type hypersensitivity (DTH) mediated by T cells

It should, however, be noted that many immune-mediated diseases are complex and may involve more than one of these mechanisms.

17.3 TYPE I HYPERSENSITIVITY

The type I hypersensitivity reaction has two phases; the first phase is a period of sensitization of the immune system to the antigen/allergen (**Figure 17.1**) and the second phase is

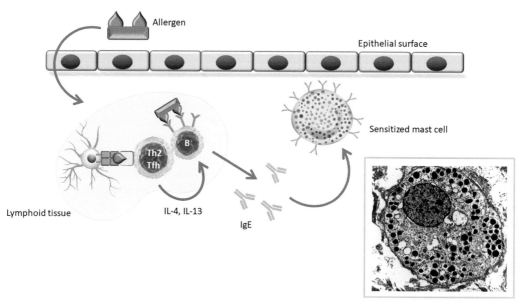

Figure 17.1 Sensitization phase of type I hypersensitivity. Repeated exposure to allergen absorbed across an epithelial barrier leads to Th2 and Tfh-regulated production of allergen-specific IgE antibodies that occupy FcεRI receptors on tissue mast cells or circulating basophils. Inset: transmission electron microscope image of a feline mast cell showing numerous cytoplasmic granules.

re-exposure of the sensitized individual to antigen with clinical manifestation of the hypersensitivity response.

The antigens that induce type I hypersensitivity are generally encountered at the **mucosal and cutaneous epithelial surfaces** of the body. For example, a dendritic cell within the bronchial epithelium may encounter an inhaled antigen or a Langerhans cell within the epidermis may capture a percutaneously absorbed antigen. Injured epithelial cells secrete cytokines such as IL-33 and thymic stromal lymphopoetin (TSLP) that instructs dendritic cells to generate cytokine signals (signal 3) to antigen-specific naïve CD4$^+$ T cells, which differentiate into **Th2 and Tfh cells**. The Tfh cell, in particular will promote a humoral immune response, with expansion of antigen-specific B cells that class-switch towards **IgE production**. The net effect of this sensitization phase is the generation of high concentrations of **antigen-specific IgE**,

which may be detected in the blood and, more importantly, which bind to **FcεRI receptors** on the surface of circulating **basophils** and tissue **mast cells**. Many of these IgE-coated mast cells will reside beneath the epithelium of the body surface where the antigen was originally encountered. Additionally, there is likely to be some class switching towards production of a restricted subclass of IgG that is able to bind to Fcγ receptors on the same cells (termed '**homocytotropic IgG**'). The individual with allergen-specific IgE-coated mast cells is now sensitized, but at this stage displays no tissue pathology or clinical signs.

The second ('**elicitation**') phase of the type I hypersensitivity reaction occurs when a sensitized individual re-encounters the antigen/allergen (**Figure 17.2**). On this occasion, the antigen binds to the IgE-coated mast cells or basophils after penetrating the particular mucosal or cutaneous barrier. When antigen

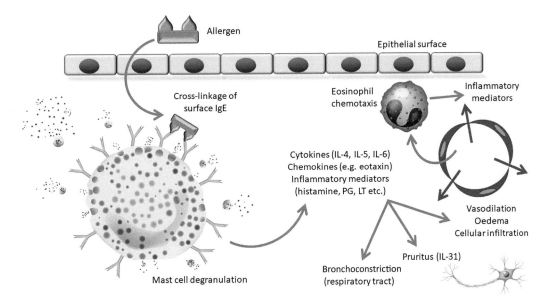

Figure 17.2 Type I hypersensitivity. When a sensitized individual re-encounters allergen, this cross-links mast cell-bound IgE, leading to degranulation of that cell and release of potent bioactive molecules that mediate the rapid local vascular/inflammatory changes of vasodilation, oedema and cellular infiltration, as well as bronchoconstriction (respiratory tract exposure) and pruritus. Infiltration of inflammatory cells (especially eosinophils) may lead to the 'late phase response' some hours later as a result of production of further pro-inflammatory mediators.

is bound by two or more IgE molecules, this **cross-linking** of FcεRI-receptors initiates a signalling pathway within that mast cell. The consequence of mast cell activation is the rapid release (within seconds) of cytoplasmic granules (**degranulation**) containing numerous potent, pre-formed, biologically active mediators including histamine, heparin, serotonin, tryptase, chymase and neutral proteases. In addition, the synthesis of a range of lipid mediators and cytokines is initiated within minutes. The most important of these derive from arachidonic acid, which forms by cleavage of membrane phospholipids by phospholipase A2. Arachidonic acid is subsequently modified by cyclooxygenase (COX) enzymes to gives rise to prostaglandins and thromboxanes and, if modified by lipoxygenase enzymes, to leukotrienes. Mast cells may also synthesize and secrete a range of pro-inflammatory and immunoregulatory cytokines (e.g. IL-4, IL-5, IL-6, IL-13, TNF-α), which are generated over several hours following initial degranulation.

Mast cell degranulation and release of pre-formed mediators occurs very rapidly after antigen exposure and the speed of this response (with clinical effects seen within **15–20 minutes**) is why the reaction is termed '**immediate-type hypersensitivity**'. The consequences of mast cell degranulation include:

- **Vasodilation** with resulting tissue **oedema**, leakage of serum proteins and extravasation of **inflammatory cells**.
- Contraction of smooth muscle which leads to **bronchoconstriction** in the airways.
- Interaction of mediators with local nerve endings (pruriceptors) that are linked to specific centres in the central nervous system mediating itch (**pruritus**). This has greatest relevance in the skin and pruritus is a major part of allergic skin disease. The mediators responsible for this effect include histamine, some prostaglandins and leukotrienes, and neuropeptides.

A key mediator of pruritus is **IL-31**, which is secreted by Th2 cells, mast cells and eosinophils.

Although type I hypersensitivity is generally equated with this range of allergic diseases, the same mechanism likely evolved to form a protective immune response to **parasitic (particularly helminth) infestation**.

A second stage of this reaction, known as the **late-phase response**, occurs **4–24 hours** after antigen exposure. During this late-phase response there is active **recruitment of eosinophils and monocyte/macrophages** into the tissue. The presence of eosinophils in particular is regarded as a hallmark of the type I hypersensitivity reaction (**Figure 17.3**). Under the influence of IL-4 and IL-13, infiltrating monocytes undergo differentiation into M2 macrophages, which are instrumental in tissue repair. Type I hypersensitivity reactions may become chronic in nature following recurrent episodes of allergen exposure. Chronicity can alter the nature of the tissue immune response, which may subsequently become dominated by Th1 cell-mediated immunity stimulated by secondary infection (e.g. staphylococcal bacteria or *Malassezia* yeasts in the skin), or involve tissue remodelling, mediated by cytokines (e.g. fibroblast or epithelial growth factors) and enzymes (e.g. matrix metalloproteinases).

Some type I hypersensitivity reactions are **systemic** and involve degranulation of circulating basophils or tissue mast cells throughout the body. These may be severe and life-threatening events and are known as **anaphylactic reactions** (the process is **anaphylaxis**). The sensitizing allergens in this instance are often ingested (e.g. dietary components such as peanuts in humans or drugs such as penicillin) or injected (e.g. insect sting). However, the majority of type I hypersensitivity reactions are **localized** and affect the tissue that comes into contact with the inciting antigen. The multitude of allergic diseases that afflict man and domestic animals

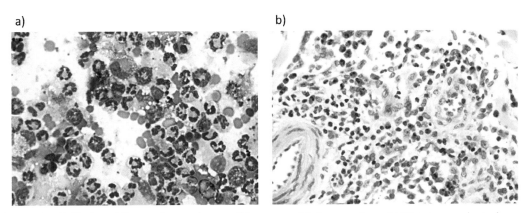

Figure 17.3 Eosinophil recruitment in type I hypersensitivity reactions. (a) This is a cytological preparation of bronchoalveolar lavage fluid from a dog with pulmonary disease (eosinophilic bronchopneumopathy) presumptively caused by a type I hypersensitivity reaction. Numerous eosinophils are recruited into the airways of affected dogs. (b) This skin biopsy is from a horse with sweet itch (Culicoides sensitivity) and shows a perivascular infiltration of mononuclear cells and eosinophils typical of the more chronic stages of disease.

are excellent examples of localized allergic responses affecting the **skin** (e.g. atopic dermatitis, flea allergic dermatitis, equine 'sweet itch' or 'insect bite hypersensitivity'), **respiratory tract** (allergic rhinitis, asthma) or **intestinal tract** (food allergy).

17.4 TYPE II HYPERSENSITIVITY

Type II hypersensitivity (or **antibody-mediated cytotoxicity**) involves sensitization to an antigen displayed on the surface of a **target cell**. Inappropriate production of **IgG** (mostly) or **IgM** antibodies may bind to target cells and cause their destruction. The cytotoxic reaction may involve activation of the classical pathway of complement activation, opsonization and phagocytosis by macrophages or NK cell-mediated destruction via ADCC (**Figure 17.4**). Again, it should be recalled that antibody-mediated cytotoxicity has an important role in many protective immune responses such as in the **clearance of virus-infected cells** and that this immunopathological mechanism evolved for that primary purpose.

A classic example of a type II hypersensitivity reaction is that of an **incompatible blood transfusion**, where a recipient animal carries **alloantibodies** specific for **blood group antigens** expressed on the donor red blood cells (RBCs). Such an incompatible transfusion might lead to a **haemolytic reaction**, involving destruction of transfused erythrocytes within days of transfusion (whereas such cells would normally be expected to survive for several weeks). In the case of whole blood transfusions, acute **anaphylactic reactions** may also occur, but these generally involve type I hypersensitivity with sensitization to serum proteins within the donor blood.

Blood transfusion reactions are relatively uncommon in the dog and are mostly related to expression of the blood group antigen termed **dog erythrocyte antigen 1** (DEA1.1). Most dogs do not carry this blood group antigen (i.e. are DEA1.1⁻) and DEA1.1⁻ dogs do not have spontaneously arising alloantibodies to the DEA1.1 antigen. However, a DEA1.1⁻ dog that has been previously transfused with DEA1.1⁺ blood will generally develop IgG antibodies

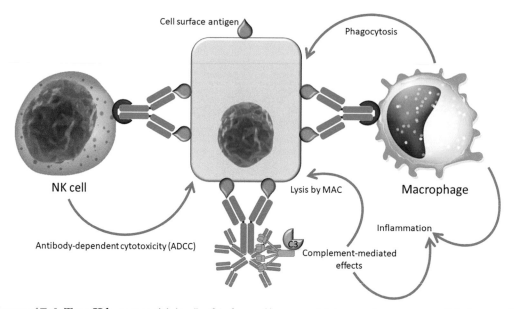

Figure 17.4 Type II hypersensitivity. In this form of hypersensitivity reaction, a target cell is destroyed following binding of IgG or IgM antibody to the cell-surface antigen. Cytolysis may be due to the effects of complement or mediated by NK cells that express Fcγ receptors. Alternatively, the entire target may be phagocytosed by macrophages. Tissue damage may also occur through a bystander effect of the inflammatory reaction.

to that blood group antigen (i.e. become sensitized) so that second and subsequent incompatible transfusions may result in a hypersensitivity reaction. Other canine blood group antigens rarely cause transfusion reactions. Although it is often said that the first transfusion is relatively safe in the dog, best clinical practice would dictate that **blood typing and cross-matching** must be performed before administering any blood transfusion (**Figure 17.5**).

A much greater risk of incompatible blood transfusion occurs in the cat. Cats may phenotypically be of blood groups A, B or AB. Type AB cats have no alloantibodies, **type A cats** may sometimes have **low-titre anti-B** alloantibodies, but **type B cats** invariably have **high-titre anti-A** alloantibodies. Although the majority of cats are type A, the prevalence of blood group B is relatively high in some pure breeds, and there is a high risk of transfusion reaction occurring when a type B cat receives type A blood.

Therefore, blood typing and cross-matching is of great importance in feline transfusion medicine. Kittens born to a type B queen who have inherited type A from their father, will ingest maternal anti-A antibodies via colostrum resulting in **neonatal isoerythrolysis** (Chapter 14), which also represents a type II hypersensitivity reaction. In feline infectious anaemia (caused by *Mycoplasma hemofelis*) red blood cell destruction can be exacerbated by antibodies binding to the infected cells, whereby this immune response to infection is detrimental to the animal.

The type II hypersensitivity reaction also forms the basis for a range of **autoimmune diseases** mediated by autoantibodies, and alloantibody formation has a role in graft rejection. Autoimmune diseases such as **autoimmune haemolytic anaemia** or **thrombocytopenia** utilize this immunopathological mechanism, as do those autoimmune diseases in which autoantibody binds to specific **cell-surface molecules to**

Figure 17.5 Blood transfusion. This dog affected with immune-mediated haemolytic anaemia is receiving a whole blood transfusion after blood typing and cross-matching to ensure that the donor blood is compatible and that the recipient does not have pre-formed alloantibodies to blood group antigens on the surface of the donor cells. (Photograph courtesy S. Warman)

interfere with their normal function. This is seen in autoimmune skin diseases of the pemphigus complex, whereby autoantibodies directed against keratinocyte cell adhesion molecules (e.g. components of the desmosomes or hemidesmosomes) leads to loss of epithelial integrity and formation of vesicles and erosions of the skin and mucocutaneous junctions (**Figure 17.6**). This also occurs in **myasthenia gravis** (recognized in dogs and cats) in which autoantibody specific for the acetylcholine (nicotinic) receptor of the neuromuscular junction causes blockade or destruction of these receptors, leading to episodic skeletal muscle weakness and collapse (**Figure 17.7**).

17.5 TYPE III HYPERSENSITIVITY

Type III hypersensitivity (or **immune-complex hypersensitivity**) involves the formation and tissue deposition of immune complexes with subsequent activation of complement system. The end products of complement activation subsequently cause cellular damage (membrane attack complex; MAC) and inflammation (C3a and C5a). Complement proteins C3a and C5a were originally described as '**anaphylatoxins**' owing to their potent effect in stimulating mast cell degranulation. Immune complexes may form locally, at the site of antigen deposition in the presence of high levels of circulating IgG antibodies. **Complement activation** at that site leads to a **localized tissue inflammatory reaction** (**Figure 17.8**). This localized type III hypersensitivity reaction with immune complex vasculitis in the subcutis is known as an **Arthus reaction** and may occur following allergen-specific immunotherapy and vaccination, but is rarely seen clinically. Pulmonary reactions (hypersensitivity pneumonitis) are recognized in man and generally relate to occupational exposure to allergen. The classical reaction occurs in 'farmer's lung' in which repeated sensitization to the spores of the thermophilic actinomycete *Saccharopolyspora rectivirgula* in bales of mouldy hay (stored indoors in warm temperatures while still damp) leads to IgG production

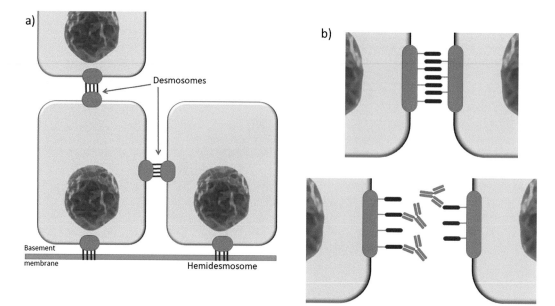

Figure 17.6 Type II hypersensitivity can affect keratinocyte cell adhesion. (a) Keratinocytes are anchored to the basement membrane by hemidesmosomes and to each other by desmosomes. These cell adhesion molecules maintain the integrity of the stratified epithelium. (b) Autoantibodies directed against protein components of these cell adhesion molecules can disrupt their function, leading to a loss of cell adhesion and formation of vesicles/ulcers.

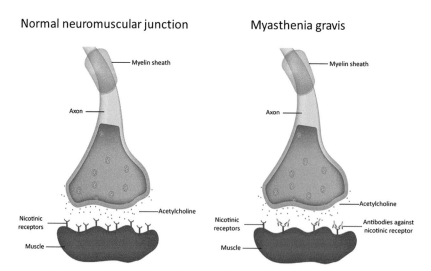

Figure 17.7 Myasthenia gravis. In this autoimmune disease, autoantibodies specific for the acetylcholine (nicotinic) receptor of the neuromuscular junction block the receptor for binding by acetylcholine. The antibodies may also cause endocytosis and depletion of the receptors on the post-synaptic membrane.

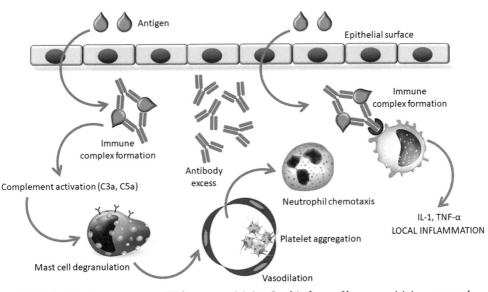

Figure 17.8 Antibody excess type III hypersensitivity. In this form of hypersensitivity, repeated exposure to antigen leads to high concentrations of IgG antibody. Re-exposure to antigen results in the formation of immune complexes with an excess of antibody at the site of antigen penetration. Activation of the classical pathway of complement triggers a local tissue inflammatory reaction involving vasodilation, recruitment of granulocytes and monocytes, mast cell degranulation and platelet aggregation within local capillaries. Activated macrophages phagocytose immune complexes and release pro-inflammatory cytokines.

and the potential for a hypersensitivity response on inhalation of the allergen. Clinical signs of pulmonary disease usually become manifest **within 24 hours** of re-exposure. The same disease may occur in **cattle housed over winter** and fed hay containing these spores, which are small enough (1 µm) to pass through the upper respiratory tract defences and into the lungs. A similar pathogenesis may be involved in **equine asthma/inflammatory airway disease**, which is believed to result from sensitization to moulds and other allergens in dusty stable air and to involve elements of both type I and III hypersensitivity. **Canine 'blue eye'** is a further example of a localized type III hypersensitivity reaction. This occurs in some dogs that are infected by or vaccinated with canine adenovirus-1. Immune complexes form within the anterior uvea and lead to localized inflammation (uveitis), corneal endothelial damage and corneal oedema (**Figure 17.9**).

A **systemic form of type III hypersensitivity** is triggered by a **high concentration of circulating antigen**, leading to the formation of immune complexes with antigen excess in the circulation. These complexes are generally relatively **small and soluble** and circulate in the bloodstream until they deposit in the predilection sites of the **renal glomerulus, synovium** and **the walls of small blood vessels**. Once lodged in the tissue, there is complement activation and triggering of the same inflammatory pathways described previously (**Figure 17.10**). The consequences of vasculitis may be the formation of a **thrombus** that occludes the capillary lumen, leading to **ischaemic necrosis** of the tissue supplied by that vessel (**Figure 17.11**). Vasculitis can be observed clinically within the skin, where discrete round foci of necrosis may develop (**Figure 17.12a**) or there may be necrosis and loss of the tips of the ears or tail tip (**Figure 17.12b**).

Figure 17.9 Antibody excess type III hypersensitivity. This dog has 'blue eye', which is corneal oedema that occurs secondary to local uveal tract inflammation triggered by the formation of immune complexes of canine adenovirus-1 antigen with antibodies formed in infected or vaccinated dogs.

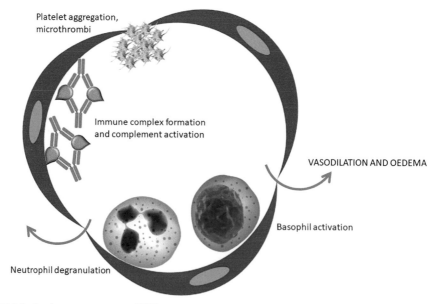

Figure 17.10 Antigen excess type III hypersensitivity causes vasculitis. In this form of hypersensitivity, a high concentration of antigen leads to the formation of small, soluble immune complexes. In areas of turbulent blood flow, these are forced against the walls of capillaries and may lodge at these sites. Complement activation results in recruitment of neutrophils with subsequent inflammation within the vessel wall (vasculitis). Local platelet aggregation may result in the formation of microthrombi and ischaemic necrosis of tissue supplied by the vessel. Platelets and basophils release vasoactive amines to cause vasodilation and increased vessel permeability.

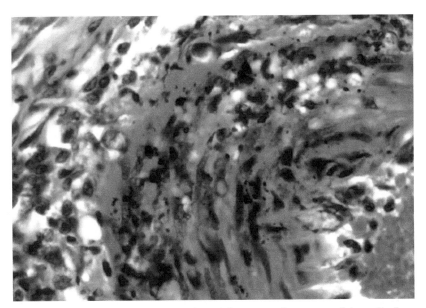

Figure 17.11 Immune complex induced vasculitis. This image shows part of the wall of an arterial blood vessel in the kidney of a dog with multisystemic immune complex disease secondary to *Leishmania* infection. There is neutrophilic inflammation, necrosis and fibrin deposition within the smooth muscle wall of the vessel.

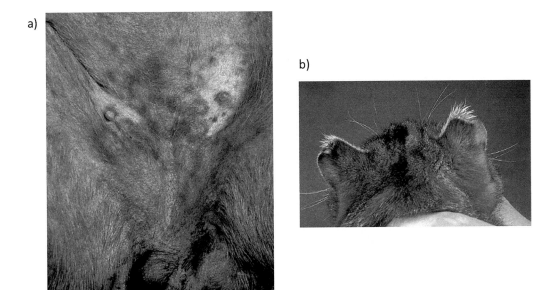

Figure 17.12 Vasculitis. (a) The ventral abdomen of this dog has numerous discrete round 'target' lesions consistent with vasculitis. This dog had septicaemia and it is likely that immune complexes of bacterial antigen and antibody had formed and deposited within the walls of cutaneous capillaries. (b) The tips of the ears of this cat have become necrotic and detached. The underlying cause was vasculitis with thrombosis of small capillaries and ischaemic necrosis of the skin.

Whether circulating immune complexes will induce such a reaction is determined by a number of factors:

- The **biochemical nature** of the antigen and antibody.
- Whether the normal immune complex **clearance mechanisms** (e.g. macrophage phagocytosis in spleen or liver) are impaired.
- Whether there are pre-existing **lesions in the vascular wall** resulting in contraction of endothelial cells and exposure of subendothelial collagen.
- The **nature of blood flow** in the capillary bed. As the renal glomerulus has a turbulent and high pressure flow, more likely to force complexes against the vascular walls, this is a clear predilection site for deposition and induction of immune complex glomerulonephritis (**Figure 17.13**).

Systemic immune complex disease may include clinical signs such as lameness, joint pain or swelling (associated with non-erosive polyarthritis), visual impairment and ocular pain, cutaneous lesions and protein-losing nephropathy (associated with glomerulonephritis). Adsorption of immune complexes to red blood cells and platelets may also cause immune-mediated haemolytic anaemia and thrombocytopenia. This form of immune complex disease is certainly recognized in animals, but often the antigenic cause is not determined. In these instances the disease is often described as **'idiopathic'** or **'immune-mediated'**, but that simply reflects an inability to identify the causative antigen. In autoimmune immune complex diseases (e.g. systemic lupus erythematosus) the antigen is derived from self-tissue.

Immune complexes are formed as a means of clearing infectious agents through phagocytosis by macrophages in the spleen and liver. However, this mechanism is also a classical means of producing an immune-mediated reaction secondary to an infectious disease, where the infectious agent might be eliminated or be persistent, but complexes of residual antigen and antibody now perpetuate clinical disease by causing **secondary immunopathology**. There are several examples of this phenomenon in veterinary medicine, and canine leishmaniosis

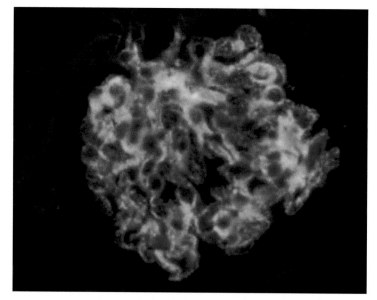

Figure 17.13 Immune complex glomerulonephritis. This section of kidney from a cat with protein-losing nephropathy has been probed with antibody specific for feline IgG and detected by immunofluorescence microscopy. The fine granular deposits around the capillary loops of the glomeruli are immune complexes that have deposited in the walls of these vessels.

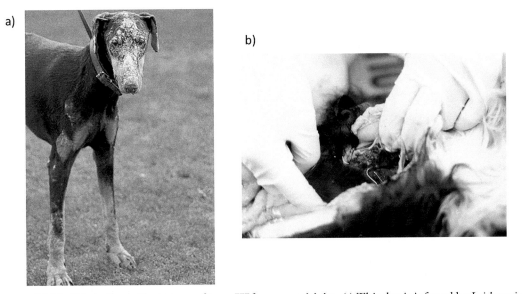

Figure 17.14 Clinical manifestations of type III hypersensitivity. (a) This dog is infected by *Leishmania infantum*. A sandfly has transmitted the infection to the skin establishing a granulomatous dermatitis apparent as areas of alopecia and scaling over the face of the dog. This dog is also lame and has protein-losing nephropathy. These two clinical signs relate to the development of circulating immune complexes and their deposition in capillaries of the synovium and renal glomerulus. (b) This cat was euthanized as a result of the wet (effusive) form of feline infectious peritonitis, whereby a type III hypersensitivity to coronavirus antigen leads to widespread vasculitis and effusion into the body cavities.

provides one such instance. Although this disease involves granulomatous inflammatory lesions in skin and other organs, some of the clinical changes arise via secondary immune complex formation, with deposition of immune complexes within the synovium, glomerulus and uveal tract (**Figure 17.14a**). The wet (effusive) form of feline infectious peritonitis occurs as a result of type III hypersensitivity to the coronavirus antigen, leading to widespread vasculitis and effusion into the body cavities (**Figure 17.14b**). Another example is **purpura hemorrhagica** in horses following infection with *Streptococcus equi* (strangles) and occasionally other respiratory infections. Horses may develop subcutaneous oedema and petechial mucosal hemorrhages due to a systemic immune complex vasculitis.

17.6 TYPE IV HYPERSENSITIVITY

Whereas types I–III hypersensitivity are all mediated by antibody, the type IV reaction is **cell-mediated**. Additionally, and because of this cellular nature, the type IV reaction has a prolonged onset (**24–72 hours**) and is therefore also known as **delayed-type hypersensitivity (DTH)**. The sensitization phase of a type IV response generates antigen-specific T lymphocytes, in particular **Th1 cells**. Subsequent exposure to the sensitizing antigen leads to antigen presentation and reactivation of those effector T cells. These cells home to the site of antigen exposure and release pro-inflammatory cytokines, including **IFN-γ and chemokines**, which, in turn, results in the upregulation of vascular addressins, and **recruitment of mononuclear inflammatory cells** including macrophages, CD4+ and CD8+ T cells, NK cells and some granulocytes. Further pro-inflammatory cytokines are released by these infiltrating cells, exacerbating the inflammatory reaction in the tissues (**Figures 17.15, 17.16**). This type of response is actually an appropriate one for dealing with intracellular (vesicular) pathogens such as mycobacteria, in which the activated

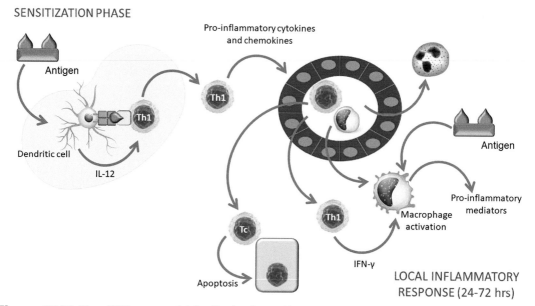

Figure 17.15 Type IV hypersensitivity. In this form of hypersensitivity, the sensitized individual develops antigen-specific Th1 cells that become activated upon re-exposure to antigen and home to the site of exposure. These cells produce cytokines (particularly IFN-γ) and chemokines that have vascular effects and enhance recruitment of a range of inflammatory cells into the tissues. These infiltrating cells (particularly macrophages) amplify the local inflammatory reaction. In some cases, there may also be some degree of cellular damage (apoptosis) mediated by CD8$^+$ killer T cells.

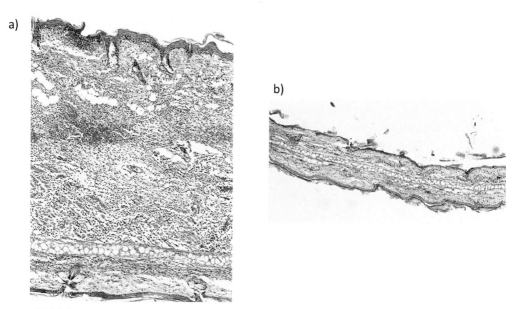

Figure 17.16 Type IV hypersensitivity reaction in tissues. (a) In this experimental demonstration of type IV hypersensitivity, a rat was sensitized by injection of a synthetic peptide in adjuvant. Following sensitization, the rat was re-exposed to peptide injected into the dermis of the ear. Within 24 hours the site of injection had become raised, reddened and firm. Biopsy reveals that this is attributed to an influx of mononuclear inflammatory cells into the dermis. (b) A control histological section from the contralateral ear injected with another peptide, to which the animal had not been sensitized.

macrophages form granulomas that wall off the infection. However, the formation of granulomas results in formation of space-occupying lesions and contributes to the clinical signs and immunopathology associated with such infectious diseases (**Figure 17.17**). This type of systemic immune response to coronavirus antigen is responsible for the pathology seen in the dry (granulomatous) form of feline infectious peritonitis.

The **tuberculin skin test** is a classic example of a type IV hypersensitivity reaction that is used to screen cattle for infection with *Mycobacterium bovis*. It is based on the detection of memory CD4+ T cells generated in infected animals (**Figure 17.18**). The test consists of the **intradermal**

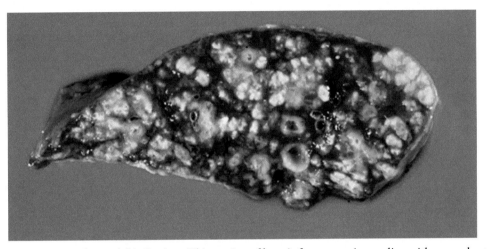

Figure 17.17 Mycobacterial infection. This section of lung is from a captive sea lion with a mycobacterial infection. The lung displays the classical coalescing granulomas that form during mycobacterial infection. The type IV immunopathological mechanism evolved to deal with obligate intracellular pathogens such as *Mycobacterium* spp.

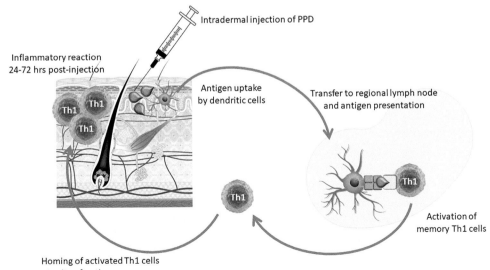

Figure 17.18 Mechanism of tuberculin testing. Injected antigen is transported to regional lymph nodes where memory Th1 cells are activated and then home back to the site of injection to mediate an inflammatory response.

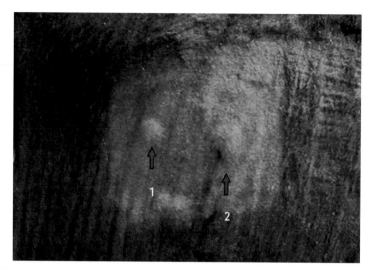

Figure 17.19 Tuberculin testing. Tuberculin testing of cattle remains a major veterinary activity worldwide. Skin sites on the lateral neck are selected, clipped and the thickness of a skin fold measured with calipers. Two intradermal injections (of avian and bovine tuberculin/PPD) are given. The test is read 72 hours later by repeating the caliper measurement. If the thickness of the skin fold at the site of injection of the bovine tuberculin is more than 4 mm greater than the fold for the avian tuberculin, the animal is considered a 'reactor'. 1, *M. avium*; 2, *M. bovis*.

injection of a protein extract (purified protein derivative, PPD) of mycobacteria in either the tail fold or the neck. The injection site is measured after **48–72 hours for swelling and induration** caused by the accumulation of inflammatory cells such as lymphocytes and macrophages (**Figure 17.19**). An alternative test is designed to measure production of IFN-γ in blood samples collected from animals following stimulation in the laboratory with mycobacterial proteins.

KEY POINTS

- Hypersensitivity requires an initial period of sensitization to antigen. On re-exposure of the sensitized individual there is an inappropriately excessive immune response to that antigen.
- Allergy is the pathological manifestation of hypersensitivity to innocuous foreign (environmental) antigens.
- An allergen is an antigen that triggers an allergic response.
- Hypersensitivity responses evolved to protect from infectious disease, but they have been subverted to cause immune-mediated disease.
- Gell and Coombs designed the classification system for types I–IV hypersensitivity.
- In type I (immediate) hypersensitivity, sensitization leads to Th2/Tfh-regulated production of IgE antibody that binds to mast cells and basophils.
- Re-exposure to antigen in type I hypersensitivity leads to IgE cross-linkage, mast cell degranulation and release of bioactive molecules.

- Mast cell degranulation leads to vasodilation, bronchoconstriction (respiratory tract exposure) and pruritus (skin exposure) within 15–20 minutes of contact with the allergen.
- The late-phase response (4–24 hours) of type I hypersensitivity involves tissue recruitment of eosinophils and macrophages and secretion of additional pro-inflammatory mediators.
- Type I hypersensitivity reactions may be localized (skin, respiratory and intestinal tracts) or generalized (anaphylaxis).
- In type II hypersensitivity (antibody-mediated cytotoxicity), sensitization leads to production of IgG or IgM antibody, which mediates destruction of a target cell.
- Blood transfusion reactions are an example of a type II hypersensitivity reaction.
- The type II hypersensitivity mechanism also underlies a range of autoimmune diseases including those in which anti-receptor antibodies form (myasthenia gravis).
- Type III hypersensitivity (immune complex hypersensitivity) may occur in antibody excess or antigen excess.
- In antibody excess type III hypersensitivity, immune complex forms at the site of tissue exposure to antigen, leading to localized tissue inflammation (the Arthus reaction).
- In antigen excess type III hypersensitivity, soluble circulating immune complexes lodge in capillary walls in the renal glomerulus, uveal tract, synovium or skin (immune complex disease).
- The basic lesion of antigen excess type III hypersensitivity is vasculitis, which may progress to thrombosis and ischaemic necrosis.
- Antigen excess immune complex formation is an important cause of post-infectious secondary immunopathology.
- Type IV or DTH involves cells rather than antibodies and has a slow onset (24–72 hours).
- Antigen-specific Th1 cells are the key elements in the type IV hypersensitivity mechanism that underlies tuberculin tests.

CHAPTER 18

HYPERSENSITIVITY DISORDERS AND ALLERGIC DISEASE

OBJECTIVES

At the end of this chapter, you should be able to:

- Define allergy.
- Review the immunopathological mechanisms involved in allergic disease.
- Discuss factors that predispose to allergic disease.
- Discuss the 'hygiene hypothesis'.
- Give examples of hypersensitivity disorders and allergic diseases that affect the skin, respiratory tract and gastrointestinal tract in animals.
- Describe the methods used in diagnosis of allergic disease.

18.1   INTRODUCTION

Allergy is the clinical manifestation of hypersensitivity reactions to innocuous environmental antigens (allergens). These are typically type I (immediate-type) hypersensitivity reactions, associated with an **IgE response to the allergen**, recruitment and activation of **mast cells and eosinophils**, and a central regulatory role of **type 2 cytokines (IL-4, IL-5 and IL-13)** secreted by **Th2, Tfh and ILC2 cells**. However, type II-IV hypersensitivities to environmental antigens may also lead to tissue pathology in some circumstances. The term '**atopy**' is used to describe the genetically mediated propensity of an individual to produce type I hypersensitivity reactions to environmental allergens by producing **allergen-specific IgE antibodies**. Such individuals are referred to as being '**atopic**' and have **an exaggerated Th2 immune response**, probably as a consequence of **impairment of Treg cell** function (**Figure 18.1**).

Allergic diseases are of major significance in human and veterinary medicine. The incidence of allergic diseases such as **hayfever**, **asthma** and **atopic dermatitis** (AD) has markedly increased in the human population over the past five decades. The speed with which this has occurred indicates that this increase is likely caused by **environmental factors** rather than genetic changes in the human population. Although allergic diseases have also been increasingly diagnosed in companion animals, it is uncertain if this is due to a truly increased incidence or because of increased recognition and improved diagnostic methods.

The **hygiene hypothesis** postulates that the increasing prevalence of allergic (and some autoimmune) diseases over the past century has been caused by changes in lifestyle and in

DOI: 10.1201/9781003310969-18

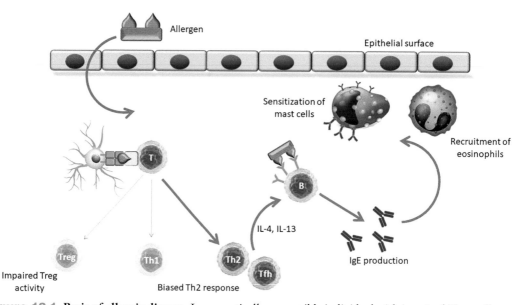

Figure 18.1 **Basis of allergic disease.** In a genetically susceptible individual with impaired Treg cell function, exposure to innocuous environmental allergen leads to an expanded Th2 immune response that (alongside Tfh cells) mediates production of allergen-specific IgE antibody and recruitment of mast cells and eosinophils. In the initial phases of such an aberrant immune response there is impairment of Th1 activity, but in chronic disease with secondary infection, Th1 cells may become an important contributor to the tissue pathology.

particular the early life experience. Children now lead a more '**sanitized' lifestyle** that fails to adequately redirect the developing immune system towards Treg cell and Th1 cell activity. Such influences include factors such as:

- Spending more time indoors.
- A centrally heated and carpeted indoor environment that may contain more allergens but fewer microorganisms than in previous decades because of widespread use of antimicrobial cleaning agents.
- Smaller families with low sibling numbers.
- Increased use of antibiotics.

Similarly, rural living and exposure to outdoor green space appears to be inversely related to the incidence of atopic dermatitis in dogs. The identified factors increase the degree and duration of exposure to allergens and can change the composition of the **intestinal and skin microbiome.** Living in an urban environment has been shown to reduce the diversity of the skin microbiome in dogs. The intestinal microbiome is important in developing and maintaining the integrity of the immune system including the **expansion of Treg cells.** Early exposure to probiotics such as *Lactobacillus* may reduce the severity of atopic dermatitis in dogs. It also appears that **helminths** have a much more potent effect on the **induction of Treg cells** than do many bacteria or viruses. For many years an inverse relationship between intestinal parasitism and allergy has been recognized, despite the fact that both situations involve a Th2-dominated immune response. The endoparasitism that typically occurs in human populations within developing countries appears to have a protective effect, as such, individuals are rarely affected by allergy or autoimmunity. Indeed, deliberate establishment of an intestinal parasitic infection in human patients with allergy has led to marked clinical improvement in some individuals.

18.2 FACTORS PREDISPOSING TO ALLERGIC DISEASE

As is characteristic of many immune-mediated diseases, allergic disease has a **multifactorial** basis. One of the strongest predisposing factors in humans and dogs is the **genetic background**. There are clear **breed predispositions** to allergic disease in dogs. Certain breeds including Labrador and Golden retrievers and West Highland White terriers are at increased odds of developing atopic dermatitis and recent studies have suggested that there are breed-specific patterns in the nature of the disease (e.g. lesion distribution). Experimental lines of beagles have been selected for their propensity to develop IgE responses ('**high IgE responder beagles**') when sensitized to a variety of dietary allergens or aeroallergens. A study in Golden and Labrador retrievers estimated the heritability of atopic dermatitis to be 0.47, meaning that genetic factors contributed to nearly 50% of the risk. Despite this strong genetic predisposition, the precise molecular basis for allergy is not yet clearly defined. Although a wide range of genetic associations has been highlighted, the results from different studies are inconsistent, likely reflecting differences in breed, geographical location (gene pool) and environment between the populations investigated. Genomic studies have shown associations with numerous polymorphic candidate genes, including those of the **MHC** and a **cytokine gene cluster** encoding a number of Th2-related molecules. These genetic associations indicate an important role for CD4[+] T cells in allergic disease and an imbalance in their differentiation into effector cells (i.e. impaired Treg cells, reduced Th1 responses and excessive Th2 activity).

There are similar genetic influences on the development of allergic disease in horses. Ponies, and particularly Icelandic ponies, are susceptible to developing hypersensitivity to the saliva of *Culicoides spp.* midges (**insect bite hypersensitivity** or '**sweet itch**'). Of note is the fact that the incidence of this allergy is greatest in adult Icelandic ponies born in Iceland (which lacks midges) and subsequently exported to countries in which the midge is present. The prevalence of disease increases with increasing time since export. There are also genetic influences on the development of **equine asthma (inflammatory airway disease)** in the horse. Although hypersensitivity disease occurs in cats, there are, in contrast, no recognized breed associations.

Age influences on the development of allergic disease are also apparent. There is a clear age predisposition for canine AD, with up to 75% of cases being diagnosed at less than three years of age. **Equine asthma** comes in two forms, a mild to moderate form that is more common in young adult horses and a severe form (also known as recurrent airway obstruction) that occurs in older animals.

Allergen exposure is an obvious factor predisposing to the development of these disorders. Antigens that can trigger allergic reactions are typically **proteins or small molecules (haptens) attached to proteins**. Many allergenic proteins are enzymes, but the role of the enzymatic activity in inducing a type 2 immune response is unclear. Humans, dogs and cats share reactivity to common indoor environmental allergens (e.g. house dust mites, human dander) and pollens, although there may be differences in the precise epitopes leading to sensitization in different species. There are often geographical differences in the relative significance of particular allergens. Exposure to haemophagous arthropods (e.g. fleas, flies, mosquitoes) is also clearly required for the onset of diseases such as **flea allergy dermatitis** (FAD). Repeated (but possibly intermittent) ingestion of dietary antigen is generally required in order to develop a **dietary hypersensitivity**. Coupled with allergen exposure, there may be a **reduction in mucocutaneous barrier function** in susceptible individuals due to a genetic defect or injury. Such biomechanical defects may allow increased penetrance of inhaled, ingested or percutaneously absorbed allergens. Injury to epithelial cells can trigger the release of epithelial cytokines, IL-25, IL-33 and thymic stromal lymphopoietin (TSLP), which activate ILC2 and instruct dendritic cells to direct the differentiation of CD4+ T cells to Th2 cells.

18.3 SYSTEMIC ALLERGIC REACTIONS

A relatively uncommon manifestation of type I hypersensitivity in animals is the presence of the related conditions of **urticaria**, **angioedema** and **anaphylaxis**. These all involve sensitization to a range of possible allergens including drugs, vaccines, incompatible erythrocytes, food, plants and biting or stinging insects. All have an acute onset, as is characteristic of a type I hypersensitivity reaction. Urticarial lesions are most often cutaneous and involve the formation of **localized or generalized wheals** that may be **pruritic** (**Figure 18.2**). In angioedema there is involvement of subcutaneous tissue and **facial swelling** is a common manifestation (**Figure 18.3**). These lesions are

generally **transient** and resolve spontaneously, but angioedema may become life-threatening if the process extends to involve the larynx and upper respiratory tract. The most severe of this group of disorders is **anaphylaxis** in which there is **acute systemic vasodilation** leading to generalized oedema, a drop in blood pressure and shock that may lead to death. In the dog this process particularly affects the liver which becomes engorged with blood, whereas in most other species the '**shock organs**' are the lung and intestine.

In contrast to companion animal species, allergic disease is rarely documented in domestic livestock. This may relate to factors such as lifespan, environment and endogenous parasitism, all of which differ significantly, compared with companion animal populations. **Milk allergy** (directed against the casein protein of milk) is recognized in cows during periods of milk retention and generally relates to the drying off period. The increased intramammary pressure forces milk proteins into the circulation, allowing sensitization to occur. This disease may recur in the same cow during subsequent drying off periods and is thought to have a genetic basis in Channel Island breeds. The clinical presentation of milk allergy may include

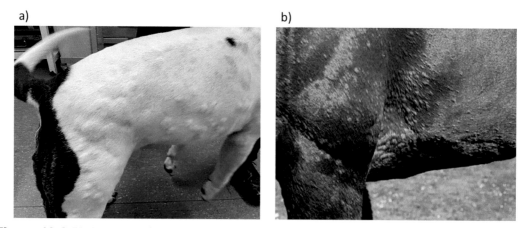

a)   b)

Figure 18.2 Urticaria. (a) This dog is showing signs of urticaria following exposure to an (unknown) allergen against which they must have previously been sensitized (photograph courtesy of Jelena Ristić). (b) Within minutes of being exposed to an allergen this horse has developed multiple oedematous wheals over most of its body. These lesions are mild and transient and resolved spontaneously.

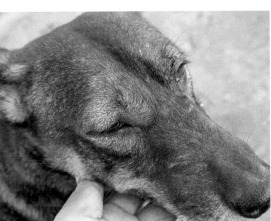

Figure 18.3 Angioedema. These two dogs are showing signs of angioedema (facial swelling) following exposure to an allergen against which they have previously been sensitized.

cutaneous urticaria (facial or generalized), muscle tremor, dyspnoea, restlessness, self-licking or even extreme behavioural changes such as charging or bellowing.

18.4 ALLERGIC SKIN DISEASE

The most common canine allergic disease in many parts of the world is **flea allergic dermatitis (FAD)**. The causative allergens are proteins within the saliva of *Ctenocephalides felis*, some of which have now been identified. The immune response to these antigens can be a classical type I hypersensitivity with a late-phase response, or a mixed type I and type IV hypersensitivity reaction. The cutaneous lesions may become chronic due to **self-trauma** and **secondary infection** with bacteria and/or yeasts and typically involve the skin of the tail base and hindlimbs (**Figure 18.4**). Intriguingly, some dogs that are chronically exposed to fleas appear to develop tolerance rather than hypersensitivity. Dogs may also develop allergic reactions following sensitization to the bites of other arthropods (e.g. ticks) or as part of the patho-genesis of mite infestation/mange (e.g. *Sarcoptes scabei*).

Canine **atopic dermatitis (AD)** has many similarities to the disease in man. A wide spectrum of causative allergens may be involved, but these are most often '**indoor allergens**', particularly those derived from **house dust mites** (*Dermatophagoides pteronyssinus* and *D. farinae*) or '**outdoor allergens**', particularly those derived from tree or grass pollens. The best characterized allergen is a molecule known as Der f15, which is a 98 kDa chitinase enzyme found within the cells lining the digestive tract of the house dust mite. In contrast, the major mite allergens in human AD are of lower molecular weight and are typically components of mite faecal particles.

The immunology of canine AD has been well studied and involves numerous elements such as:

- **Type 2** cytokine association with increased expression of genes encoding IL-4, IL-5 and IL-13 in lesional skin.
- Production of **IgE**, IgG1 and IgG4 allergen-specific antibodies. However, the

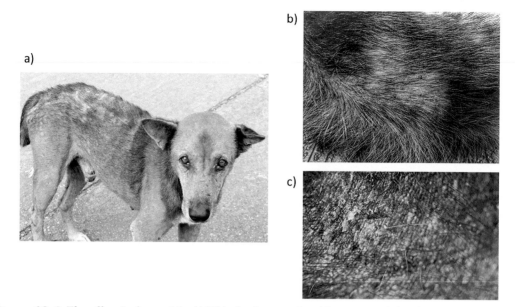

Figure 18.4 Flea allergic dermatitis. (a) This dog has acutely self-traumatized the skin over the base of the tail. This area is classically involved in the pruritic response triggered by flea allergy. (b) Self-trauma leads to inflammation of the skin. (c) In this more chronic case of FAD there has been prolonged self-trauma leading to alopecia and thickening and hyperpigmentation of the skin.

role of these antibodies in the pathogenesis of AD remains uncertain.

- **Impairment of Treg cell function** with reduction in IL-10 production.
- A role for epidermal Langerhans cells in allergen capture.
- Increased expression of MHC class II by Langerhans cells and dermal dendritic cells.
- CD4+ and CD8+ T cell infiltration of lesional dermis.
- A **late-phase response**.
- Progression to a Th1 response with IFN-γ production, associated **with chronicity and secondary infection**.

Canine AD has **breed and age predispositions** and typically presents as erythematous and pruritic lesions that particularly affect the **face and feet**, but which may become more generalized. The lesions progress following self-trauma and **secondary bacterial and yeast infection** (**Figure 18.5**). The diagnosis of AD requires exclusion of other causes of pruritic skin disease, including ectoparasites, bacterial infections and food allergy.

Dietary hypersensitivity or **food allergy** is poorly characterized in the dog with different clinical manifestations, which can be indistinguishable from AD. In this case, affected animals develop IgE antibodies to dietary components, most commonly beef, chicken, milk, eggs, corn, wheat and soya. The disease may be reproduced experimentally and colonies of spontaneously affected dogs have also been studied. Most dogs with dietary hypersensitivity present with a pruritic cutaneous disease that may mimic AD, but animals may also present with primary gastrointestinal disease (vomiting, diarrhoea and weight loss).

Allergic skin disorders also occur in the **cat**, but their immunopathogenesis is less well investigated and there are some species differences. The clinical presentation of cutaneous type I hypersensitivity disease (FAD, AD or

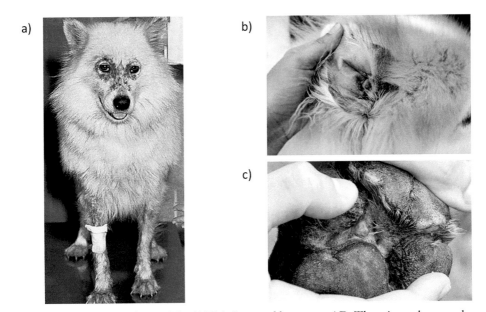

Figure 18.5 Canine atopic dermatitis. (a) This Samoyed has severe AD. There is erythema and pruritus affecting the face and forelimbs. (b) Otitis externa is a common finding in dogs affected with AD. (c) In this patient there is evidence of secondary infection between the footpads.

food allergy) is more variable, with a series of 'cutaneous reaction patterns' including **ulcerative facial dermatitis, symmetrical alopecia, 'miliary' dermatitis or lesions indicative of the 'eosinophilic granuloma complex'** (**Figure 18.6**). Some immunological features of feline allergic skin disease are similar to those described for the dog; for example, lesional skin is infiltrated by IL-4-producing T cells and dendritic cells upregulate MHC class II expression. The role of allergen-specific IgE in feline AD is not clear, as clinically normal cats may have significant levels of such antibodies. Cats may also develop sensitization to the bites of arthropods other than fleas, and mosquito bite hypersensitivity is well documented.

The most significant allergic skin disease of the horse is **insect bite hypersensitivity (IBH)**, in which there is sensitization to salivary proteins of the biting midges of the genus *Culicoides*. This leads to pruritic skin disease with self-trauma of the mane and tail regions and secondary infection (**Figure 18.7**). The disease is most prevalent during the summer months, when midges are most active. The immunology of this disease is becoming well documented. The causative salivary antigens are currently being characterized and the serum IgE response to these proteins has been measured. Lesional skin from affected Icelandic ponies demonstrates an increased number of CD4$^+$ T cells, increased IL-13 gene expression and decreased Foxp3 and IL-10 gene expression compared to normal or non-lesional skin. These findings are consistent with elevated Th2 and depressed Treg cell activity. Icelandic ponies with IBH also have fewer CD4$^+$CD25$^+$Foxp3$^+$ Treg cells in their blood than unaffected animals.

Contact allergic dermatitis is caused by a type IV (delayed type) hypersensitivity reaction to chemicals that come in contact with the skin. These chemicals are mostly small lipophilic molecules that can penetrate the epidermis. They are not immunogenic by themselves, but act as **haptens** that bind to skin proteins and possibly directly to MHC molecules. The initial

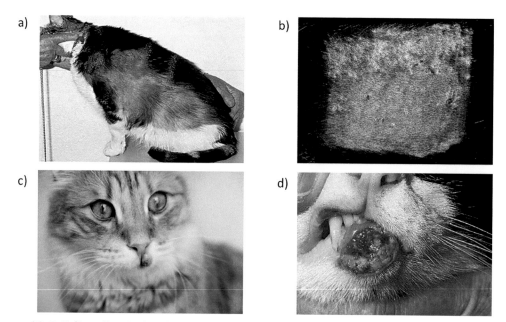

Figure 18.6 Feline allergic skin disease. The cat may develop one of a series of 'cutaneous reaction patterns' in response to allergic skin disease. (a) Symmetrical alopecia in a cat with AD. (b) Miliary dermatitis in a cat with flea allergy. (c) and (d) Two examples of cats demonstrating lesions consistent with 'eosinophilic granuloma complex'.

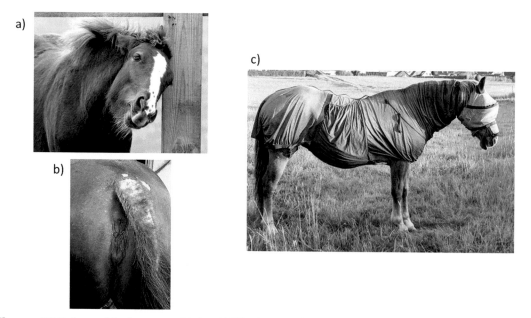

Figure 18.7 Insect bite hypersensitivity. (a) This horse is showing signs of pruritus associated with IBH. (b) This horse has lesions related to self-trauma over the tail base region. The lesions develop in summer with exposure to bites of *Culicoides* midges. (Photograph courtesy of D. Wilson.) (c) Prevention of IBH in this horse through avoidance of contact with midge bites using a rug.

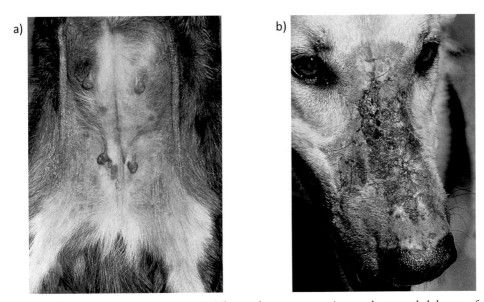

Figure 18.8 Contact allergic dermatitis. (a) The erythematous reaction on the ventral abdomen of this dog developed within 72 hours of coming into contact with a particular plant. (b) This dog had been treated with a topical aural medication containing neomycin against which it had become sensitized. Subsequently, the owner applied the same medication to a small lesion on the nose, leading to development of this extensive alopecic and crusting lesion that resolved after the medication was withdrawn.

exposure leads to the activation of **CD4⁺ and CD8⁺ T cells**, and subsequent exposure causes the accumulation of T cells that secrete cytokines and chemokines that recruit other cells into the skin (**Figure 18.8**). Clinically, skin lesions develop at the site of exposure, initially characterized by edema and erythema, and progressing to scaling and lichenification over time.

Erythema multiforme and **toxic epidermal necrolysis** are systemic skin diseases caused by the destruction of keratinocytes in the skin and mucosal tissues by CD8⁺ T cells and NK cells. The cause of these diseases is often unknown (idiopathic) or they may be associated with neoplasms, infections or drug treatment. Toxic epidermal necrolysis is believed to involve a type IV hypersensitivity to drugs including some antibiotics. Erythema multiforme is a more benign disease which affects less than 10% of the skin, whereas toxic epidermal necrolysis affects more than 30% of the skin, is often associated with systemic clinical signs, and car-

ries a poor prognosis. The lesions of erythema multiforme often involve raised erythematous papules and plaques of the axillary and groin skin, ears, footpads and mucocutaneous junctions. Epidermal detachment, erosions and ulcers are common manifestations of toxic epidermal necrolysis.

18.5 RESPIRATORY DISEASE

Just as asthma in human patients is not a single entity, but rather a collection of diseases with common clinical manifestations, but different underlying pathogenesis, **equine asthma** refers to a group of non-infectious, chronic inflammatory diseases of the lower airways in horses. It can be subclassified into a **mild to moderate type** that occurs mainly in young adult racehorses (also referred to as **inflammatory airway disease**) and a **severe type** in older horses (also referred to as **recurrent airway obstruction**). As more research is conducted, it is likely

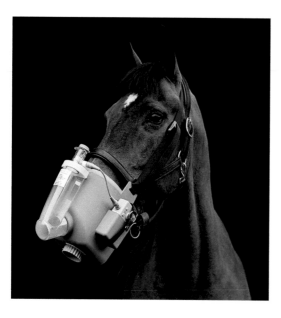

Figure 18.9 **Horse receiving bronchodilator therapy via a nebulizer.** (Image courtesy of B. Hogan, Creative Design Lead, Haygain.)

that additional subclassifications will emerge. Equine asthma is characterized by broncho-constriction, inflammation and remodelling with increased mucus production and thickening of the bronchiolar smooth muscle. The pathogenesis is not well understood, but equine asthma is associated with **respiratory exposure to dust**, **endotoxins**, **mould and particulate matter**. Horses housed in stables and fed hay are at increased risk. Eosinophils are increased in bronchoalveolar lavage fluid of some horses with mild asthma, while inflammation in horses with severe asthma is dominated by neutrophils. The condition may be treated using bronchodilator drugs (e.g. beta-2 adrenoreceptor agonists), which can be administered to horses by use of a nebulizer (**Figure 18.9**). **Asthma is also a common condition in cats**. Affected cats may present in severe respiratory distress with open mouth breathing and increased respiratory rate (status asthmaticus), or develop a chronic bronchitis with wheezing, coughing and dyspnea. There is an **increase of eosinophils** in bronchoalveolar lavage fluid. The diagnosis includes ruling out parasitic diseases, in particular heartworm infection and aelurostrongylosis (feline lungworm).

18.6 GASTROINTESTINAL DISEASE

The gastrointestinal tract is routinely exposed to environmental antigens from proteins in the diet as well as the huge amount of foreign antigen associated with the gut microbiome. Under normal conditions, the immune response is able to manage this situation at a local level, both in terms of innate immunity (gamma-delta T cells, tissue macrophages and mast cells) and adaptive immunity (intraepithelial lymphocytes, Th2/Th3/Treg and B cells) that permits mucosal tolerance to innocuous antigens and a primarily antibody-mediated (IgA and IgE) response to foreign antigens associated with pathogens. However, there are occasions where mucosal immunity goes awry, leading to gut pathology and chronic diarrhoea. **Gluten-sensitive enteropathy** can occur in Irish Setter dogs. Although the trigger (wheat gluten) is similar to that seen in human coeliac disease, the underlying pathogenesis is likely to be different and difficult to classify in terms of the Gell and Coombs classification system, with a mixed type of hypersensitivity reactions likely contributing to the tissue pathology. Affected dogs present with chronic diarrhoea and weight loss that responds to a gluten-free diet.

Inflammatory bowel disease (IBD) is a relatively common condition in dogs. A histological diagnosis, based on the type of cellular infiltration seen in the mucosa/submucosa is important (**Figure 18.10**) and helps to differentiate this condition from intestinal lymphoma. The various histological features of IBD indicate that this is a heterogeneous disease and likely involves a mixture of type III and type IV hypersensitivities. Clinically, IBD can be divided into food-responsive enteropathy (i.e. animals respond positively to a hypoallergenic diet) or antibiotic-responsive enteropathy (i.e. animals respond positively to antimicrobial drug therapy). The latter probably represents a hypersensitivity reaction to antigens associated with the microbial flora. There is evidence of dysbiosis in these animals, which may predispose to a hypersensitivity reaction and although antibiotic therapy may go some way to correcting this situation, an alternative approach of 'faecal transplantation' has shown promise for addressing the underlying dysbiosis, moderat-

ing the immune reaction and improving the clinical signs.

IBD is common in German shepherd dogs and research into the disease in this particular breed has improved our understanding of the immunopathology associated with this condition. It appears that IBD involves a complex series of defective mucosal mechanisms that impact on barrier function (reduced production of antimicrobial peptides, increased permeability), innate immunity (pattern recognition receptor defects in NOD2 and/or TLR-5) and adaptive immunity (secretory IgA deficiency). These deficiencies may lead to 'antigen overload' from dietary or microbiome (bacterial overgrowth/dysbiosis) sources that induces a switch from mucosal immunity/tolerance (mediated by relatively anti-inflammatory cells such as Th2/Th3/Treg cells and M2-type macrophages) to a more pro-inflammatory response (mediated by Th1/Th17 cells and M1 macrophages) (**Figure 18.11**). **Anal furunculosis/ perianal fistula** in German shepherd dogs (**Figure 18.12**) probably has a similar immunopathogenesis to IBD, but is possibly more

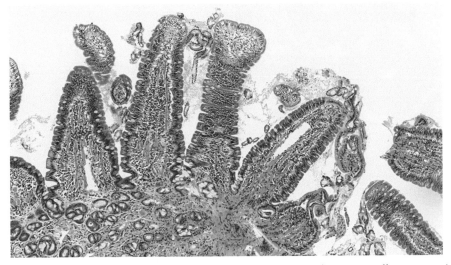

Figure 18.10 Histopathology of IBD. A histopathological diagnosis of IBD is usually accompanied by a description of the predominant cell type infiltrating the mucosa/submucosa. In this example there is a moderate lymphoplasmacytic infiltrate in the lamina propria. (Image courtesy of J. Williams, RVC.)

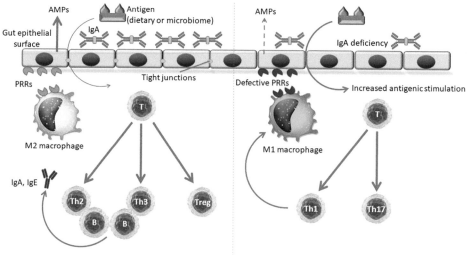

MUCOSAL 'HOMEOSTASIS'   INFLAMMATORY BOWEL DISEASE

Figure 18.11 Pathogenesis of IBD. Mucosal 'homeostasis' involves a protective barrier (production of antimicrobial peptides [AMPs], tight junctions between enterocytes, secretory IgA), innate (pattern recognition receptors, submucosal M2 macrophages) and adaptive immunity (mediated by Th2, Th3 and Treg cells and production of IgA/IgE antibodies from B cells). Susceptibility to IBD involves a number of factors that can impact on the barrier function, including reduced secretion of AMPs, selective IgA deficiency and increased permeability, leading to 'antigen overload'. There may also be innate immune deficiencies resulting from mutations in PRRs such as TLR-5 and NOD2. This leads to a shift in the immune response to a more pro-inflammatory phenotype, induced by Th1 cells, Th17 cells and M1 macrophages of cell-mediated immunity. In essence, this represents development of a type IV hypersensitivity reaction to either dietary or microbial antigens in the gut.

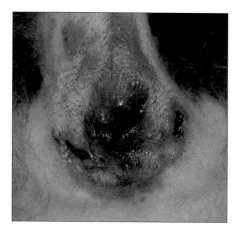

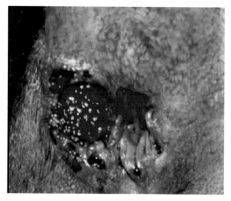

Figure 18.12 Anal furunculosis (AF)/perianal fistula in German shepherd dogs. The picture on the left is from a dog affected with AF showing fistulous lesions that extend deep into the perianal tissues. The picture on the right is from an AF patient showing primarily ulcerative lesions of the perineum, potentially associated with the production of matrix metalloproteinases from M1 macrophages within the lesional tissues. Both dogs responded well to ciclosporin therapy. (Photographs courtesy of Arthur House.)

of a type IV hypersensitivity directed towards antigens associated with the faecal microflora, rather than the intestinal microbiome. The immune-mediated nature of the pathology is evidenced by the clinical response to ciclosporin therapy in affected animals.

18.7 DIAGNOSIS OF ALLERGIC DISEASE

Diagnosis of allergic disease in animals requires the clinician to evaluate the patient carefully. Most important is a detailed **clinical history** taking into account the breed and age of the animal, its environment, diet and the management of ecto- and endoparasites. The initial stages of diagnosis may involve excluding contact with known allergens to determine whether there is any clinical improvement. A response to a simple course of animal and environmental **flea control** would be consistent with a diagnosis of FAD. Placing the animal on a **restricted protein home-cooked diet** or a **commercial**

hypoallergenic or hydrolyzed protein diet should lead to alleviation of the clinical signs of dietary hypersensitivity. These diets work on the principle of feeding a source of protein to which the animal has not been previously exposed (restricted protein) or protein fragments that are too small to enable cross-linking of mast cell-associated IgE (hydrolyzed protein). Gradual reintroduction of elements of the previous diet may then identify the causative allergen/s, but once disease is in remission such dietary challenge is often not performed. Similarly, for diseases such as contact allergy, removal from the home environment often leads to remission and phased reintroduction to areas of the environment can potentially identify the causative allergen.

Diagnosis of AD involves undertaking these initial steps in order to rule out FAD or food allergy. It is necessary to identify the causative allergen(s) responsible for disease in an animal when avoidance or **allergen-specific immunotherapy (ASIT)** are being considered as a

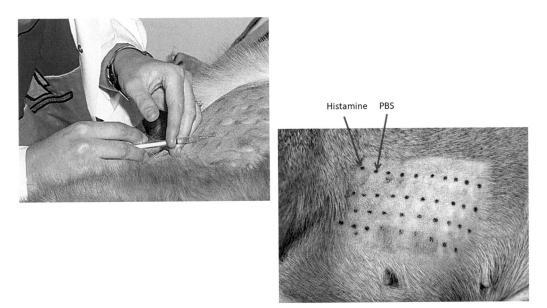

Figure 18.13 Intradermal skin test. This atopic dog is receiving a series of intradermal injections of allergens into the clipped skin over the flank. The development of a wheal within 20 minutes of injection of similar magnitude to that induced by local injection of histamine (as positive control) indicates that the dog is reacting to that allergen and has cutaneous mast cells coated by allergen-specific IgE. Phosphate buffered saline (PBS) is injected as a negative control.

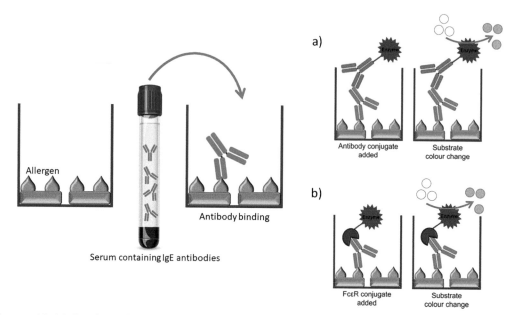

Figure 18.14 Serological testing for allergen-specific IgE. Various allergens are applied to individual wells of the test plate. Serum is added, and any IgE present will bind. The presence of IgE is detected either using (a) an anti-IgE:enzyme conjugate or (b) enzyme-labelled recombinant FcεR. Addition of a substrate leads to a colour change in the wells containing those allergens that are reactive.

treatment option. This may be achieved by the use of an **intradermal skin test** (**IDST**), where a small quantity of a panel of test allergens is injected intradermally into the skin of the patient. Injections of saline and histamine are used as negative and positive controls, respectively. The presence of dermal mast cells with bound allergen-specific IgE will be indicated by the development of an oedematous **wheal** at the site of injection within **20 minutes** of the injection being given (**Figure 18.13**). An alternative approach involves the detection of allergen-specific IgE in a blood sample, most often by **ELISA** (**Figure 18.14**). A wide range of commercial serological tests are now marketed for the diagnosis of allergic diseases in dogs, cats and horses. Studies have shown a poor correlation between the IDST and some of these serologic tests. The latter often give positive results even in normal dogs. The IDST is considered the gold standard, but serology testing is more convenient, safer and less likely to be affected by therapeutic drugs. Particular care should be taken with using serological tests that claim to be able to identify dietary hypersensitivity/food allergy as there is poor correlation between results of these tests and the response to a dietary trial.

KEY POINTS

- Allergy is a state of immunological sensitization to an innocuous environmental allergen that leads to inappropriate immune responses on re-exposure to the allergen.

- Atopy is the genetically mediated propensity of an individual to become sensitized to environmental allergens by producing allergen-specific IgE antibodies mediating type I hypersensitivity reactions.
- Atopy involves excessive production of type 2 cytokines (IL-4, IL-5 and IL-13) and insufficient Treg cell activity.
- There are breed predispositions for allergic disease in dogs and horses, indicating a genetic basis, but the specific genes have not been fully identified.
- The onset of clinical signs in canine AD is typically at a young age.
- Secondary infection may exacerbate the pathology induced by the allergic reaction.
- Major allergic diseases of animals include urticaria, angioedema, anaphylaxis, FAD, AD and food allergy of dogs and cats, and urticaria, IBH and asthma of horses.
- Inflammatory bowel disease and anal furunculosis are complex inflammatory conditions that likely represent a hypersensitivity towards dietary proteins and/or microbial antigens associated with the gut microbiome.
- Diagnosis of allergic disease involves clinical history, ruling-out ectoparasite or food allergy by exclusion. IDST and serology can be used to identify the allergens that induce the allergic reaction in some cases.

CHAPTER 19

AUTOIMMUNITY AND AUTOIMMUNE DISEASES

OBJECTIVES

At the end of this chapter, you should be able to:

- Discuss the immunological basis for autoimmunity.
- Discuss how genetic factors can predispose to autoimmune disease.
- Discuss the environmental factors that might predispose to autoimmune disease.
- Describe the spectrum of antibody-mediated and cell-mediated autoimmune diseases in animals.
- Describe the principle and interpretation of the Coombs test, antinuclear antibody (ANA) test and thyroglobulin autoantibody test.

19.1 INTRODUCTION

Autoimmunity is defined as the presence of **autoantibodies** and/or **autoreactive T cells** that react with self-antigens expressed in normal tissues. The presence of such autoantibodies and autoreactive T cells is relatively common in healthy animals, indicating that **central tolerance** mechanisms are not perfect and regulatory mechanisms associated with **peripheral tolerance** are important. **Autoimmune disease** represents the progression of tissue pathology and development of clinical signs that results from a **failure of self-tolerance and regulatory processes** allowing activation of self-reactive B and T lymphocytes (**Figure 19.1**). In this chapter we will explore the immunological and clinical basis for autoimmune diseases in domestic animals.

The pathogenesis of autoimmune disease is a **multifactorial** process involving a number of interlinked **predisposing and triggering factors** (**Figure 19.2**). Several of these factors are required before autoimmune disease occurs. The key elements of the autoimmune response are:

- A susceptible **genetic background**.
- **Predisposing factors** including age, sex, lifestyle and diet.
- **Environmental factors** such as exposure to UV irradiation (for some diseases), chemicals (including drugs) or infectious agents that may act as triggers for the autoimmune process.
- A **breakdown of immunological control mechanisms** permitting the activation of autoreactive lymphocytes and failure to properly regulate their responses.

A wide spectrum of autoimmune diseases exists, as there are many self-antigens that may

DOI: 10.1201/9781003310969-19

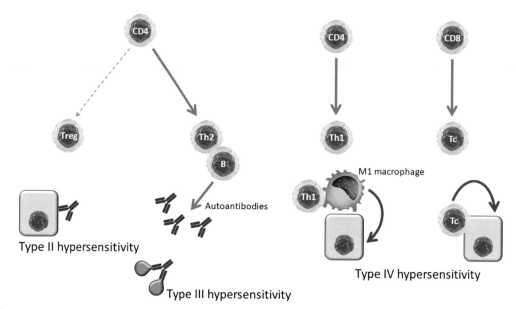

Figure 19.1 Mechanisms of autoimmunity. Autoimmunity occurs when inappropriate activation of autoreactive T and B lymphocytes occurs alongside deficient Treg cell activity. Autoreactive Th2 cells may amplify autoantibody production and direct cellular damage (type II hypersensitivity) or immune complex disease (type III hypersensitivity). Alternatively, autoreactivity involving cell-mediated immunity (CD4+ Th1 or CD8+ Tc cells) can lead to cellular injury and tissue damage.

Figure 19.2 Factors contributing to risk of autoimmune disease.

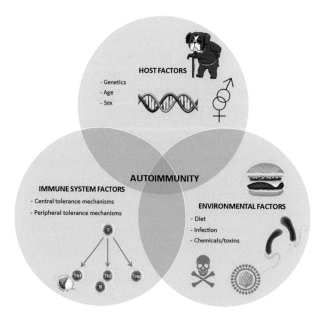

potentially become targets for an autoimmune response. Virtually any body system can be affected by autoimmune disease. Most autoimmune diseases are '**organ specific**' and target autoantigens related to a single body system

or organ, but some are '**multisystemic**' and involve two or more body systems. Human autoimmune diseases are of major medical and economic significance with a global prevalence of 3 to 5% of the population. Examples of human autoimmune diseases include type 1 diabetes, rheumatoid arthritis, systemic lupus erythematosus (SLE), myasthenia gravis, Hashimoto's thyroiditis, Graves' disease, pemphigus and multiple sclerosis. In veterinary medicine, autoimmune diseases are **most prevalent in the dog**. They occur with less frequency in cats and horses and are only sporadically in production animals, most likely due to the fact that these latter species generally do not live to an age at which autoimmunity becomes apparent.

19.2 THE GENETIC BASIS FOR AUTOIMMUNITY

There is a clear **genetic predisposition** to autoimmune disease in people and animals. Particular **breeds of dog** are susceptible to certain autoimmune diseases and these often occur within pedigree lines (**Figure 19.3**). This breed susceptibility is often recognized internationally, due to the '**founder effect**' whereby the entire breed can be traced back to relatively few (often related) animals. These 'founders' may have been selected for their physical characteristics and attributes, but may also carry disease-associated genetic variants that are then passed down to subsequent generations. Thus, selection for a particular phenotype (such as coat colour, ear conformation etc.) tends to drive the genes responsible towards homozygosity, but if

there is a deleterious mutation in '**linkage disequilibrium**' with that particular gene, it also becomes highly prevalent in the breed population. A classic example of a predisposed breed is the cocker spaniel, which is recognized internationally as highly susceptible to various types of autoimmune disease.

While a few autoimmune syndromes have been identified in people caused by a single gene mutation (e.g. *AIRE* in human autoimmune polyendocrine syndrome), genetic susceptibility to autoimmune diseases is typically polygenic and complex, with individual genes having a relatively small effect, but which is cumulative. The genes most strongly linked to susceptibility towards autoimmunity are those of the MHC. **Disease associations with MHC class II haplotypes** have been established in humans and dogs and in some cases, it is estimated that these may represent around 50% of the genetic risk (e.g. for HLA-DR3 and HLA-DR4 in human type 1 diabetes). The reason for such a close association between autoimmunity and the MHC might be explained by their important role in positive and negative selection in primary lymphoid tissues and antigen presentation in secondary lymphoid tissues. Alternatively, the MHC region contains many other immune response genes (including, for example, *TNF*) that are in linkage disequilibrium and might also contribute to genetic risk of autoimmunity.

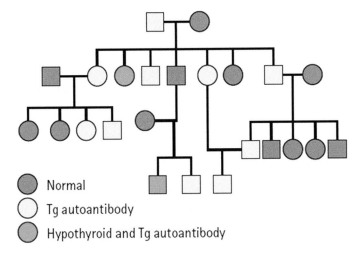

Figure 19.3 Inheritance of canine autoimmunity. This pedigree shows the inheritance of hypothyroidism associated with the presence of serum autoantibody to thyroglobulin through three generations of a family of Great Danes. (Data from Haines DM, Lording PM, Penhale WJ (1984) Survey of thyroglobulin autoantibodies in dogs. *American Journal of Veterinary Research* **45**:1493–1497.)

⬤ Normal

◯ Tg autoantibody

◓ Hypothyroid and Tg autoantibody

Polymorphisms in other immune response genes are likely to contribute to the overall genetic risk of autoimmunity, such as *AIRE*, which is involved in self-antigen expression in the thymus, and *CTLA4*, which is associated with downregulation of T cell responses and which is constitutively expressed by Treg cells. A genetic association with autoimmunity has also been made with **IgA deficiency**. An inherited inability to produce sufficient IgA results in weakened mucosal defence and greater susceptibility to infection. As will be described later, infectious agents may trigger autoimmunity, thus potentially explaining the association between IgA deficiency and autoimmune disease. Whereas genetic variants in immune response genes may increase the overall risk of autoimmunity, other more tissue specific genes may be important in determining the nature of the autoimmune response and the target tissue. This is certainly the case in type 1 diabetes in people, whereby polymorphisms in the insulin gene are major genetic risk factors, after taking into account the effect mediated via the MHC genotype.

Although the genetic background strongly determines the inherent susceptibility to autoimmune disease, it is not the sole determinant and there is incomplete penetrance, even in those individuals that are carrying multiple genetic risk factors. This is well documented in twin concordance studies in humans, whereby identical twins that are separated at birth, often only one twin will develop autoimmune disease, indicating that environmental influences have a major role to play.

19.3 INTRINSIC/HOST FACTORS IN SUSCEPTIBILITY TO AUTOIMMUNE DISEASE

In both humans and dogs, autoimmune disease most commonly occurs in **middle to older age**. The reason likely relates to age-associated changes in immune function (**immunosenescence**). Older animals retain the ability to make humoral and recall immune responses, often having elevated IgG production. In contrast, there is a reduced capacity for primary immune responses, associated with a shift in the balance of circulating T cells, with relatively more CD8$^+$ cells and fewer CD4$^+$ T cells, more memory T cells and fewer naïve T cells. If the declining population of CD4$^+$ cells includes Tregs, this could potentially account for an increased propensity towards autoimmune reactivity.

Many human autoimmune diseases are more common in females than in males. However, this does not appear to be the case for most canine autoimmune diseases, where there is little evidence to support a sex bias. This may, in part, relate to the fact that many companion animals are neutered. Epidemiological studies suggest that neutered dogs are more likely to be affected with immune-mediated diseases than sexually intact dogs. The hormonal influence on autoimmunity has been studied in experimental rodent models. For example, particular strains of laboratory rats will develop lymphocytic thyroiditis and diabetes mellitus if they are thymectomized early in neonatal life. These rats have a disturbed balance of T cells (particularly lacking in nTregs), which is likely to be related to the onset of these autoimmune syndromes. The incidence of thyroiditis is much greater in female compared with male animals. If female rats are ovariectomized and then injected with oestrogen, the incidence of thyroiditis significantly decreases, but when ovariectomized rats are injected with progesterone the incidence of disease becomes even greater than in an intact animal, highlighting the role of this latter steroid hormone as a risk factor.

There are several examples of situations in which an autoimmune disease might arise concurrently with or subsequent to neoplasia or the presence of a chronic inflammatory disease. For example, immune-mediated cytopenias might arise secondary to a range of tumours (lymphoma, splenic haemangiosarcoma, myeloproliferative disease of the bone marrow) or to chronic pancreatitis or inflammatory bowel disease. The precise mechanisms underlying these associations are, however, poorly defined.

19.4 EXTRINSIC/ENVIRONMENTAL FACTORS IN SUSCEPTIBILITY TO AUTOIMMUNITY

Exposure to environmental risk factors may be associated with an increased risk of autoimmunity in animals that have other genetic and host risk factors. **UV light** is linked to the development of canine nasal cutaneous lupus erythematosus (CLE) (**Figure 19.4**). It is proposed that UV light might in some way modify the antigenic structure of target autoantigens within the skin. Chemical agents that act as haptens may have a similar effect when they bind to self-antigens (**Figure 19.5**).

19.4.1 Drugs and Vaccines as Triggers for Autoimmunity

There is little doubt that some **drugs** may be able to precipitate an autoimmune reaction. Antimicrobial drugs are most implicated and

Figure 19.4 Environmental triggers of autoimmunity. This dog has nasal CLE, a complex immune-mediated disease targeting the skin of the planum nasale. Immunological mechanisms involved in this disease include a lymphocytic infiltration of the junction between epidermis and dermis and immune complex deposition at the basement membrane zone of the epidermis. UV light is thought to act as one trigger for the occurrence of this disease.

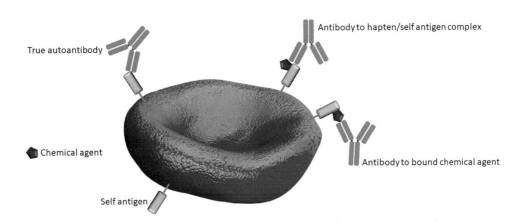

Figure 19.5 The 'innocent bystander' effect. A hapten that attaches to protein(s) on the surface of a host cell (such as an erythrocyte) directly provides a target for the adaptive immune response, or can combine with a self-antigen to form a novel target ('modified self'). The immune response, in this case antibody and classical complement, will destroy the host cell after binding to the specific target, in a process known as 'bystander destruction'.

trimethoprim–sulphonamides (TMS) are particularly recognized as triggers of IMHA, thrombocytopenia and neutropenia in dogs. In these cases, the drug may be acting as a **hapten**, which by binding to the surface of a host cell, directly forms a target for an immune response. Alternatively, by modifying the structure of a self-protein the drug forms a novel drug–protein combination that is also capable of inducing an immune response (**Figure 19.5**). Dobermann dogs seem to be particularly predisposed to this effect, but TMS drugs may induce immune-mediated cytopenia in a range of other dog breeds. Antibiotics and topical ectoparasitic drugs have also been associated with pemphigus foliaceus (an autoimmune skin disease) in dogs. In feline medicine the drugs used for medical management of hyperthyroidism (i.e. carbimazole/methimazole) have been recognized as triggers for IMHA, IMTP and serum ANA production.

There is a great deal of interest in both human and veterinary medicine as to whether vaccines might also trigger autoimmune disease. Although some genetically susceptible dogs might develop immune-mediated cytopenias or polyarthropathy one to two weeks post vaccination, such events are extremely rare and a causal relationship is unclear. A more common occurrence is the induction of serum autoantibodies to a range of connective tissue or epithelial proteins following vaccination, but there is little evidence that these antibodies are pathogenic and a cause of clinical disease. This phenomenon probably arises because of the incorporation of small quantities of serum proteins or cellular components used in vaccine manufacture. This may be particularly relevant with viral vaccines, where the virus has been propagated in cell culture, prior to being processed to formulate the vaccine.

19.4.2 Infections as Triggers of Autoimmunity

There is reasonable experimental evidence to support the notion that some infectious agents can trigger an autoimmune response. Several rodent models illustrate the situation whereby infection, or challenge with an antigen derived from an infectious agent, is able to induce an autoimmune reaction in a genetically susceptible strain. However, direct evidence of autoimmune disease in people or domestic animals directly linked to an infection is somewhat limited. A variety of mechanisms have been postulated to explain the possible association of infectious agents with autoimmune disease.

- The infection may result in activation of the innate immune system, inflammation and cellular damage/destruction. This may result in presentation of self as well as foreign antigens, increased expression of costimulatory molecules by dendritic cells and secretion of cytokines that can activate autoreactive T cells. This probably happens quite commonly during infection, but the activated autoreactive T cells are normally downregulated by Tregs after recovery from the infection. A decreased number or function of Tregs may allow uncontrolled propagation and activation of such autoreactive T cells, resulting in autoimmune disease, postinfection. Inflammation may also cause the release of intracellular antigens that are normally sequestered and the production of proteolytic enzymes that generate novel peptides capable of binding to MHC molecules for recognition by T cells. As these peptides were not generated in the thymus, T cells reactive with such '**cryptic epitopes**' are not necessarily deleted. While the autoreactive immune response may initially be directed against a single or small number of epitopes on a self-protein, the number of antigens and antigenic epitopes and, hence, the number of activated autoreactive T and B cells can increase over time (**epitope spreading**). This contributes to the progressive nature of autoimmune diseases (**Figure 19.6**).

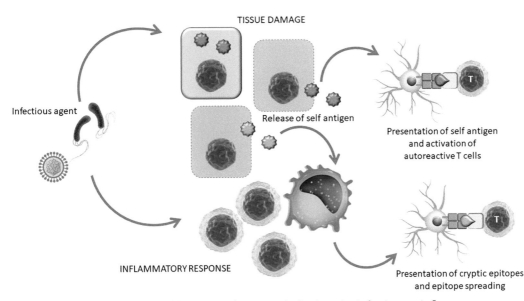

TISSUE DAMAGE

Infectious agent

Release of self antigen

Presentation of self antigen
and activation of
autoreactive T cells

INFLAMMATORY RESPONSE

Presentation of cryptic epitopes
and epitope spreading

Figure 19.6 Presentation of self-antigens during an infection. An infectious or inflammatory response within a tissue involving tissue damage and infiltration of pathogen-specific lymphocytes and local cytokine production may lead to secondary autoimmunity. Damaged tissue may release self-antigen, which can be processed and presented for activation of autoreactive lymphocytes. Alternatively, the inflammatory milieu may alter the activity of APCs to allow them to present previously 'cryptic' (hidden) self-antigenic peptides. The immune response to pathogens might extend to self-antigens via the process of 'epitope spreading'.

- Polyclonal activation of T and B lymphocytes may result in the activation of autoreactive T cells and generation of autoantibodies. Some microbial proteins act as B cell mitogens and cause polyclonal hypergammaglobulinaemia and peripheral lymphadenopathy. Superantigens are produced by certain staphylococci and streptococci, and a range of other infectious agents. A superantigen is able to non-specifically activate a subset of T cells by cross-linking the variable region of the TCR β chain with MHC class II outside the peptide-binding groove.
- **Molecular mimicry** refers to the presence of shared or similar B cell and T cell epitopes between self-antigens and microbial molecules. An appropriate immune response to the pathogen may result in inadvertent activation of autoreactive B and T cells. This process is involved in rheumatic fever in human

patients following streptococcal infections. Antibodies against streptococcal antigens cross-react with myocardial proteins and can cause myocarditis. Molecular mimicry may also be involved in **equine recurrent uveitis** following leptospirosis. Antibodies against *L. interrogans* have been found to crossreact with lens, uveal and retinal proteins (see also Chapter 15).
- Another possible mechanism involves microorganisms that attach to the membrane of host cells or infects that cell, leading to expression of microbial antigen on the cell surface. An appropriate adaptive immune response is made to the infectious agent, but this also destroys the host cell that carries the organism, leading to exacerbation of the clinical signs. This effect may be important in diseases such as feline infectious anaemia (**Figure 19.7a**), where circulating erythrocytes are parasitized by the surface membrane-dwelling

a)   b)

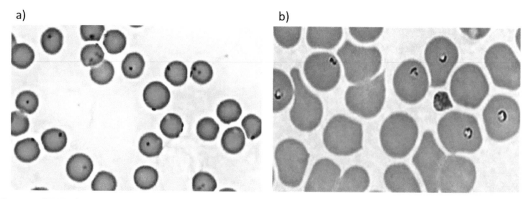

Figure 19.7 **Autoimmunity associated with infection.** An infectious agent that attaches to the surface of a host cell (e.g. an RBC) directly provides a target for the adaptive immune response or combines with a self-antigen to form a novel target ('modified self'). The immune response, in this case antibody and classical complement, will destroy the host cell in addition to the specific target in a process known as 'bystander destruction'. (a) An example of such an infection is feline infectious anaemia in which *Mycoplasma haemofelis* parasitizes the surface of circulating erythrocytes. Cats with this infection become Coombs' test positive. The red cell-bound antibody might be specific for the parasite, an erythrocyte membrane protein (autoantigen) or a novel epitope formed through the infection. (b) The erythrocytes from this dog are parasitized by one of the small forms of Babesia. Such dogs may become Coombs' test positive and it is suggested that the antibodies associated with erythrocytes are true autoantibodies with specificity for red cell membrane antigens. It is conceivable that a Babesia antigen carries an epitope that acts as a molecular mimic for an epitope within a red cell membrane protein.

organism *Mycoplasma haemofelis*. Cats with this disease are generally Coombs' test positive (i.e. have antibody associated with the surface of RBCs) and the infection and anaemia precede the development of these erythrocyte-bound antibodies. Although it is possible that the antibodies target the infectious agent, there are two alternative explanations. The first is that the infection triggers the production of true autoantibodies and the second suggests that by binding to an autoantigen, the infectious agent changes the structure of that antigen, making it a target for an immune response (the **'modified self'** hypothesis, as may also occur with the binding of a drug as discussed previously).

In addition to feline haemoplasmosis, there are a number of other examples of companion animal autoimmune disease that might be trig-gered by infectious agents. The best examples relate to the group of arthropod-transmitted infectious agents (e.g. *Babesia, Ehrlichia, Leish-mania, Borrelia, Anaplasma* and *Rickettsia*). The complex interplay between these infectious agents, the arthropods that transmit them and the host immune system often leads to **second-ary immunopathology** in the chronic stages of infection. These infections are often associated with clinical signs such as polyarthritis, uveitis, glomerulonephritis, anaemia and thrombocyto-penia and the presence of autoantibodies spe-cific for RBCs or platelets, or the presence of serum circulating immune complexes or ANA (**Figure 19.7b**).

19.5   AUTOIMMUNE DISEASES MEDIATED BY AUTOANTIBODIES

Some canine autoimmune diseases are primarily mediated by autoantibodies and include those representative of a type II hypersensitivity, namely immune-mediated haemolytic anaemia (IMHA), immune-mediated thrombocytopenia (IMTP), autoimmune skin diseases of the pemphigus complex and myasthenia gravis. In addition, there are other autoimmune diseases that are mediated by a type III hypersensitivity, namely autoimmune non-erosive polyarthritis and SLE. These canine autoimmune diseases are **analogous to the equivalent human disorders**, with a similar immunological basis, tissue pathology and clinical presentation. For this reason, there is currently great interest in the study of these canine autoimmune disease as spontaneous models of the same human disorder.

IMHA involves the destruction of erythrocytes mediated by IgM or IgG autoantibody and the classical pathway of complement (type II hypersensitivity) (**Figure 19.8**). IMHA may also be associated with type III hypersensitivity to non-erythrocyte antigens, whereby immune complexes can become deposited on the surface of the red blood cells, triggering the complement system with subsequent cellular damage and destruction. In IMTP, autoantibody destruction of platelets leads to thrombocytopenia and a bleeding disorder characterized by the presence of petechial haemorrhages, ecchymoses (**Figure 19.9**), melaena and/or epistaxis.

There is a range of (somewhat similar) autoimmune skin diseases that come under the general classification of the **pemphigus complex**. In these disorders, autoantibodies bind to targets associated with keratinocyte cell adhesion (e.g. hemidesmosomes and desmosomes) leading to failure to maintain the integrity of the stratified squamous epithelium of the epidermis and resulting in formation of vesicles, ulcers, erosions and crusting lesions (**Figure 19.10**).

Multisystemic autoimmune disease (also known as systemic lupus erythematosus; SLE) is a type III hypersensitivity resulting from generation of autoantibodies against nuclear

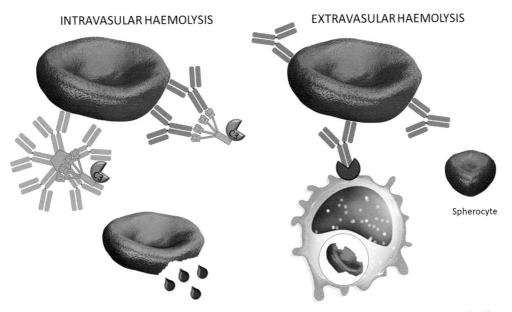

INTRAVASULAR HAEMOLYSIS   EXTRAVASULAR HAEMOLYSIS

Spherocyte

Figure 19.8 Immune-mediated haemolytic anaemia. Production of autoantibodies against red cell surface antigens (type II hypersensitivity) leads to intravascular haemolysis, mediated by complement (IgM and IgG) and extravascular haemolysis mediated by opsonization (IgG) and phagocytosis by splenic and hepatic macrophages. Incomplete phagocytosis results in formation of 'spherocytes', which are an important indicator of IMHA on haematology.

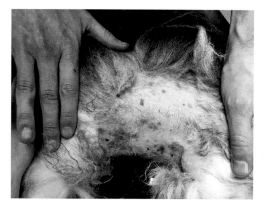

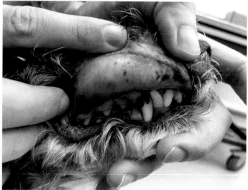

Figure 19.9 Clinical presentation of IMTP. This dog has a very low platelet count associated with the presence of anti-platelet antibodies that lead to their destruction and the appearance of ecchymoses and petechial haemorrhages. (Images courtesy of Dr Barbara Glanemann, Royal Veterinary College)

Figure 19.10 Autoimmune skin disease. This dog is showing typical mucocutaneous ulcerative/crusting lesions associated with autoimmune skin disease of the pemphigus complex.

antigens, such as DNA and histone proteins. Immune complex formation and their intravascular deposition lead to pathology, particularly in the kidney, where glomerulonephritis occurs. Immune complex deposition and vasculitis in the skin can lead to clinical signs and immune complex deposition in the synovial capillaries can lead to a non-erosive polyarthropathy. These immune complexes may also attach to the surface of erythrocytes and platelets and SLE patients can also show signs consistent with IMHA and/or IMTP.

19.6   AUTOIMMUNE DISEASES MEDIATED BY AUTOREACTIVE T CELLS

Autoimmune diseases involving cell-mediated immunity (type IV hypersensitivity) include rheumatoid arthritis/erosive polyarthritis, hypothyroidism, hypoadrenocorticism (Addison's disease), exocrine pancreatic insufficiency and keratoconjunctivitis sicca. These disorders are typically associated with lymphocytic (primarily T cell) and monocyte/macrophage infiltration of the tissues and an inflammatory reaction that leads to cellular damage and destruction. In rheumatoid arthritis (**Figure 19.11**), Th1 cell release of IFN-gamma leads to activation of synovial macrophages and production of pro-inflammatory mediators (including TNF) and matrix metalloproteinases, the latter associated with development of erosive lesions of the articular cartilage.

Lymphocytic thyroiditis involves the destruction of thyroid follicular epithelium by a cell-mediated immune reaction (Th1 and cytotoxic T lymphocytes) (**Figure 19.12a, b**). Subclinical lymphocytic thyroiditis (evidenced by the presence of thyroglobulin autoantibodies) probably occurs some time before there is clinical evidence of hypothyroidism associated with a deficiency in production of thyroxine (T4). Thyroglobulin autoantibodies are likely to represent a consequence of tissue destruction and release of self-antigens from the colloid

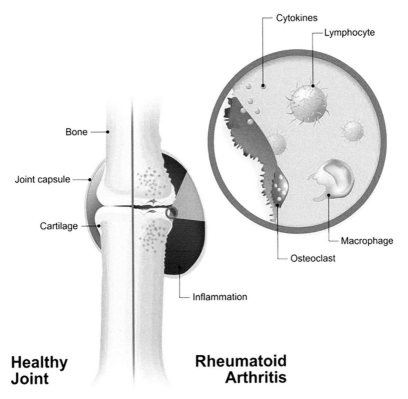

Figure 19.11 Rheumatoid arthritis. Lymphocytic infiltration of the synovium and subsequent 'pannus' formation leads to production of pro-inflammatory mediators and matrix metalloproteinases that are responsible for degradation of type II collagen and an erosive arthropathy.

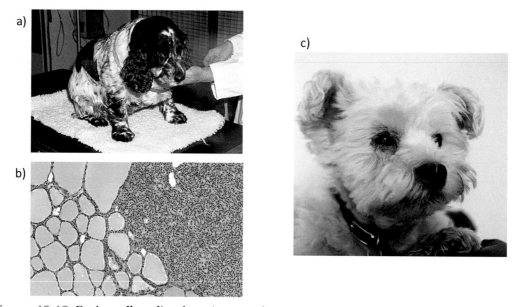

Figure 19.12 Canine cell-mediated autoimmune disease. (a) Cocker spaniel dog affected with hypothyroidism. (b) Histopathology of the thyroid gland showing lymphocytic thyroiditis. As more tissue is destroyed by the cell-mediated autoimmune reaction, this eventually leads to thyroid atrophy. (c) West Highland white terrier affected with keratoconjunctivitis sicca ('dry eye') caused by lymphocytic infiltration and pathology of the lacrimal gland.

(i.e. epitope spreading), rather than the cause of the thyroid pathology itself (which is mediated by T cells). Similarly, in hypoadrenocorticism (Addison's disease) the pathology is indicative of lymphocytic adrenalitis, leading to end stage adrenocortical atrophy, with evidence for a 'bystander' autoantibody response (e.g. against P450 side-chain cleavage enzyme and 21-hydroxylase). Lymphocytic infiltration and an inflammatory process in the lacrimal glands leads to clinical signs associated with keratoconjunctivitis sicca ('dry eye'), which can be treated with artificial tears and topical ciclosporin eye drops (**Figure 19.12c**). Exocrine pancreatic insufficiency is another example, whereby organ function is compromised as a result of a cell-mediated autoimmune pathology in the target tissue, which in this case results in a failure of the pancreatic exocrine tissue to produce digestive enzymes, leading to chronic diarrhoea and weight loss.

19.7 DIAGNOSTIC TESTS FOR AUTOIMMUNE DISEASE

A clinical suspicion of autoimmune disease should be supported by immunodiagnostic testing where possible, to demonstrate the presence of an autoimmune response with specificity for self-antigen within the target tissue or organ. In veterinary medicine there are relatively few commercially available tests available for the diagnosis of autoimmune disease and these are mostly designed to detect circulating (serum) or cell-bound autoantibody. These tests include:

- The Coombs' test (or direct antiglobulin test, DAT) for the detection of erythrocyte-bound autoantibodies.
- Tests for the detection of anti-platelet or anti-neutrophil antibodies.
- Radioimmunoassay for the detection of antibody to the neuromuscular

acetylcholine receptor (AChR) in
myasthenia gravis.

- ELISA for the detection of thyroglobulin,
  T3 or T4 autoantibodies.
- Indirect immunofluorescence,
  immunoprecipitation or ELISA for the
  detection of serum ANA.
- Indirect agglutination or ELISA for
  the detection of serum or synovial fluid
  rheumatoid factor (RF).
- Tissue biopsy immunofluorescence or
  immunohistochemistry for the detection
  of cell-associated autoantibody or immune
  complex *in situ*; for example, interepithelial
  adhesion molecule autoantibody in
  pemphigus foliaceus or glomerular
  immune complex in membranous
  glomerulonephritis.

Although a range of other autoantibody detection methods has been documented in the scientific literature, these tests are not widely available in clinical practice. Examples of the most commonly performed tests are discussed next.

19.7.1 Direct Agglutination (Coombs') Test

The Direct Agglutination Test (DAT) is also named Coombs' test in recognition of Robin Coombs, a veterinarian, who was instrumental in developing this technique. The test aims to demonstrate the presence of antibody and/or complement attached to the surface of circulating erythrocytes, indicative of IMHA. The DAT relies on the principle of **agglutination** discussed in Chapter 10. When erythrocytes coated by antibody are incubated with antiserum, this will cause cross-linkage, resulting in the suspended red cells being held in a lattice-like arrangement, which is seen visually as agglutination (**Figure 19.13**).

An EDTA anticoagulated blood sample is collected from the patient that is suspected of having IMHA. The sample is centrifuged and the plasma and buffy coat removed by pipette. The pelleted erythrocytes are resuspended in phosphate buffered saline (PBS) and re-centrifuged. This process is repeated to wash the red cells of plasma protein. First, a **saline agglutination test** is performed to determine whether spontaneous agglutination occurs. The blood is diluted with saline 1:2 to 1:4 and evaluated microscopically. If no agglutination is observed, a DAT is performed.

The **antiserum** generally used in the DAT has multiple specificities for **IgG, IgM** and **complement C3**. The test may also be performed with individual antisera specific for these three immunoreactants. The Coombs' test is performed in the round-bottomed wells of a plastic microtitration plate. Each antiserum is two-fold diluted (from 1/2 or 1/5 starting dilution) across the rows of the plate. A set of negative control wells containing only PBS is also included. An equal volume of the red cell suspension is added to the wells of serially diluted antiserum and PBS control. The reactants are then mixed and the plate incubated at an appropriate temperature, most often 37°C, although a duplicate plate may be incubated at 4°C. Following incubation, the plate is examined for the presence of agglutination and the titre of reaction with each antiserum is determined. Negative results at high concentration of the antiserum may occur due to the **prozone effect** (excess of antibodies). The titre obtained does not necessarily correlate with the severity of anaemia, but the pattern and temperature of reactions may correlate with the clinical presentation and pathogenesis of haemolysis. When performed correctly, false-negative results are uncommon. Flow cytometry is an alternative methodology whereby red cell-bound antibody is detected by antibodies conjugated with a fluorescent label.

19.7.2 Antinuclear Antibody (ANA) Test

The ANA test aims to demonstrate the presence of serum autoantibodies specific for one or more

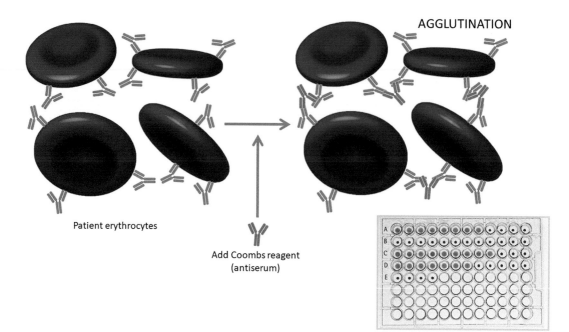

AGGLUTINATION

Patient erythrocytes

Add Coombs reagent
(antiserum)

Figure 19.13 Coombs' test. Erythrocytes from an animal affected with IMHA are coated by antibody and/or complement *in vivo*. Incubation of these cells with antiserum specific for the immunoreactants leads to agglutination of the cells, which may be identified visually. Inset: a canine Coombs' test performed in a microtitration system. Four antisera are used in this test: Row A, polyvalent canine Coombs' reagent; Row B, anti-dog IgG; Row C, anti-dog IgM; and Row D, anti-dog complement C3. Each antiserum is double-diluted across the plate from a starting dilution of 1/5 in well number 1 to a final dilution of 1/10,240 in well number 12. Row E contains PBS as a negative control. An equal volume of a 2.5% suspension of washed patient red cells is added to all the wells and the plate is mixed and incubated at either 4° or 37°C. Agglutination is seen as a 'shield' of cells spread over the base of the well, whereas in the absence of agglutination, the cells form a 'button' at the bottom of the round-bottomed well. This dog is Coombs' test positive, with red cells coated by IgM (titre 1,280) and complement C3 (titre 320).

of the constituents of the nucleus, including single and double-stranded DNA, histone proteins and chromatin. Although it is possible to determine the fine specificity of the reaction by other techniques such as ELISA, the ANA test performed most widely simply seeks evidence for the presence of any autoantibodies reactive against nuclear antigens. A proportion of normal animals or animals with chronic disease may demonstrate low-titre serum ANA (usually IgM) as a reflection of cell turnover in the body. In the context of autoimmune disease, only a high-titre (IgG) ANA is regarded as significant.

The presence of **high-titre serum ANA** is one of the diagnostic criteria for the rare multisystemic autoimmune disease **SLE**, but might also be present in animals with a range of other autoimmune disorders.

The detection of serum ANA is generally performed by the **indirect immunofluorescence assay** (see Chapter 10). A nucleated cell line is grown in the wells of a sectored glass microscope slide and fixed to the glass surface (such slides are prepared commercially). Serial dilutions of patient serum are added to individual wells of the slide and if ANA is present, it

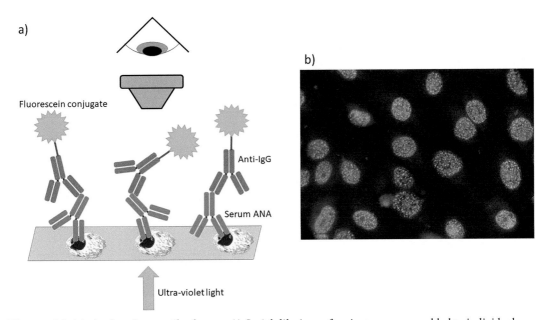

Figure 19.14 Antinuclear antibody test. (a) Serial dilutions of patient serum are added to individual wells of a sectored glass microscope slide on which a nucleated cell line has been grown. ANA binds to the nucleus and in turn a secondary antiserum binds the patient's autoantibody. The antiserum is conjugated to a fluorochrome, which emits light when the slide is viewed with a microscope that produces UV light of appropriate wavelength. (b) The appearance of a positive ANA test. The nuclei of these cells emit apple-green fluorescence in a speckled pattern when viewed under UV light of appropriate wavelength.

will bind to the nucleus of the cells. This immunological reaction is then visualized by the use of a secondary antiserum (e.g. anti-canine IgG), which is conjugated to a fluorochrome (**Figure 19.14a**). When the slide is viewed under UV light of appropriate wavelength, the cellular nuclei appear apple-green on a black background (**Figure 19.14b**) indicating a positive result. The titre of autoantibody can be determined from assessment of the dilution series and the pattern of nuclear labelling (diffuse, rim, speckled or nucleolar) may also be reported.

19.7.3 Thyroglobulin Autoantibody (TGAA) Assay

Hypothyroidism (resulting from **lymphocytic thyroiditis**) is a relatively common disease in dogs that is similar to human Hashimoto's thyroiditis. Although the pathology is largely due to cell-mediated destruction of thyroid tissue, a range of serum autoantibodies to components of the colloid can also be identified and provide a useful diagnostic tool. The most widely available immunoassay is that for detection of autoantibodies specific for thyroglobulin (Tg). **Tg autoantibodies** are most readily identified by use of an ELISA. The wells of these microtitre trays come pre-coated with purified canine Tg protein. Appropriately diluted patient serum is then added and if autoantibody is present, it will bind to the Tg antigen. This reaction is then visualized by use of a secondary antiserum (anti-canine IgG) conjugated to an enzyme (in this case horseradish peroxidase). Addition of the substrate will lead to a colour change in positive wells, which may be measured spectrophotometrically to allow interpretation of the result.

KEY POINTS

- Autoimmunity results from a failure of self-tolerance in genetically susceptible individuals.
- Autoimmunity involves an immunological imbalance, particularly with respect to insufficient Treg cell activity.
- Autoimmunity may lead to clinical signs of autoimmune disease.
- Autoimmune disease may be organ-specific or multisystemic.
- The immunopathology of autoimmunity may involve type II–IV hypersensitivity mechanisms.
- Autoimmune disease has a genetic basis and occurs in particular breeds and pedigrees of dogs.
- MHC genes are most closely associated with susceptibility or resistance to autoimmune disease.
- Autoimmune disease typically occurs in middle to older age, possibly due to immunological changes associated with immunosenescence.
- Autoimmune disease has a female gender predisposition in humans, but not always in dogs.
- Potential triggers for autoimmune disease include cancer, stress, UV light, administration of drugs and exposure to infectious agents.
- Microorganisms may trigger autoimmunity by inducing non-specific polyclonal activation of lymphocytes or non-specific superantigen stimulation of autoreactive cells.
- An infectious agent attached to a host cell may cause bystander destruction of that cell.
- Epitopes of infectious agents may act as molecular mimics of self-epitopes.
- There are several examples of antibody-mediated (type II hypersensitivity) autoimmune disease in veterinary practice, including IMHA, IMTP and autoimmune skin diseases of the pemphigus complex.
- There are some examples of antibody-mediated (type III hypersensitivity) autoimmune disease in veterinary practice, including SLE.
- There are several examples of cell-mediated (type IV hypersensitivity) autoimmune diseases seen in dogs, including rheumatoid arthritis, hypothyroidism, Addison's disease, exocrine pancreatic insufficiency and keratoconjunctivitis sicca.
- A range of commercial immunodiagnostic tests are available for the detection of autoantibodies in blood samples.

CHAPTER 20

CANCER IMMUNOLOGY AND IMMUNE SYSTEM NEOPLASIA

OBJECTIVES

At the end of this chapter, you should be able to:

- Define what is meant by 'tumour antigen'.
- Discuss how the adaptive immune system responds to the presence of a tumour.
- Discuss how tumours can evade the immune response.
- Give examples of current approaches to immunotherapy for cancer.
- Discuss the anatomical, histological, cytological, immunophenotypic and molecular classification of lymphoma and understand why such a classification is important.
- Briefly discuss the prevalence and distribution of lymphoma in various domestic animal species.
- Distinguish lymphoid leukaemia from lymphoma.
- Distinguish plasmacytoma from multiple myeloma.
- Describe the clinical signs associated with multiple myeloma.
- Distinguish canine histiocytoma from histiocytic sarcoma.

20.1 INTRODUCTION

Cancer refers to disease caused by malignant tumours. The disease is the result of the uncontrolled proliferation and spread of neoplastic cells that damage normal tissues. Tumour cells can derive from any normal body cell through **neoplastic transformation**, which is the result of genetic mutations in proto-oncogenes, tumour suppressor genes and genes that encode proteins involved in apoptosis or DNA repair. These mutations are called **driver mutations** that are directly involved in tumour development as opposed to **passenger mutations** that are the result of the overall **genetic instability** of neoplastic cells. Malignant tumours have the ability to **invade local tissues and metastasize** to other organs via the lymph and/or blood. In addition, they can evade recognition and killing by the immune system. **Immune surveillance** refers to the critical role that the immune system plays in preventing cancer by detecting and killing neoplastic cells before they can cause disease. This is indicated by the increased occurrence of certain types of tumours in immunosuppressed/immunodeficient animals. The greater incidence of cancer associated with aging is caused by the lifetime accumulation of genetic mutations and probably also by a decreased function of the immune system with age (**immunosenescence**). Although the frequent occurrence of cancer in animals

DOI: 10.1201/9781003310969-20

and people indicates that the immune system is often ineffective in preventing the growth of tumour cells, recent advances have shown that it is possible to utilize the immune system to kill neoplastic cells through **immunotherapy**. This chapter considers aspects of how the immune system recognizes and responds to neoplastic cells and mechanisms that malignant tumours may develop to escape host defence mechanisms. The chapter will end with a discussion of the range of tumours of cells of the immune system itself.

20.2   TUMOUR ANTIGENS

Tumour antigens are molecules expressed by tumour cells that may be recognized by the immune system. The specific nature of tumour antigens is unknown for most tumours. They can be classified into different categories according to their origin.

- **Neoantigens** are proteins with an altered amino acid sequence as a result of gene mutations. The majority of neoantigens are encoded by passenger mutations and fewer by driver mutations. Increased genetic instability of neoplastic cells correlates with more mutations and more neoantigens that may lead to activation of T cells.
- Some tumour cells produce proteins that are not expressed by the cells from which the tumour originated or are expressed at an increased level. These **normal proteins** may stimulate immune responses because they are usually only expressed during embryonic development, in restricted tissues such as the testis (e.g. cancer-testis antigens such as the MAGE family), or at low very levels resulting in lack of tolerance. An example is tyrosinase which is expressed at low levels by normal melanocytes, but at increased levels in melanoma cells. Some normal antigens expressed by tumour cells, e.g. CD20 on B cell lymphoma cells, do not

induce an immune response, but can be targets of monoclonal antibody therapy.

- In the case of **virus-induced neoplasia**, viral proteins can elicit immune responses that can kill tumour cells. Expression of the feline oncornavirus-associated cell membrane antigen (FOCMA) by cells infected with feline leukemia virus (FeLV) is a good example of this phenomenon.

20.3   THE ANTI-TUMOUR IMMUNE RESPONSE

The immune response to cancer typically proceeds through three stages. The first of these is **elimination**, during which the immune system is able to completely destroy tumour cells. The second stage is **equilibrium**, during which genetic mutations give rise to immune-resistant tumour cells, but the immune system continues to destroy susceptible cancer cells. Over time and under selection pressure by the immune system, the tumour becomes less immunogenic. This has been referred to as **immunoediting**. Equilibrium may last for many years in individual patients. The final stage is **escape**, at which point the immune system is overwhelmed by the strategies the tumour has developed to evade immune detection and destruction.

The immune response to tumours involves both the innate and adaptive immune systems. Experimental studies in laboratory rodents and clinical studies in human patients have demonstrated that **CD8⁺ cytotoxic T cells** are probably the most important effector mechanism in killing tumour cells. As discussed previously (Chapter 7), activation of CD8⁺ T cells as well as CD4⁺ T cells requires antigen presentation and expression of costimulatory molecules by dendritic cells in the tumour-draining lymph node(s). Dendritic cells take up apoptotic tumour cells or tumour cell fragments and generate MHC class I/tumour peptide complexes via **cross-presentation**. The activation and maturation of dendritic cells in the tumour environment and lymph node probably requires the release of danger-associated molecular patterns from dying tumour cells. Activated cytotoxic T lymphocytes will recirculate to the location of the tumour and it is widely recognized that most neoplasms have a peripheral infiltrate of lymphoid cells. Indeed, immunohistochemical studies have readily demonstrated such T cell infiltration (**Figure 20.1**). **CD4⁺ Th1 cells** also seem to play an important role in the anti-tumour immune response by enhancing the differentiation of naïve CD8⁺ T cells into cytotoxic T cells and by secreting **IFN-γ**, which induces macrophages to become M1 (pro-inflammatory) macrophages (**Figure 20.2**). Unfortunately, this immune response is often unable to contain a tumour, once the neoplastic lesion has reached the stage of clinical disease. The rapid division of tumour cells readily overwhelms the ability of the immune system to destroy individual targets. One rare exception to this is the example of the **canine cutaneous histiocytoma** presented in Chapter 7 (**Figure 7.15**). The spontaneous regression that characterizes these tumours is known to be associated with the action of infiltrating CD8⁺ cytotoxic T cells. Although **antibodies** against tumour antigens can often be detected in blood, these are not thought to have a particularly important role in the body's natural defence against tumours. However, therapy with tumour-specific monoclonal antibodies has become more commonly used in human medicine, and similar strategies are being developed for companion animals. Antibodies can potentially kill tumour cells by activation of the complement cascade and by antibody-dependent cellular cytotoxicity, mediated by NK cells.

Natural Killer (NK) cells have the ability to kill a variety of tumour cells and may play a role in immune surveillance. The mode of action of these cells is discussed in Chapter 3. Activation of NK cells is inhibited by receptors that recognize MHC class I on healthy cells. Decreased expression of MHC class I avoids killing of tumour cells by cytotoxic T cells, but makes them more susceptible to killing by NK cells. Tumour cells also often express stress molecules that are recognized by activating receptors on NK cells. In addition to killing of tumour cells directly, secretion of IFN-γ by activated NK cells supports the activation of M1 macrophages. Cytokine therapy with IL-2 or IL-15 is aimed at enhancing NK cell activity.

Tumour-associated macrophages (TAMs) can inhibit as well as promote the growth of tumours (see the following section) depending

a)

b)

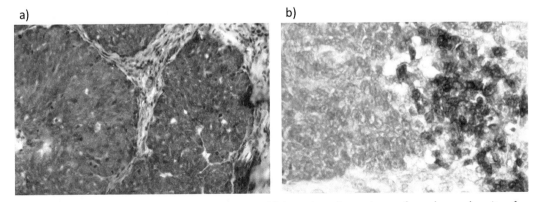

Figure 20.1 The immune response to tumours. (a) A section of a carcinoma from the nasal cavity of a cat. (b) A tissue section from the tumour, labelled immunohistochemically, shows infiltration of CD3⁺ T cells at the margins of the lobules of neoplastic epithelial cells.

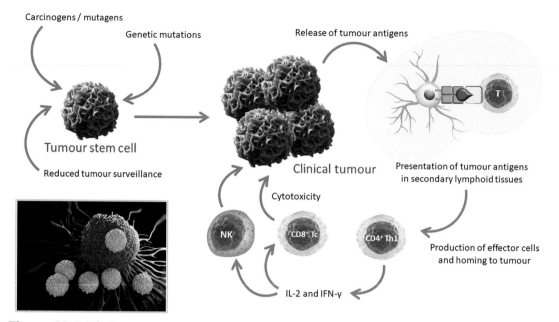

Figure 20.2 The immune response to tumours. A series of tumour initiating and promoting events leads to transformation of a cell or tumour stem cell and proliferation of that cell to the stage of induction of tissue pathology and clinical signs (clinical neoplasia). The tumour cells display neo-antigens that are processed and presented by APCs within the regional lymphoid tissue. The anti-tumour immune response comprises CD4+ Th1 and CD8+ Tc cells that home to the periphery of the tumour to attempt cytotoxic destruction of the neoplastic cells. There is an additional role for NK cells, whose action may also be amplified by Th1-derived IFN-γ. Inset: rendering of a tumour cell being attacked by CD8+ Tc cells.

on their activation state. M1 macrophages activated by IFN-γ secreted by CD4+ Th1 and NK cells can kill tumour cells using nitric oxide, reactive oxygen species and lysosomal enzymes.

20.4 TUMOUR-PROMOTING IMMUNE ACTIVITY

Malignant tumours are often associated with a chronic inflammatory reaction that can become so widespread that it contributes (e.g. via release of cytokines such as TNF) to systemic clinical signs of cancer including cachexia, anaemia and fatigue. In addition, the inflammatory response may create a local environment that supports the proliferation and spread of neoplastic cells. Indeed, chronic inflammation itself is a risk factor for the development of tumours. Examples include sarcomas in feline eyes following

injury, injection site sarcomas in cats following vaccination (see Chapter 13) and gastrointestinal lymphoma associated with colonization by proinflammatory bacteria and inflammation. The exact mechanisms by which chronic inflammation enhances tumour growth are poorly understood (**Figure 20.3**). The tumour-promoting effects seem to be mostly associated with molecules secreted by myeloid cells such as alternatively activated (M2) macrophages and poorly defined myeloid-derived suppressor cells (MDSCs). Secreted factors may promote cell proliferation and inhibit cell death. Macrophages can secrete matrix metalloproteinases that degrade and remodel the extracellular matrix and facilitate invasion of tissues, and vascular endothelial growth factor that stimulates angiogenesis. Dendritic cells and MDSCs may drive the differentiation of CD4+ T cells to

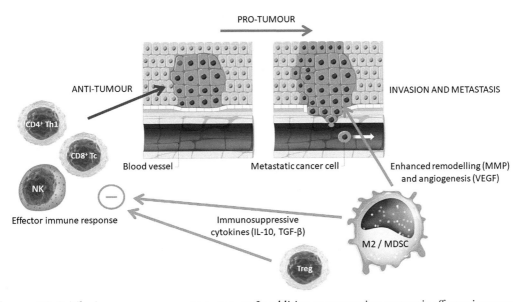

Figure 20.3 The immune response to tumours. In addition to appropriate cytotoxic effector immune responses to tumours, there are counteractive responses that may enhance the malignant potential of the cancer. These include the presence of Treg cells within the population of tumour-infiltrating lymphocytes and the presence of TAMs of the M2 phenotype that may encourage the growth, invasion and metastasis of the tumour.

Foxp3⁺ Tregs while tumour cells and inflammatory cells release chemokines that recruit Tregs into the tumours. MDSCs and Tregs suppress the immune response by various mechanisms including the secretion of immunosuppressive cytokines such as IL-10 and TGF-β. Dogs with cancer have elevated numbers of CD4⁺ Treg cells in blood and tumour-draining lymph nodes. Tregs, MDSCs and M2 macrophages would appear to be appropriate targets for tumour therapy, as inhibiting or deleting these populations will allow more effective anti-tumour immune responses and reduce their metastatic potential. In this regard, a recent study of low-dose cyclophosphamide therapy of dogs with soft tissue sarcomas has shown a reduced number of circulating Treg cells in treated patients.

20.5 IMMUNE EVASION BY TUMOURS

The high prevalence of tumours in companion animals, most of which are not immunodefi-

cient, indicates that tumour cells have developed **strategies to avoid recognition and/or killing by immune cells**, including reduced expression and presentation of tumour antigens and inhibition of T cell responses. Understanding the mechanisms by which tumours evade the immune response offers potential avenues for therapeutic intervention and mobilization of an effective immune response.

- Genetic instability of tumour cells results in random mutations, some of which may **reduce expression of tumour-specific antigens**. This offers a **selective advantage** to these tumour cells in terms of avoiding immune detection, if the proteins are not needed for cell growth and survival.
- **Reduced presentation of antigenic peptides** makes tumour cells less susceptible to killing by cytotoxic T cells. This may be result from decreased expression of MHC class I molecules or proteins associated with the processing of

intracellular antigen, such as components of the proteasome and TAP1 and TAP2 molecules.

- Chronic exposure to tumour cells may result in **increased expression of inhibitory costimulatory molecules on T cells, in particular CTLA-4 and PD-1**. The ligands of CTLA-4 are CD80 and CD86 expressed on antigen presenting cells in the draining lymph node, and the ligands of PD-1 are PD-L1 and PD-L2, which can be expressed by tumour cells and myeloid cells in the tumour microenvironment. Binding of CTLA-4 and PD-1 to their ligands inhibits the effector function of T cells.

- Tumour cells can **recruit Tregs and MDSCs** that suppress immune responses by secreting immunosuppressive molecules such as **IL-10 and TGF-β**. In addition, tumour cells themselves often secrete TGF-β, which further contributes to the immunosuppressive environment.

20.6   IMMUNOTHERAPY TO ENHANCE THE ANTI-TUMOUR IMMUNE RESPONSE

The conventional approach to tumour therapy currently relies on use of cytotoxic drugs and radiation therapy that destroy tumour cells. Studies in experimental animals have demonstrated that the therapeutic effect of some treatments such as doxycyclin and radiation is enhanced by the immune response. This is attributed to the release of damage-associated molecular patterns (DAMPs) from dying tumour cells which activate dendritic cells. Adjunctive therapies that might enhance the systemic cellular immune response also have a role in clinical management. These will be discussed further in Chapter 21. Immunotherapy of tumours can be passive, in which tumour-specific antibodies or T cells are administered

to the patient, or active, in which the immune system is mobilized to eliminate tumour cells. Tumour immunotherapy has the potential to be highly specific and potentially avoid detrimental effects on normal cells and can induce a durable response. There are various examples of such approaches, some of which have been adopted in companion animal oncology and others are actively being developed for clinical use.

- Administration of tumour-specific monoclonal antibodies. These antibodies target antigens expressed by tumour cells (although these are often also normal cells). An example is CD20-specific antibodies which kill normal and neoplastic B cells through ADCC and complement-mediated lysis. Such antibodies can also be engineered to target cytotoxic drugs specifically to tumour cells and thereby reduce the toxicity associated with systemic chemotherapy.

- One of the most promising strategies is the use of monoclonal antibodies against those molecules that act as a 'brake' on T cell responses including CTLA-4 and PD-1/ PD-L1/2. These monoclonal antibodies act as **'checkpoint inhibitors'** and have been successfully employed in human cancer patients (**Figure 20.4**). A potential side effect is the development of autoimmune disease, because of the lack of inhibition of T cells.

- T cells can be isolated from tumours or blood and their numbers expanded *in vitro* by activating them with IL-2. The **autologous** lymphocytes ('**lymphokine activated killer' (LAK) cells**) are transferred back into the patient to more effectively target tumour cells. A recent development is to transfect T cells from the patient with an engineered **chimeric antigen receptor (CAR)** composed of the variable domain of an antibody that is specific for a tumour antigen and cytoplasmic domains derived from CD3 and

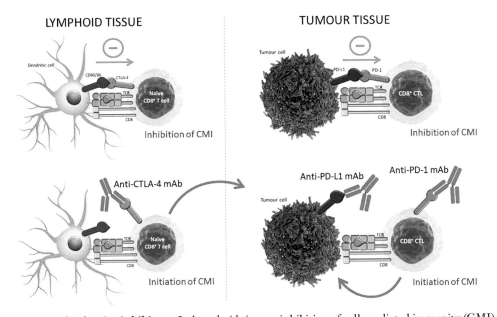

Figure 20.4 Checkpoint inhibitors. In lymphoid tissues, inhibition of cell-mediated immunity (CMI) can occur when tumour antigens are presented in the context of inhibitory (CTLA-4), rather than co-stimulatory (CD28), signals. Administration of anti-CTLA-4 antibody can inhibit this effect and enhance activation of tumour-specific naïve CD8⁺ T cells. Within the tumour microenvironment, expression of inhibitory molecules on the tumour cell surface (PD-L1, PD-L2) can inhibit the activity of CD8⁺ cytotoxic T lymphocytes (CTLs). Use of monoclonal antibodies directed against either PD-L1 or alternatively PD-1, expressed on the T cell, can abrogate this effect and permit their normal function of cytotoxicity.

costimulatory receptors to allow for a potent activation signal once the receptor engages with the tumour cell. These **CAR T cells** are transferred back into the patient to kill the tumour cells (**Figure 20.5**).

- Vaccines against cancer-inducing viruses such as **Marek's disease virus in poultry** and **feline leukemia virus in cats** can be considered cancer vaccines, similar to the human papilloma virus vaccine for prevention of cervical cancer. These vaccines have been used successfully for many years to reduce the prevalence of tumours in animals.

- The development of vaccines against other forms of cancer that have no known infectious etiology has been more elusive. **Autologous cancer vaccines** are prepared by removing tumour tissue from a patient, separating the tissue into individual cells, killing the cells and injecting them in combination with an adjuvant into the patient. The efficacy of this approach is often uncertain and anecdotal. A **DNA vaccine against oral melanoma** in dogs is commercially available. This vaccine consists of a plasmid engineered with a gene that encodes human tyrosinase. The use of the human (rather than canine) tyrosinase is thought to enhance the immunogenicity of the vaccine. Transfection of dendritic cells following injection of the vaccine is expected to generate a T cell response against melanoma cells (**Figure 20.6**). Some patients may develop vitiligo as a result of 'off-target' effects on normal melanocytes. The efficacy of this therapeutic vaccine seems to be quite variable, which may be

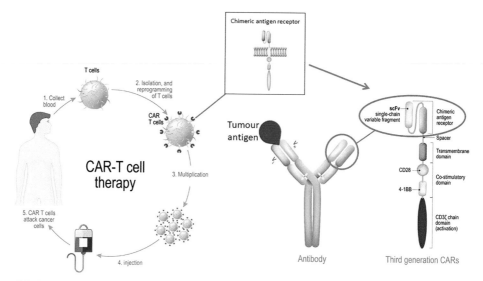

Figure 20.5 CAR T cell therapy. T cells can be isolated from the patient's blood and engineered to express a recombinant chimeric antigen receptor containing a single-chain Fv, derived from a monoclonal antibody that recognizes a tumour antigen target. This is combined with a transmembrane domain to anchor the molecule to the T cell surface and intracellular signalling domains that mimic signal 1 and signal 2 that would normally be delivered via the TCR/CD3 complex and CD28. The 'reprogrammed' CAR T cells are expanded *in vitro* then infused back into the patient, where they may target and destroy cancer cells.

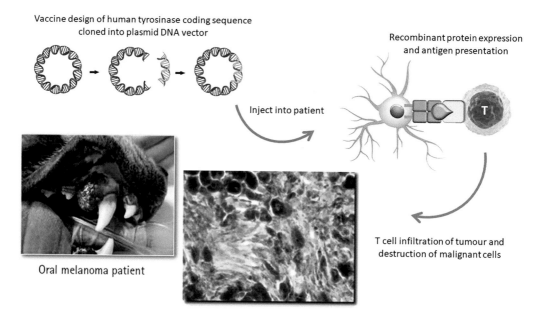

Figure 20.6 DNA vaccine against canine melanoma. This plasmid DNA vaccine incorporates the human tyrosinase gene. When injected into a dog with oral malignant melanoma, the gene transfects dendritic cells, and tyrosinase peptides are presented. This amplifies the cell-mediated immune response against this target antigen, expressed by the malignant melanoma cells.

due to either a failure of cross-reactivity with the canine antigen expressed by the cancer cells, or downregulation of this antigen in the primary tumour or its metastases. Another vaccine approach is the use of *Listeria monocytogenes* **bacteria genetically modified to express a growth factor receptor HER2/neu construct.** A subset (about 40%) of osteosarcomas in dogs overexpress HER2/neu. The immune response to HER2 was associated with an increased number of IFN-γ secreting cells and may increase the survival of dogs with osteosarcoma.

- Other efforts are aimed at boosting the immune response with systemic or localized treatment with cytokines such as IL-2. A commercial product is available for adjunctive therapy of feline fibrosarcoma. The mechanisms of action of this approach may be to enhance NK cell activation and to counteract tolerance towards tumour antigens by stimulating T cells that would otherwise become anergic. This may also be the case when adjuvants are used in tumour immunotherapy, whereby 'danger signals' are elicited from the innate immune system to potentiate the adaptive immune response to the tumour antigens.

20.7 TUMOURS OF THE IMMUNE SYSTEM

The cells of the immune system may undergo neoplastic transformation leading to the formation of a range of different tumour types. Tumours of lymphocytes (lymphoma, lymphoid leukaemia), plasma cells (plasmacytoma, multiple myeloma) and dendritic cells (histiocytoma, localized and disseminated histiocytic sarcoma) will be discussed here, although there are a number of other less common variants. A detailed discussion of oncogenesis is beyond the scope of this chapter, but development of these neo-plasms invariably involves genetic mutations compounded by co-factors such as age, failure of immune surveillance and the action of carcinogens or oncogenic retroviruses.

20.7.1 Lymphoma

Lymphoma (sometimes referred to as lymphosarcoma) is probably the **most common tumour recognized in domestic animals.** The disease is most frequent in **dogs, cats and horses** because these animals generally have a longer lifespan and this is most often a disease of middle to older age. Sporadic cases of lymphoma are recognized in production animals and the disease is significant in poultry. Lymphoma is a solid tumour that causes nodular or diffuse enlargement of the viscera in which it grows.

Lymphoma may be classified on the basis of (1) **anatomical distribution**, (2) **histological appearance**, (3) **cytological appearance**, (4) **immunophenotype** and (5) **gene rearrangement**. In human medicine, the classification scheme developed by the World Health Organization (WHO) takes all of these elements into account when defining the numerous subtypes of lymphoma. The latest **WHO classification scheme** has now been successfully adapted for canine and equine disease. The aim of such classification is to allow formulation of a prognosis and to select the optimum treatment approach for that specific tumour type.

Anatomically, lymphoma may be classified as (1) **multicentric**, (2) **thymic**, (3) **alimentary**, (4) **cutaneous** or (5) **solitary**. Multicentric lymphoma involves widespread organ involvement, presumably starting in lymphoid tissue and possibly spreading to other sites. Any organ may be involved, but those most commonly affected are lymph nodes, spleen, liver, kidney, heart, lungs and gastrointestinal tract (**Figure 20.7**). Thymic lymphoma involves the thymus and possibly also the mediastinal or cervical lymph nodes (**Figure 20.8**). Alimentary lymphoma may arise at any level of the gastrointestinal

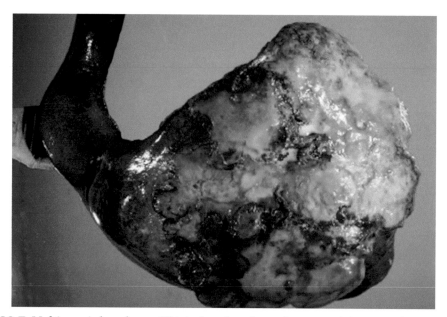

Figure 20.7 Multicentric lymphoma. This is the spleen from a horse containing a very large (note the relative size of the hand on the left) cream-coloured mass with the microscopic appearance of a lymphoma. Multiple lymph nodes were similarly enlarged and similar neoplastic tissue was present surrounding the pelvic–femoral articulation and arising from the intercostal area of the thoracic cavity.

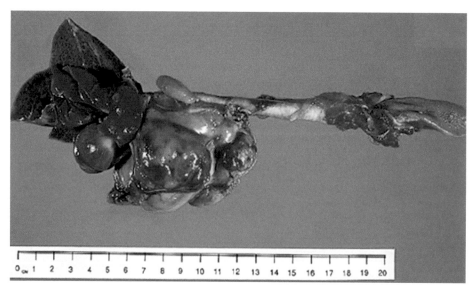

Figure 20.8 Thymic lymphoma. This is the thoracic content from a cat infected with FeLV. The heart and lungs are to the left of the photograph and anterior to the heart is a large cream-coloured mass arising from the thymus. The mass is several times larger than the heart. The cat presented with dyspnoea, dysphagia and local subcutaneous oedema.

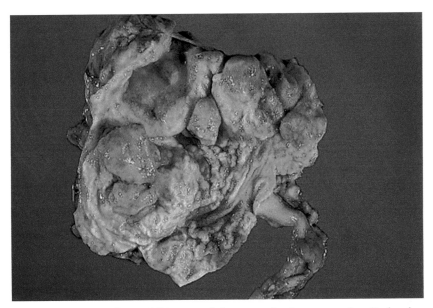

Figure 20.9 Alimentary lymphoma. This stomach from a cat has been opened to display the mucosal surface, which is entirely replaced by a series of nodular masses. The cat had a history of chronic vomiting (sometimes haematemesis) and weight loss. It has been suggested that this form of lymphoma may be associated with the presence of gastric *Helicobacter* spp.

tract and extend to the mesenteric lymph nodes and possibly the liver and other abdominal viscera (**Figure 20.9**). Alimentary lymphoma may be solitary (i.e. a single location in the stomach or intestinal tract) or there may be multiple foci of disease. Cutaneous lymphoma arises in the skin and may extend to draining lymph nodes. Two forms of cutaneous lymphoma are recognized. **Dermal lymphoma** arises within the dermis and does not extend to the epidermis and may be of T or B cell phenotype (see more on this later). In contrast, **epitheliotropic lymphoma** extends into the epidermis to form small intraepidermal aggregates known as 'Pautrier's microabscesses'. Epitheliotropic lymphoma is always of T cell origin and the cells express integrins that are responsible for the epithelial localization (**Figure 20.10**). Finally, solitary lymphoma is a single tumour in any body organ.

A histological classification of lymphoma can be made after examination of tissue biopsies and this area has recently become increasingly complex as new subtle variations of tumour growth are identified. Simplistically, neoplastic lymphocytes generally grow in one of two basic patterns. The most common is '**diffuse lymphoma**' where normal tissue structure is entirely obliterated by an infiltrating sheet of closely packed tumour cells (**Figure 20.11a**). In contrast, in '**follicular lymphoma**' the neoplastic cells (generally within lymphoid tissue) grow in a pattern reminiscent of lymphoid follicles, but these aggregates are grossly expanded and have abnormal microanatomy (**Figure 20.11b**).

The cytological classification of lymphoma relates to examination of aspirates from a tumour mass or tumour cells exfoliated into body fluids. Neoplastic lymphocytes satisfy the fundamental criteria for malignancy (e.g. cellular pleomorphism, increased mitotic rate) and are also often categorized in terms of size (e.g. small, medium or large lymphocytes), the relative proportions of nucleus and cytoplasm and the appearance of nuclear chromatin and nucleoli (**Figure 20.12a**).

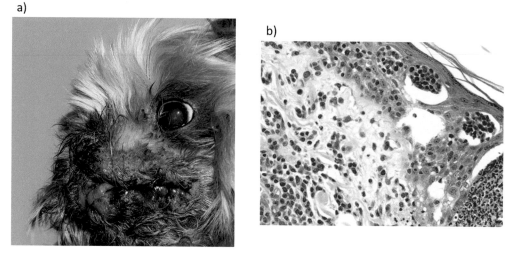

Figure 20.10 **Epitheliotropic lymphoma.** (a) A dog affected with epitheliotropic lymphoma. (b) Skin biopsy showing clusters of neoplastic lymphocytes within the epidermis (Pautrier's microabscesses). These are always T cell tumours and the neoplastic cells express adhesion molecules that interact with ligands on the surface of keratinocytes.

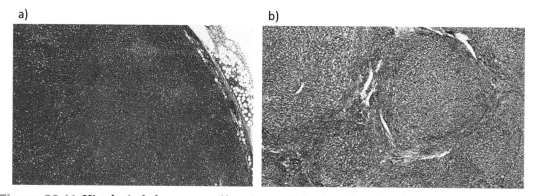

Figure 20.11 **Histological phenotypes of lymphoma.** (a) Section of lymph node from a dog with diffuse lymphoma. The normal corticomedullary structure is completely obliterated by a diffuse sheet of neoplastic lymphocytes that extends focally into the perinodal adipose tissue. (b) Section of lymph node from a dog with follicular lymphoma. The neoplastic cells have a nodular arrangement, which is sometimes difficult to distinguish from hyperplastic cortical follicles. These are generally tumours of B cell origin.

Lymphoma **immunophenotype** refers to the cell of origin of the tumour. **Immunohistochemical**, **immunocytochemical** or **flow cytometric** analyses can be used to determine the nature of surface molecule expression by the tumour cells. In a research setting, this may be taken to a high level of phenotypic classification, but in general practice it is now fairly common to determine whether a lymphoma is at least of T or B cell origin (**Figure 20.12b**). In dogs, this has prognostic significance when considered in conjunction with histological features. Studies have shown the longest median survival time (or relapse-free survival time) for low-grade T

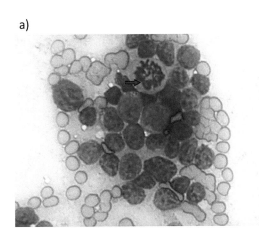

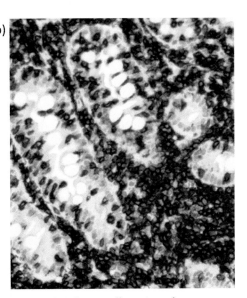

Figure 20.12 Lymphoma cytology and immunophenotyping. (a) A fine-needle aspirate from the lymph node of a cat with multicentric lymphoma. There is a population of lymphoblastic cells with minimal cytoplasm surrounding a nucleus with aggregated chromatin. A bizarre mitotic figure is present in one cell (arrowed). (b) It is now routine to determine whether a lymphoma is of T or B cell origin by immunophenotyping. This biopsy sample is from a cat with alimentary lymphoma labelled with antibody to CD3. The neoplastic lymphocytes infiltrating the lamina propria are small T lymphocytes. T cells are also present within the crypt epithelial layer, suggesting a tendency to epitheliotropism.

cell lymphoma, the shortest survival time for high-grade T cell lymphoma and intermediate survival for dogs with B cell lymphoma.

It is sometimes difficult to distinguish between increased numbers of lymphocytes as a result of an inflammatory reaction versus infiltration/proliferation of neoplastic lymphoid cells, based on routine histology. This is particularly problematic when trying to differentiate between inflammatory bowel disease and intestinal lymphoma, based on histopathology. **Clonality testing** utilizes PCR technology to determine whether a population of lymphocytes carries **rearrangements** in the variable or joining region elements of genes encoding specific **T or B cell receptors** (TCR β or γ chains; BCR heavy chain). Whereas in a normal or reactive lymphocyte population one would expect to see a polyclonal population of cells, in a neoplastic population there is restricted receptor usage

consistent with clonal expansion of the tumour cell (**Figure 20.13**). A polyclonal lymphocytosis of B cells has been reported in English bulldogs. This appears to be a non-neoplastic process and is associated with hypergammaglobulinemia due to an increase of polyclonal IgA and/or IgM.

As stated previously, lymphoma may arise in any species, but is particularly common in dogs and cats. **Canine lymphoma** is a disease of middle to older age and the most frequently recognized anatomical form is multicentric. The most common form is diffuse large B cell lymphoma, which appears to have a predilection for Golden and Labrador retrievers and German shepherd dogs. T cell lymphomas make up about 25–30% of all lymphomas, with the two main types being the peripheral T cell lymphoma and cutaneous T cell lymphoma. Boxer dogs are overrepresented among dogs affected with peripheral T cell lymphoma. Breed

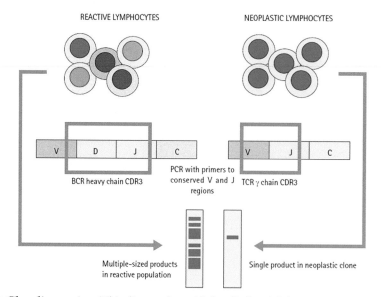

Figure 20.13 Clonality testing. This diagnostic tool helps distinguish between a reactive population of lymphocytes of mixed antigenic specificity and a clonal population of neoplastic lymphocytes with identical receptor specificity. The complementarity determining region 3 (CDR3) of the BCR or TCR forms the antigen-binding site and is an area of high sequence variability due to differential recombination between VDJ genes and further somatic mutation in B cells. In clonality testing, PCR amplification of the CDR3 region of the BCR heavy chain (VDJ) or the TCR γ chain (VJ) is performed. The PCR employs primers that anneal at conserved regions of the V and J genes in each case. The amplicons can then be separated by size (e.g. by polyacrylamide gel electrophoresis) and visualized. Products of multiple sizes indicate a polyclonal reactive population, while a single-sized product (monoclonal) or a relatively few products (oligoclonal) is more consistent with neoplastic transformation.

predilections suggest a genetic basis for these tumours. Numerous investigations have failed to provide any evidence for a canine oncogenic retrovirus. **Feline lymphoma** may be associated with **FeLV** infection and there is some evidence that co-infection with feline immunodeficiency virus (**FIV**) may also contribute to lymphomagenesis. However, since the advent of vaccines for FeLV, the prevalence of this virus in the cat population has decreased and the epidemiologic pattern of feline lymphoma has altered. Thirty years ago, feline lymphoma was almost invariably FeLV-associated and most commonly presented as mediastinal lymphoma in young animals of pure breed. Currently, the most common presentation of feline lymphoma is alimentary in older cats that are often negative on retrovirus testing.

Bovine lymphoma may also be retrovirus associated. Bovine leukaemia virus (**BLV**) may be responsible for multicentric lymphoma in adult cattle, but multicentric lymphoma in young calves or thymic lymphoma in animals under 30 months of age is not usually BLV associated. Lymphoma is relatively rare in pigs, sheep and goats, but is not an uncommon disease in the horse, where all anatomical variants are recognized. Avian lymphoma is of major significance in domestic poultry and is a virus-associated disease. **Marek's disease (neurolymphomatosis)** is caused by a DNA virus (**herpesvirus**) and presents in young birds as paralysis relating to the tropism of the tumour for growth around nerve trunks. In contrast, **lymphoid leukosis (visceral lymphomatosis)** is triggered by an RNA virus (**retrovirus**) and generally affects the viscera of older birds. **Vaccines** are available that successfully control the occurrence of Marek's disease.

20.7.2   Lymphoid Leukaemia, Plasmacytoma and Multiple Myeloma

Lymphoid leukaemia is a form of lymphoid neoplasia that arises from the **bone marrow**, extends to the **peripheral blood** and from there colonizes vascular organs such as the spleen and liver. The diagnosis is generally made from evaluation of a blood smear in an animal with profound lymphocytosis (**Figure 20.14a**). The disease is relatively **rare in animals**, but is most often recognized in dogs and cats. Feline lymphoid leukaemia may be FeLV associated. Two subtypes of lymphoid leukaemia are reported. **Acute lymphoblastic leukaemia** (ALL) is a severe disease with acute onset and a poor prognosis. Affected animals have large lymphoblastic cells within the circulation. In contrast, **chronic lymphoid leukaemia** (CLL) involves a better differentiated population of small lymphocytes and has a slower clinical course and a better prognosis.

Plasmacytoma is a **benign** plasma cell tumour that appears as a slow-growing nodular mass in the skin or mucosal tissues, most often affecting the lips, oral cavity or distal extremities of **dogs** and **cats**. These relatively rare tumours usually do not progress to malignancy and are not associated with monoclonal gammopathy. Multiple myeloma is a **malignant plasma cell tumour** that is again relatively uncommon, but is most often documented in dogs and, to a lesser extent, in cats. Classically, the tumour arises in the **bone marrow** of long bones or vertebral bodies, where it may cause localized osteolysis. However, recent studies suggest that **feline myeloma** may be a primary disease of the **abdominal viscera**, particularly the liver and spleen (**Figure 20.14b**). The neoplastic plasma cells often secrete immunoglobulins, resulting in a very high serum protein concentration, which on electrophoresis presents as a sharply defined band in the gamma globulin fraction, correlating with this clonally derived immunoglobulin (**Figure 20.15**). This situation is referred to as a **monoclonal gammopathy** and the abnormal immunoglobulin as a **paraprotein**. Paraproteins are most commonly of the IgG or IgA class. IgM paraproteins are uncommon and to date IgD or IgE paraproteins have not been identified in animals. The abnormal nature of the serum paraproteins may be demonstrated by **immunoelectrophoresis** or

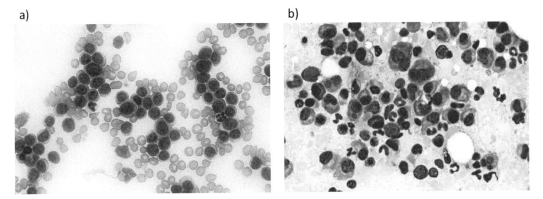

a)   b)

Figure 20.14 **Lymphoid leukaemia and multiple myeloma.** (a) This is a blood smear from a dog with marked lymphocytosis consistent with lymphoid leukaemia. These tumours arise from the bone marrow and spill over into the circulating blood. Neoplastic cells may then colonize vascular organs such as the spleen and liver. (b) This aspirate from the bone marrow of a dog with multiple myeloma reveals a dominance of abnormal plasma cells, some of which are binucleate.

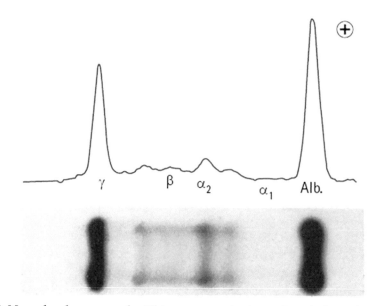

Figure 20.15 Monoclonal gammopathy. This serum protein electrophoresis shows the presence of a restricted band of high concentration within the gamma globulin region. Such monoclonal gammopathy is typical of multiple myeloma, but may occasionally occur also in some infectious diseases.

western blotting, which are used diagnostically in this disease. **Paraproteinaemia**, particularly of the IgM or IgA class, may be associated with **serum hyperviscosity**, which may cause congestive heart failure, renal disease, and ocular and neurologic abnormalities. In addition, the cells often produce an excess of light chain proteins which are referred to as **Bence-Jones proteins**. The low molecular weight of the light chains allow them to pass the glomerular basement membrane which causes **proteinuria** and may result in **renal disease**. Multiple myelomas are also often associated with **recurrent infections** as a result of general immunosuppression, **increased tendency to bleed** and **hypercalcemia**. The bleeding diathesis may be caused by interference of the paraproteins with platelet function and by decreased production of platelets. Hypercalcemia may be caused by direct destruction of bone (**osteolysis**) or the propensity of these tumours (and, incidentally, some T cell lymphomas) to produce a range of **parathyroid hormone-related peptides** (**PTHrp**) that mimic the effect of PTH and inappropriately and excessively elevate the serum calcium concentration.

20.7.3　Histiocytoma and Histiocytic Sarcoma

There is a spectrum of neoplastic disease of dendritic cells in dogs and cats that includes benign canine histiocytoma and the malignant histiocytic sarcomas. Cutaneous histiocytoma is a relatively **common skin tumour of the dog** and has been extensively studied because of its propensity for **spontaneous regression** (see Chapter 7, **Figure 7.15**). The most common clinical presentation is of a rapidly growing, solitary nodular mass arising anywhere in the haired skin, usually in young animals. Although histologically having the appearance of macrophages, immunohistochemistry has identified this as a tumour of **epidermal Langerhans cells**. The spontaneous regression that occurs over several weeks is mediated by infiltrating CD8+ T cells.

Localized histiocytic sarcoma most often presents as a solitary skin mass arising

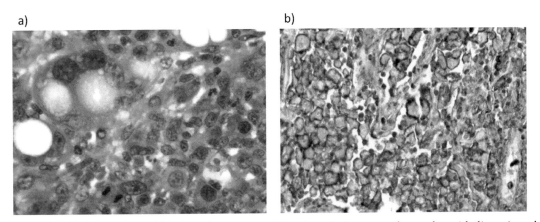

Figure 20.16 Disseminated histiocytic sarcoma. (a) Section of a lung mass from a dog with disseminated histiocytic sarcoma. The neoplastic cells are highly pleomorphic and mitotic. Note the presence of large multinucleate cells with cytoplasmic vacuolation. (b) Canine disseminated histiocytic sarcoma showing immunohistochemical expression of CD18. This is one of the cell-surface molecules used to phenotype the dendritic cell origin of these tumours.

particularly on the limbs and around the joints of dogs. These tumours may metastasize to regional lymph nodes. There are breed predispositions for the Bernese mountain dog, flat-coated retriever, Golden and Labrador retriever and Rottweiler. **Disseminated histiocytic sarcoma** shares the same breed predispositions, but is a multicentric disease involving a wide range of viscera, but rarely the skin (**Figure 20.16**). Both tumour types are of dendritic cell origin. Disseminated histiocytic sarcoma carries a poor prognosis, but non-metastatic localized histiocytic sarcoma may be treated by surgical excision and/or radiotherapy.

KEY POINTS

- Carcinogenesis is a complex process that involves genetic mutations, exposure to carcinogens and reduced immune surveillance.
- Tumour antigens are targets of the immune response and may be of a variety of origins.
- The anti-tumour immune response is cell-mediated and involves NK cells, CD8$^+$ cytotoxic T lymphocytes, M1 macrophages and CD4$^+$ Th1 cells.
- Some elements of the immune system (M2 macrophages and Treg cells) promote tumour growth, invasiveness and metastasis.
- Various immunotherapeutic approaches to treating cancer are being evaluated, including monoclonal antibodies, autologous T cells, vaccines and cytokines.
- Lymphoma may be classified on the basis of anatomical distribution, histological and cytological appearance, immunophenotype and gene rearrangement.
- Lymphoma classification helps determine prognosis and therapeutic management.
- Anatomically, lymphoma may be multicentric, thymic, alimentary, cutaneous or solitary.
- Cutaneous lymphoma may be dermal or epitheliotropic.

- Animal lymphoma may currently be classified as T or B cell in origin. Canine high-grade T cell lymphoma carries a less favourable prognosis than B cell lymphoma.
- Clonality testing determines whether a population of lymphocytes in a sample expresses multiple different TCRs or BCRs (normal or reactive) or a restricted range of receptors (lymphoma).
- Feline and bovine lymphoma may be caused by retrovirus infection.
- FeLV testing and vaccination have altered the pattern of feline lymphoma, which is now most commonly seen in the alimentary tract of older cats.
- Lymphoma in poultry is induced by a herpesvirus (Marek's disease) or retrovirus (lymphoid leukosis). Vaccines are available for the control of Marek's disease.
- Lymphoid leukaemia is a rare neoplasm that arises in the bone marrow and extends to the blood.
- Plasmacytoma is a benign plasma cell tumour of the skin or oral cavity.
- Multiple myeloma is a malignant plasma cell tumour of the bone or viscera, most often reported in dogs and associated with paraproteinemia, immunosuppression, bleeding, proteinuria and hypercalcaemia.
- Multiple myeloma may be diagnosed by serum protein electrophoresis, immunoelectrophoresis or western blotting.
- Histiocytoma is the most common canine skin tumour. It is of Langerhans cell origin and shows spontaneous regression, mediated by $CD8^+$ cytotoxic T cells.
- Localized or disseminated histiocytic sarcomas are malignant tumours of dendritic cell origin, most often recognized in certain breeds of dog.

IMMUNOTHERAPY

OBJECTIVES

At the end of this chapter, you should be able to:

- Discuss the mode of action of glucocorticoids, cytotoxic adjunct immunosuppressive drugs and ciclosporin in the management of immune-mediated disease.
- Discuss the mode of action of allergen-specific immunotherapy in the management of allergic disease.
- Discuss the utility and mode of action of intravenous immunoglobulin (IVIG) therapy in immune-mediated disease.
- Discuss the generation and mode of action of therapeutic monoclonal antibodies.
- Appreciate the utility of substances designed to stimulate innate immune responses to enhance protection against infectious diseases.

21.1 INTRODUCTION

A wide range of immune-mediated diseases may affect domestic animals. These are of greatest significance in companion animal species, which are generally longer lived. There is a demand for medical management of these diseases and a requirement for products that might either suppress (e.g. in autoimmunity or allergy) or enhance (e.g. in immunodeficiency, infectious or neoplastic disease) immune function. Unfortunately, this requirement is not always met by the current availability of safe and effective medicines. Immunosuppressive drugs currently used in animals have mostly been adopted from human medicine. These drugs are often used empirically without solid pharmacokinetic data and many are not specifically licensed veterinary products. There are even fewer options

available as a means of stimulating immune function and again most of such agents are unlicensed. This chapter reviews some of these products and also describes recent advances in immunotherapy that may potentially be available in veterinary medicine in the near future.

21.2 IMMUNOSUPPRESSIVE AGENTS

21.2.1 Glucocorticoids

The most widely used immunosuppressive drugs in veterinary medicine are glucocorticoids. Glucocorticoids are naturally produced by the adrenal cortex in response to stimulation by adrenocorticotropic hormone (ACTH). The secretion of ACTH in the anterior pituitary is stimulated by corticotropin-releasing hormone (CRH) from the hypothalamus.

DOI: 10.1201/9781003310969-21

The **hypothalamic-pituitary-adrenal (HPA) axis** is inhibited by glucocorticoids in a **negative feedback loop**. The inflammatory cytokines **IL-1, IL-6 and TNF stimulate the HPA axis**, and inhibition of their production by glucocorticoids forms another negative feedback loop (**Figure 21.1**). Their lipophilic nature and small molecular weight allows glucocorticoid molecules to diffuse readily through the plasma membrane of cells, where they bind to **cytoplasmic receptors**. The hormone/receptor complex then moves into the nucleus, where it acts as a transcription factor, **binding to steroid response elements (SRE) within the promoter of certain genes** leading to either **activation or repression of gene transcription** (**Figure 21.2**). All nucleated cells have glucocorticoid receptors and glucocorticoids can affect the transcription of many genes, causing a wide range of anti-inflammatory, metabolic and immunologic effects. Several synthetic glucocorticoids are available for systemic or local administration, which differ in terms of their **formulation, relative potency** and **duration of action** (Table 21.1). Glucocorticoids have both **anti-inflammatory** and **immunosuppressive** capacity, dependent upon the dose administered. Glucocorticoids have effects on a range of leucocytes that can be receptor- and gene transcription-independent at high doses. They cause **stabilization of the cell membrane** of granulocytes, mast cells and macrophages, thereby inhibiting production of inflammatory mediators such as prostaglandins by these cells. Glucocorticoids preferentially inhibit cell-mediated, rather than antibody-mediated immune responses. Even long-term treatment with glucocorticoids does not appear to interfere substantially with the antibody response to vaccines. Evidence from rodent models suggests that glucocorticoids primarily suppress CD4+ Th1 and Th17 cell responses

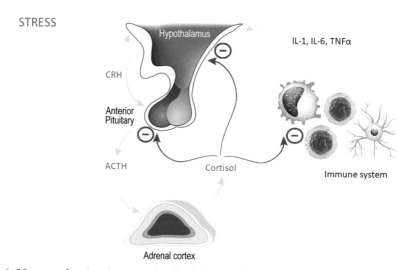

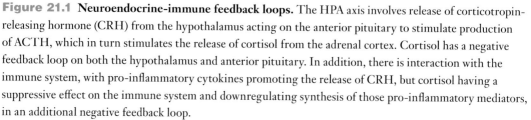

Figure 21.1 Neuroendocrine-immune feedback loops. The HPA axis involves release of corticotropin-releasing hormone (CRH) from the hypothalamus acting on the anterior pituitary to stimulate production of ACTH, which in turn stimulates the release of cortisol from the adrenal cortex. Cortisol has a negative feedback loop on both the hypothalamus and anterior pituitary. In addition, there is interaction with the immune system, with pro-inflammatory cytokines promoting the release of CRH, but cortisol having a suppressive effect on the immune system and downregulating synthesis of those pro-inflammatory mediators, in an additional negative feedback loop.

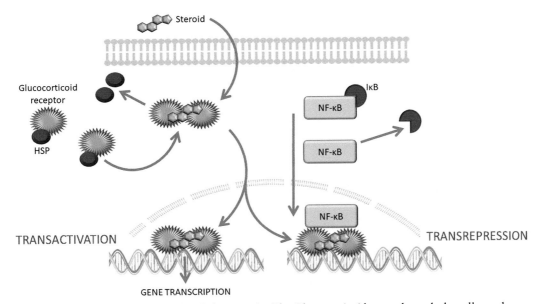

Figure 21.2 Mechanism of action of glucocorticoids. Glucocorticoids pass through the cell membrane and bind to cytoplasmic glucocorticoid receptors, displacing heat shock protein (HSP) molecules. The combination of steroid and receptor then passes into the nucleus where binding to steroid response elements (SRE) within the promoter either initiates gene expression (transactivation) or blocks the ability of pro-inflammatory signalling molecules (such as NF-κB) from binding to their target site, thus inhibiting gene transcription (transrepression).

Table 21.1 Relative Anti-inflammatory Potency of Glucocorticoids

GLUCOCORTICOID	RELATIVE POTENCY	DURATION OF ACTION
Cortisone	0.8	Short acting
Hydrocortisone	1.0	Short acting
Prednisolone/prednisone	4.0	Intermediate
Methylprednisolone	5.0	Intermediate
Triamcinolone	5.0	Intermediate*
Flumethasone	15.0	Long acting
Dexamethasone	30.0	Long acting
Betamethasone	35.0	Long acting

Short acting, <12 hours; intermediate acting, 12–36 hours; long acting, >48 hours.

* Triamcinolone may act for up to 48 hours.

Prednisone and cortisone must first be activated in the liver to prednisolone and cortisol, respectively.

and have relatively little effect on Th2 and Treg cells. Glucocorticoids may, however, interfere with immunoglobulin function, by **downregulating Fc receptor expression** on phagocytic cells, thereby reducing opsonization.

Much of the information regarding glucocorticoid function in veterinary species has been extrapolated from human medicine and studies in experimental rodent systems. There have been surprisingly few investigations specifically addressing the immunological effects of this class of drug in other animal species. In general, treatment with glucocorticoids causes lymphopenia and eosinopenia, and may induce a modest reduction of serum immunoglobulins. There may be specific species differences with respect to glucocorticoid action; for example, **cats** are regarded as being relatively '**steroid-resistant**' and require higher doses of these drugs to cause immunosuppression compared with the dose rate for dogs. This difference is thought to relate to a **lower expression of glucocorticoid receptors** in the target tissues in feline species.

Glucocorticoids have a **wide clinical application** in companion animal medicine. They are used in chemotherapeutic protocols or by intravenous administration in the management of shock. Glucocorticoids administered at **anti-inflammatory** doses are used to manage the clinical signs of **allergic skin diseases** or allergic conditions of the respiratory tract (e.g. feline and equine asthma or canine eosinophilic bronchopneumopathy). In these examples, **topical** (i.e. skin cream or inhaled drug) administration may be used as an alternative to **systemic** dosing. **Immunosuppressive doses** of glucocorticoid are widely used in the management of **autoimmune or idiopathic inflammatory** disorders.

Although they are very effective at suppressing inflammation and immune responses, glucocorticoids have well-recognized **adverse effects**. They suppress the function of all T lymphocytes and not just those cells participating in the pathological immune response. This 'blanket immunosuppression' means that the recipient may be susceptible to **secondary/opportunistic infection**. Glucocorticoids mimic the metabolic effects of endogenous corticosteroid, produced by the adrenal gland (e.g. gluconeogenesis, protein catabolism and lipolysis) and with long-term use, **iatrogenic hyperadrenocorticism** (Cushing's-like syndrome) may develop, involving polyuria/polydipsia, polyphagia/weight gain, accumulation of fat/glycogen within hepatocytes leading to liver dysfunction ('steroid hepatopathy') and atrophy of the skin. Because **abrupt withdrawal** of glucocorticoid therapy may lead to clinical signs of **adrenal insufficiency** (iatrogenic hypoadrenocorticism; Addison's-like disease), the dose of glucocorticoids must always be **slowly tapered**, so that the drug is gradually withdrawn (Table 21.2) and the HPA axis allowed to return to normal function.

Some animals may be resistant to the effects of glucocorticoids. In dogs with IBD, the mechanism of such resistance has been proposed to relate to expression of the drug efflux pump p-glycoprotein (encoded by the multidrug resistance 1 gene, *MDR1*) by mucosal lymphocytes and enterocytes.

21.2.2   Cytotoxic Drugs

Cytotoxic drugs are typically used in chemotherapeutic protocols for management of neoplastic disease. In addition, cytotoxic drugs can be used in combination with glucocorticoids to provide an **additive immunosuppressive effect** in animals. These agents have a mode of action that involves **interference with the cell cycle** and they therefore impact on the division and replication of lymphocytes, particularly those that are rapidly dividing. In dogs, the mostly widely used cytotoxic agents in immunosuppressive protocols are **azathioprine, chlorambucil** and **cyclophosphamide**. **Leflunomide** (an inhibitor of pyrimidine synthesis) and **mycophenolate mofetil** (mode of action similar to azathioprine) are also sometimes employed (but are more expensive drugs), with the latter currently being evaluated for its effect in the management of a range of immune-mediated diseases.

The combination of **glucocorticoid and azathioprine** is used to treat various **immune-mediated (autoimmune) diseases in the dog**. The drugs have an additive immunosuppressive effect, so combination therapy is generally used for disease of greater clinical severity or that does not respond to glucocorticoid alone. The combination also has a '**steroid-sparing**' effect, in that it allows a lower dose of glucocorticoid to be used in the maintenance phase of therapy, thus lowering the risk of adverse effects associated with prolonged steroid use.

Azathioprine is a pro-drug of 6-mercaptopurine, which in turn is an analogue of the natural purine hypoxanthine. Hepatic metabolism of azathioprine leads to the formation of nucleotides containing mercaptopurine, which interfere with purine synthesis. Generation of

Table 21.2 Tapering Protocol for Use of Immunosuppressive Oral Prednisolone in a Dog with Primary Immune-mediated Disease

WEEK	CLINICAL STATUS	DOSE OF ORAL PREDNISOLONE
0	First diagnosis of immune-mediated disease	1 mg/kg q12h (induction dose)
2–4	Disease controlled with resolution of clinical signs	0.75 mg/kg q12h
6	Clinical resolution maintained	0.5 mg/kg q12h
8	Clinical resolution maintained	0.25 mg/kg q12h
10	Clinical resolution maintained	0.25 mg/kg q24h
12	Clinical resolution maintained	0.25 mg/kg q48h
14	Clinical resolution maintained	Discontinue prednisolone. Continue to monitor clinical progress for the next 12 months

Typical protocol for management of primary immune-mediated (autoimmune) disease in which there is good clinical response to treatment within the first few weeks of treatment. If disease relapses during the tapering protocol, the same or a higher induction dose would be implemented with more prolonged tapering.

thioinosine monophosphate leads to further metabolism to thioguanosine monophosphate and thioguanosine triphosphate. Incorporation of these metabolites into cellular DNA leads to **DNA damage** and **blockade of mitosis and cellular metabolism**. Azathioprine therefore acts on the **S phase of the cell cycle (Figure 21.3)**. The most important immunosuppressive effects of azathioprine are on **T cell proliferation** and consequently on the provision of T cell help for B lymphocyte responses.

The pharmacokinetics of azathioprine involves degradation of the drug to inactive metabolites by enzymes including xanthine oxidase and thiopurine methyltransferase (TPMT). The activity of TPMT is genetically regulated and cats have lower activity of this enzyme than dogs. This means that cats are more sensitive to the adverse effects of the drug than are dogs. For this reason, azathioprine is generally **not recommended for use in the cat** and in this species chlorambucil is more commonly incorporated into combination immunosuppressive regimes. The most significant **adverse effect** of long-term administration of azathioprine is **bone marrow suppression**, while hepatotoxicity and pancreatitis have also been reported in dogs.

Mycophenolate mofetil interferes with purine synthesis in T and B lymphocytes. The drug is metabolized in the liver to mycophenolic acid, which inhibits inosine monophosphate dehydrogenase, an essential enzyme in the purine synthesis pathway. Use of mycophenolate has shown promise in dogs for treatment of immune-mediated haemolytic anaemia (IMHA) and immune-mediated thrombocytopenia (IMTP). There are limited studies of adverse effects, but in one study, gastrointestinal disorders were the most commonly reported.

Chlorambucil and **cyclophosphamide** are **alkylating agents** that cause breaks or formation of cross-linkages between DNA strands. These changes **inhibit DNA replication and transcription to RNA**. Alkylating agents therefore affect **dividing cells**; however, they also affect cells in the **intermitotic phase** and these drugs are thus **cell cycle non-specific** in their action (**Figure 21.3**). In the context of immune-mediated disease, alkylating agents are used for their effects on lymphocytes. The adverse effects of chlorambucil are **bone marrow suppression** and **gastrointestinal disturbance**, and those of **cyclophosphamide** include **leucopenia, haemorrhagic cystitis** and **gastrointestinal upset**. All cytotoxic drugs must be handled with care by those administering them, as they are potentially mutagenic, teratogenic and carcinogenic in nature.

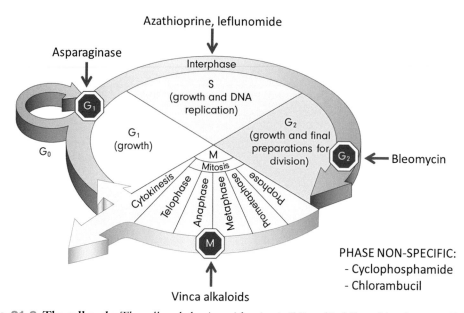

Figure 21.3 **The cell cycle.** The cell cycle begins with mitosis (M) and is followed by the gap 1 (G₁) phase during which there is RNA and protein synthesis. The G₁ phase may last for days or weeks depending on the tissue. Cells may then enter a prolonged resting state (G₀) from which they may return to G₁ to continue with the cell cycle. The S phase is relatively short (two hours) and involves DNA synthesis and there is further RNA and protein synthesis within the subsequent G₂ phase (six to eight hours in duration). The cytotoxic drug azathioprine acts during the S phase of the cell cycle, whereas cyclophosphamide and chlorambucil are not cell-cycle specific and act during the mitotic and inter-mitotic stages.

21.2.3 Ciclosporin

Ciclosporin is a potent immunosuppressive fungal polypeptide originally developed for management of rejection in human **organ transplantation**. This drug is specifically licensed for use in the dog for the management of atopic dermatitis (systemic administration) and in selected ocular immune-mediated diseases (via topical therapy). However, ciclosporin is finding increasing application in the management of a wider range of immune-mediated (autoimmune and idiopathic inflammatory) diseases in the dog and is often used in **combination with glucocorticoids** in this regard.

Ciclosporin has a more targeted mode of action than the drugs described previously and chiefly affects the function of **T lymphocytes**. When absorbed by these cells, ciclosporin binds to a cytoplasmic receptor called **cyclophilin**. The **ciclosporin–cyclophilin complex** subsequently binds to and blocks the action of the phosphatase **calcineurin**. In a normal T cell response, calcineurin is activated during the influx of Ca²⁺ into the cell that follows ligation of the TCR and stimulation of the CD3 complex and CD28 co-stimulatory receptor. Activated calcineurin (in combination with calmodulin) dephosphorylates the transcription factor **nuclear factor of activated T cells** (NF-AT), which is present in the cytoplasm of resting T cells. Upon dephosphorylation, NF-AT migrates into the nucleus and, in cooperation with other transcription factors like AP-1, induces the **transcription of cytokine and other immune response genes** by the T cell (**Figure 21.4**). The most important blocking effect is that on **IL-2 and IL-2Rα (CD25) production**, which inhibits activation

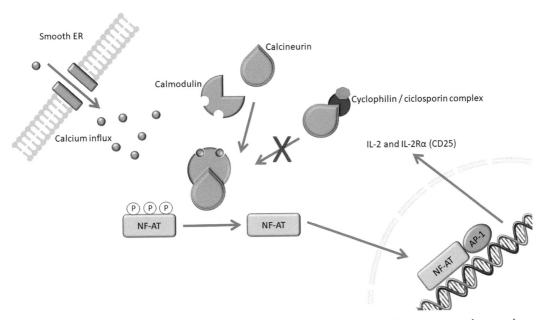

Figure 21.4 Mode of action of ciclosporin as a calcineurin inhibitor. Ciclosporin enters the cytoplasm of T cells and binds to a receptor called cyclophilin. The drug–receptor complex then binds to and blocks the action of calcineurin. Normally, when T cells are activated, calcineurin binds with calmodulin and dephosphorylates the 'nuclear factor of activated T cells' (NF-AT), which then migrates into the nucleus and binds to the transcription factor AP-1. The role of the NF-AT/AP-1 transcription factor complex is to induce the transcription of cytokine and cytokine receptor genes by the T cell. In particular, failure to produce IL-2 and the alpha subunit of the IL-2 receptor (CD25) impairs the ability of naïve T cell to become activated and undergo clonal proliferation.

and proliferation (clonal expansion) of antigen-specific naïve T cells. Although ciclosporin has a selective effect on cell-mediated immunity, the suppression of T cell help (Th2/Tfh) leads to indirect suppression of the B cell response. Incorporation of ciclosporin into cultures of mitogen-stimulated canine lymphocytes reduced transcription of genes encoding IL-2, IL-4 and IFN-γ, but not TNF. Blood lymphocytes from dogs treated with ciclosporin have been shown to have reduced cytoplasmic expression of IL-2 and IFN-γ cytokine proteins, but there was no effect on IL-4 expression. The related drug **tacrolimus** also inhibits T cell function by preventing the activation of calcineurin. Tacrolimus is often formulated into a **topical medication** and may be used to treat dogs with keratoconjunctivitis sicca, an autoimmune disease that affects the lacrimal gland, or used topically on anal furunculosis lesions in German shepherd dogs.

A range of **adverse effects** has been reported with the use of ciclosporin, including secondary infections or recrudescence of subclinical infection, gastrointestinal signs, gingival hyperplasia, hirsutism or hair shedding, and hepatic and renal toxicity when used at high dosage.

21.3 JANUS KINASE INHIBITORS

The JAK–STAT pathway of cellular activation, following binding of a cytokine to its cognate receptor, was described in Chapter 6. Novel JAK inhibitors have been developed to block this pathway and thereby inhibit the activity of a range of different cytokines. In human

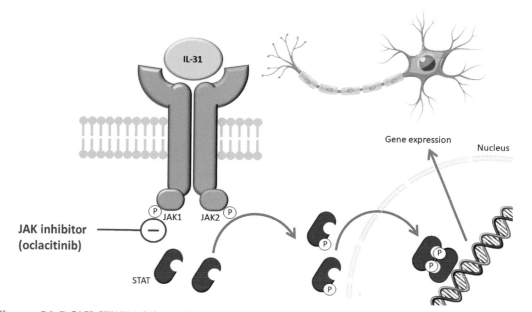

Figure 21.5 JAK-STAT inhibitor drug. The IL-31 receptor is expressed on nerve endings in the dermis. Binding of IL-31 to its receptor leads to phosphorylation of the JAK molecules and subsequent phosphorylation of STATs. When these dimerize, they migrate into the nucleus and act as transcription factors to activate the nerve cell. A JAK inhibitor drug (such as oclacitinib) prevents the intracellular signalling events following receptor binding. This inhibits the 'itch' signal and breaks the 'itch-scratch cycle' in dogs affected with atopic dermatitis.

medicine these agents are being developed for the treatment of autoimmune and inflammatory disorders, particularly rheumatoid arthritis. A JAK inhibitor, oclacitinib (Apoquel®), is used for the management of pruritus associated with canine atopic dermatitis. The drug inhibits all four JAK molecules (JAK 1, 2 and 3 and Tyk2), but is most effective in inhibiting JAK1, resulting in suppression of the function of several cytokines, including IL-31, a key mediator of the itch–scratch cycle (**Figure 21.5**).

21.4 INTRAVENOUS IMMUNOGLOBULIN THERAPY

Intravenous immunoglobulin (IVIG) therapy has been adopted from human medicine for the management of an increasing array of immune-mediated diseases. The process involves the **intravenous injection** of a high concentration of **purified human immunoglobulin**. This product is prepared from donated blood and in human medicine is used for the management of primary immunodeficiency diseases as well as inflammatory and autoimmune conditions. IVIG therapy has been successfully employed in the management of refractory **canine IMHA and IMTP**. The mechanisms by which IVIG suppresses these diseases are complex and not completely understood, but likely include the **binding** of human immunoglobulins to the **macrophage FcγR**, thereby blocking these receptors and preventing interaction with antibody-coated (i.e. opsonized) erythrocytes or platelets. IVIG therapy has also been demonstrated to be of benefit in the management of challenging cases of **autoimmune skin diseases** (particularly erythema multiforme), where the effect may relate to **inhibition of apoptosis** of target keratinocytes by blocking the **CD95 (FAS):CD95 ligand (FASL)** interaction. Human immunoglobulin

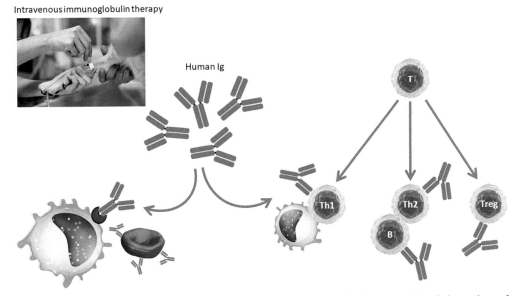

Intravenous immunoglobulin therapy

Human Ig

Figure 21.6 **Mode of action of intravenous human immunoglobulin therapy.** A single large dose of human immunoglobulin allows these molecules to bind to and block Fc receptors expressed by phagocytic cells. In patients with IMHA or IMTP this prevents antibody-coated red cells or platelets from being removed from the circulation. Some antibodies within the immunoglobulin preparation have the potential to bind to surface molecules expressed by lymphocytes and thus alter the function of these cells.

is also thought to contain a range of antibodies that can bind to various lymphocyte surface molecules, leading to inhibition of lymphocyte function (**Figure 21.6**), and recent studies have suggested that this approach might also induce activation of Treg cells in human patients with autoimmune diseases. Although purified canine or feline immunoglobulin would be preferrable (in order to avoid the potential for sensitization and hypersensitivity to human proteins), such preparations are not commercially available. Repeated administration of IVIG should not be performed to avoid allergic reactions. However, a single administration of human IVIG has not been associated with any major adverse effects.

21.5 ALLERGEN-SPECIFIC IMMUNOTHERAPY

Allergen-specific immunotherapy (ASIT) or 'desensitization' is widely practiced for the management of canine (and, less frequently,

feline) **atopic dermatitis**. In a referral setting, this process may be efficacious in up to 80% of patients. This approach requires identification of reactive allergens (see Chapter 18) followed by repeated administration of these allergens in increasing dose increments over a number of weeks, until a maintenance phase is reached. Allergens are formulated in **aqueous suspension** or may be **adjuvanted with aluminum salts** (**Figure 21.7**) and there are several different protocols by which each may be administered. The traditional method consists of subcutaneous injections of allergens. Alternative methods of delivery are sublingual administration and direct injection into lymph nodes under ultrasound guidance. The precise mode of action of ASIT is not completely understood, but it appears to involve the generation of **'blocking' IgG antibodies** that compete with mast cell-bound IgE for allergen, induction of T cell anergy and/or the **generation of Treg cells** that inhibit the function of allergen-reactive

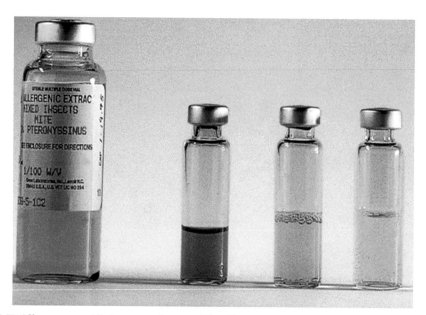

Figure 21.7 Allergen-specific immunotherapy. The allergens to which an individual patient is reactive are formulated into aqueous solutions of differing concentration (the darker vials are more concentrated). ASIT involves repeated administration of these allergens in increasing dose increments over an induction period of many weeks. Maintenance therapy requires less frequent administration. Allergens may also be alum adjuvanted and given by alternative protocols.

Th2 and Th1 effector cells. There is evidence for such mechanisms in canine atopic patients treated by ASIT, in which elevation in allergen-specific IgG and IL-10 gene transcription has been reported. Adverse reactions to ASIT are not common, although local swelling due to a type III hypersensitivity reaction to the allergen may occur following subcutaneous injections.

21.6 MONOCLONAL ANTIBODY THERAPY

The production of monoclonal antibodies has been described in Chapter 9. In addition to their use in diagnostic tests, they have found wide application in human medicine for the treatment of cancer, immune-mediated disorders and infectious diseases. The monoclonal antibody, **lokivetmab** (Cytopoint®), binds to **IL-31** and prevents it from binding to its receptor (**Figure 21.8**). IL-31 is secreted by Th2 cells in the skin of dogs with atopic dermatitis and causes pruritus by binding to receptors expressed on cutaneous nerves and neurons in the dorsal root ganglia. Subcutaneous injection of lokivetmab prevents IL-31 from binding to its receptor which reduces scratching and ameliorates the clinical disease. In addition, the blocking of IL-31 may also directly improve the barrier function and reduce overall cytokine secretion in the skin. The anti-canine IL-31 monoclonal antibody was first produced in mice. Repeated administration of a mouse immunoglobulin to dogs would induce an immune response to the mouse protein, which will subsequently impair the antibody activity and lead to the risk of an allergic reaction. To overcome this, scientists first created a **chimeric antibody** in which the Fc portion of the mouse monoclonal antibody was replaced with a canine Fc, while the antigen-binding Fab region remained the same. Subsequently, the constant framework regions of the

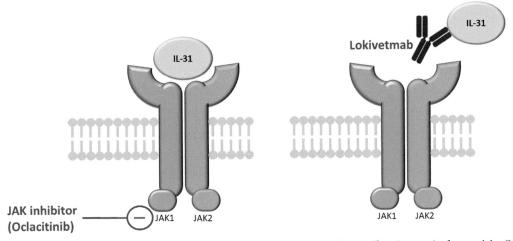

Figure 21.8 Monoclonal antibody approach to neutralizing the IL-31 effect in atopic dermatitis. On the left is shown the JAK-STAT inhibitor drug (oclacitinib) approach to blocking the intracellular signalling of the IL-31 receptor. On the right is shown an alternative approach, using a monoclonal antibody (lokivetmab) to neutralize the cytokine, blocking its ability to bind to the receptor.

V domains were replaced by canine sequences resulting in a **'caninized' monoclonal antibody** in which the entire protein is of dog origin, except for the three hypervariable regions in each of the light and heavy chains, which make up the antigen binding site. Although this greatly reduces the risk of an immune response to the therapeutic antibody, it does not completely eliminate it, as immune responses may develop that recognize the mouse sequences (anti-idiotype) or allotypic determinants on the canine portion of the protein. Such immune responses may cause a sudden or gradual loss of effectiveness of the antibody therapy.

Other monoclonal antibodies, bedinvetmab (Librela®) and frunevetmab (Solensia®) that target and neutralize nerve growth factor (NGF), can be used for the treatment of chronic pain associated with osteoarthritis in dogs and cats, respectively. The clinical success and economic feasibility of monoclonal antibody therapy in companion animals is likely to stimulate the development of other products. Monoclonal antibodies designed to treat cancer and other diseases of companion animals are under development. However, monoclonal antibod-

ies designed for treatment of canine lymphoma were only briefly available commercially before being withdrawn.

21.7 CYTOKINE THERAPY

A major advance in human medicine has been the commercial production of recombinant versions of a range of immunoregulatory cytokines. These can be directly injected into patients to manipulate the immune response for clinical benefit. For example, **recombinant human IFN-γ (rHuIFN-γ)** mimics the effect of this Th1-derived cytokine and thus be beneficial in a wide range of **infectious or neoplastic disorders**. The benefits of recombinant cytokine therapy have also been applied in veterinary medicine. One of the first applications was the use of **recombinant human granulocyte colony-stimulating factor** (rHuGCSF) or **recombinant human granulocyte–monocyte colony-stimulating factor** (rHuGMCSF) as a means of increasing bone marrow production of these leucocytes in canine chemotherapy patients. These products (e.g. Neupogen®) are able to restore subnormal blood neutrophil numbers, but if used repeatedly they will induce an immune response to the foreign human protein. The successful application of recombinant cytokine therapy to animals would therefore necessitate production of species-specific molecules. In fact, **recombinant canine GCSF and GMCSF** have both been produced, but are not commercially available. **Recombinant canine IFN-γ (Interdog®)** has been developed and is licensed in Japan for the treatment of canine atopic dermatitis.

As discussed in Chapter 3, **type I interferons** provide some degree of antiviral defence. Oral administration of **human interferon-α** may reduce the severity of inflammatory airway disease (asthma) in horses and viral infections in cats. As the protein is degraded in the gastrointestinal tract, its actions are directed at immune cells in oropharyngeal lymphoid tissues. The use of the human cytokine can induce anti-IFNα antibodies which reduce the effectiveness of subsequent treatments. A **recombinant feline IFN-ω** (Virbagen® Omega) is commercially available and may be of value in treatment of cats with persistent viral infections, such as FeLV, FIV or FIP.

Another approach is to use **cytokine genes**, incorporated into **bacterial plasmid DNA** injected into the patient. This procedure is not dissimilar to the use of DNA vaccines (see Chapter 13). The cytokine gene would be expressed within host cells and local production of the cytokine may direct the nature of the immune response. These effects might occur within lymphoid tissue or at the site of an infectious or neoplastic lesion. A canarypox virus-vectored IL-2 gene (Oncept IL-2) is used as an adjunct therapy in fibrosarcoma in cats.

21.8 IMMUNOSTIMULATORY AGENTS

Vaccination is designed to protect animals against specific infectious diseases by stimulating the adaptive immune system, resulting in the production of antibodies secreted by long lived plasma cells and memory B and T lymphocytes. However, effective vaccines are only available for relatively few infectious diseases. In addition, several diseases have a **multifactorial etiology**, a combination of viral and bacterial infections alongside environmental and management factors such as stress. A good example of this type of complex disease is **shipping fever bronchopneumonia** in feedlot cattle. The disease occurs in calves brought together from different farms often following long distance transport. The stress of transport along with exposure to viruses and mycoplasma, makes the animals susceptible to bacterial infections caused by *Mannheimia haemolytica*, *Pasteurella multocida* and *Histophilus somni*. It might be possible to increase the resistance of animals to such a disease by stimulating innate immunity. Although the innate immune system does not have the capacity for long term memory (the characteristic feature of the adaptive immune response) it is possible to induce a short-term boost (**trained immunity**). The underlying mechanisms are still being investigated, but trained immunity likely involves an increased number of innate immune cells in the blood, as well as increased cellular activation status. Immunostimulatory compounds may also be used to try to enhance or trigger the **anti-tumour immune response**. Such products are injected directly into the tumour or sometimes administered systemically.

A wide variety of 'health products' are on the veterinary market for preventive or therapeutic use in animals that claim to have immunostimulatory properties, although the evidence to support such claims is often lacking. These products are often relatively crude preparations, mostly **extracts from microorganisms or plants**. Unfortunately, mechanistic studies and scientific data from domestic animal species to support their use are often not available. Beta glucans are used as feed additives in aquaculture, poultry and swine and consist of linear or branched polysaccharides derived from plants, bacteria, yeast, fungi and algae. The size and structure of beta glucans varies based on the source and this affects their immunostimulatory property. Beta glucans are recognized by several pattern recognition receptors expressed on innate immune cells, including neutrophils and macrophages. These receptors include the C-type lectin receptor Dectin-1 and complement receptor 3 (CR3). Administration of beta glucans to animals increases the number of monocytes, phagocytosis and the ability of macrophages to kill microorganisms.

The innate immune system has several receptors that can recognize DNA, in particular **unmethylated CpG DNA motifs** that are more common in **bacterial DNA**. These receptors are present in endosomes (**TLR9**) and in the cytoplasm (**AIM2** and **cGAS-STING**) of macrophages. Intramuscular injection of **bacterial plasmid DNA encapsulated in lipid nanoparticles** (Zelnate®) in calves immediately before or soon after transport is aimed at reducing the incidence and severity of shipping fever bronchopneumonia. Research suggests that this product stimulates macrophages primarily via the cGAS-STING pathway.

Intravenous injections of extracts of **_Propionibacterium acnes_** (ImmunoRegulin®) and **mycobacteria** (Equimune®) may be used to treat horses with infectious diseases or prevent disease triggered by stressful events such as transport. These products can also be injected intra-lesionally into equine sarcoids in an attempt to activate the immune system and induce regression of the tumours. They include several microbial ligands for pattern recognition receptors, including DNA and bacterial cell wall components. Similarly, intramuscular injection of inactivated **_Parapox ovis_ virus** (Zylexis®) may be used to reduce the severity of respiratory disease in horses caused by equine herpes viruses and aggravated by stressful events.

KEY POINTS

- Glucocorticoids differ in formulation, potency and duration of action.
- Glucocorticoids have both anti-inflammatory and immunosuppressive properties.
- Glucocorticoids bind to cytoplasmic receptors. The drug–receptor complex moves into the nucleus and binds to steroid response elements within promoters, leading to gene activation or repression.
- Cats are relatively 'steroid resistant' compared with dogs, as they may have a lower expression of glucocorticoid receptors within cells.
- Prednisolone can decrease blood lymphocyte numbers and the concentration of serum IgG, IgM and IgA in normal dogs.
- Prolonged use of glucocorticoids may lead to iatrogenic hyperadrenocorticism.
- Glucocorticoids must be slowly tapered as abrupt withdrawal may lead to signs of iatrogenic hypoadrenocorticism.
- Azathioprine is used in conjunction with glucocorticoid for additive immunosuppressive and steroid-sparing effects.
- Azathioprine interferes with the S phase of the cell cycle.
- Cats are sensitive to azathioprine as they have lower activity of TPMT.
- Prolonged use of azathioprine may cause bone marrow suppression.

- Chlorambucil and cyclophosphamide are alkylating agents and act in a non-cell cycle-specific manner.
- Ciclosporin inhibits T cell activation by blocking NF-AT activation and cytokine (particularly IL-2) gene transcription.
- Allergen-specific immunotherapy in atopic dermatitis may work by inducing blocking IgG antibodies and IL-10-producing Treg cells.
- IVIG therapy may be effective in IMHA/IMTP by blocking macrophage Fc receptors and in other immune-mediated diseases by binding to lymphocyte surface molecules and modulating the function of these cells.
- Caninization of monoclonal antibodies reduces the chance of induction of an immune response against the therapeutic antibody.
- An anti-IL-31 monoclonal antibody is used in atopic dermatitis in dogs to reduce itching and inflammation.
- Anti-NGF monoclonal antibodies can be used for chronic pain in canine and feline osteoarthritis.
- Recombinant cytokines may be used to modulate immune responses.
- Extracts of plants or microorganisms are used for their stimulatory properties on the innate immune system.

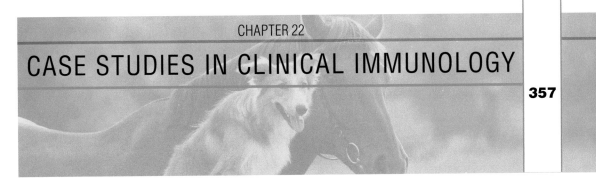

22.1 INTRODUCTION

The purpose of this final chapter is to place the fundamental knowledge of immunology, presented previously in this textbook, into a clinical context. To that end, this chapter presents a number of clinical case studies, each of which provides an example of an aspect of immunology as it relates to veterinary clinical practice. The information given on each case cannot be exhaustive, instead the major clinical and laboratory diagnostic features are presented, together with a discussion of the underlying immunological mechanism(s) and, where appropriate, the therapy employed.

22.2 THE BASIS OF IMMUNE-MEDIATED DISEASE

The four major categories of immune-mediated disease are **immunodeficiency** diseases, **hypersensitivity** (allergic) diseases, **autoimmune** diseases and **immune system tumours**. The common basis for all of these, is that they are **multifactorial** in aetiology and have a pathogenesis that reflects a **disturbance in normal immune system homeostasis**.

A helpful way of thinking about immune-mediated disease is using the **analogy of the iceberg (Figure 22.1)**. An iceberg has the main part of its structure hidden beneath the waterline with the peaks of the structure visible from the surface. In this model, the 'body' of the iceberg (beneath the waterline) represents the immune system. The immune system is surrounded by a 'sea' of triggering and predisposing factors,

which when present in appropriate combination, can impact on the immune system leading to its dysregulation. The clinical expression of immune system dysregulation, or the visible peak of the iceberg, is immune-mediated disease. The iceberg model also helps to explain the interrelationship between different immune-mediated diseases. There are several examples of concurrent or temporally distant immune-mediated diseases occurring in a single individual. These instances (e.g. the concurrence of IMHA in dogs with underlying lymphoma and the occurrence of myasthenia gravis associated with the presence of a thymoma) are situations where the surrounding factors have led to immune system dysregulation, which is expressed in two distinct fashions (as two peaks of the iceberg protruding above the waterline).

22.3 HOW COMMON IS IMMUNE-MEDIATED DISEASE?

It is important for veterinary graduates to appreciate the clinical relevance of different types of immune-mediated disease. For example, although **primary immunodeficiency** diseases are fascinating and help to explain various aspects of the immune response, they are **relatively rare** conditions and many practitioners will not encounter one of these diseases throughout their practice career. In contrast, **allergic diseases** are of **major clinical importance** and the majority of companion animal practitioners will spend a considerable amount of time dealing with patients affected by flea allergic or atopic dermatitis.

DOI: 10.1201/9781003310969-22

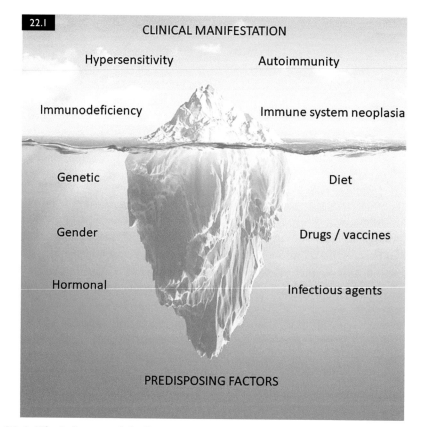

Figure 22.1 **The iceberg model of immune-mediated disease.** In this model the 'body' of the iceberg (beneath the waterline) represents the immune system. Predisposing and trigger factors (in the sea surrounding the iceberg) impact on the immune system, leading to its dysregulation. Immune system dysregulation is expressed clinically as one or more of the four types of immune-mediated disease (seen as peaks of the iceberg above the waterline) occurring contemporaneously or sequentially in the patient.

Autoimmunity and immune system neoplasia fall between these two ends of the spectrum. **Immune system neoplasia** (lymphoma, leukaemia or mast cell neoplasia) is **relatively common** in companion animals. Laboratory confirmed **autoimmune disease** is probably **less common**, but in first opinion practice, diseases such as IMHA, IMTP or pemphigus foliaceus will likely be encountered from time to time. Many animals with a vague collection of clinical signs that respond to glucocorticoid therapy are suggested to have 'immune-mediated disease', but in the absence of definitive laboratory data or a specific diagnosis, such a diagnosis is not necessarily substantiated. Having made these generalizations, it is important

to realize that there is a lack of large scale published epidemiological evidence; although more recently, there are a few studies that present the prevalence of these immune-mediated diseases with data derived from analysis of large multi-practice computerized databases.

22.4   IMMUNODIAGNOSTIC TESTING

Earlier chapters in this book (e.g. Chapter 10) have described the range of **immunodiagnostic procedures** that may be used to support a diagnosis of immune-mediated disease. The veterinary student should also be aware that several of these tests are **difficult to access** from veterinary practice as they may not be available commercially. Most clinical pathology laboratories will offer serological detection of allergen-specific IgE or some form of Coombs test. More specialist laboratories may offer molecular diagnostic tests for primary immunodeficiency, measurement of serum immunoglobulin concentration and flow cytometric phenotyping, but such tests are often difficult to source, relatively expensive and sometimes require samples that are not readily shipped over long distances. Veterinarians are cautioned against sending animal samples to local human medical diagnostic laboratories. Those immunological tests that are based on the use of antisera generally require the use of species-specific reagents and the interpretation of immunodiagnostic tests for animals is a specialized skill that should not (and legally cannot) be undertaken by unqualified individuals.

22.5 CASE 1: CANINE IMMUNE-MEDIATED HAEMOLYTIC ANAEMIA

22.5.1 Signalment
Six-year-old, neutered male English springer spaniel (**Figure 22.1a**; photo courtesy S Warman).

22.5.2 History and Physical Examination
The dog presented with a two-week history of increasing lethargy, weakness and reduced appetite. There have been no recent changes to the lifestyle of the animal and no recent administration of drugs or vaccines. The dog has lived all of its life in northern Europe in an area non-endemic for arthropod-borne infectious diseases such as babesiosis, ehrlichiosis and leishmaniasis. Clinical examination reveals pallor of the oral and conjunctival mucous membranes, mild pyrexia and mild elevation in heart and respiratory rate. Abdominal palpation reveals enlargement of the spleen.

22.5.3 Diagnostic Procedures
Routine haematology, serum biochemistry and urinalysis were performed initially. Haematological examination (see more on this later) revealed a strongly regenerative anaemia and leucocytosis due to a left shift neutrophilia and mild monocytosis. Examination of the blood smear revealed polychromasia, anisocytosis and spherocytosis with prominent nucleated erythrocytes (**Figure 22.1b**).

22.1a

PARAMETER	VALUE	NORMAL RANGE
PCV	0.13	0.35–0.55 l/l
Haemoglobin	50	120–180 g/l
Red blood cells	1.68	$5.4–8.0 \times 10^{12}$/l
MCV	80.1	65–75 fl
MCHC	373	340–370 g/l
MCH	29.8	22–25 pg
Absolute reticulocytes	280	$\leq 60 \times 10^9$/l
Platelets	251	$170–500 \times 10^9$/l
White blood cells	19.8	$5.5–17 \times 10^9$/l
Neutrophils	15.5	$3–11.5 \times 10^9$/l
Band neutrophils	1.67	$0–0.3 \times 10^9$/l
Lymphocytes	0.84	$0.7–3.6 \times 10^9$/l
Monocytes	1.68	$0.1–1.5 \times 10^9$/l
Eosinophils	0.3	$0.2–1.4 \times 10^9$/l
Basophils	0.0	$0–0.1 \times 10^9$/l
Nucleated red blood cells	2.9×10^9/l	

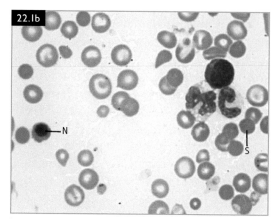

N, nucleated RBC; S, spherocyte.

Serum biochemistry and protein electrophoresis revealed the presence of a mild polyclonal hypergammaglobulinaemia, but there was no significant elevation in serum bilirubin or liver enzymes. Urinalysis was unremarkable. Survey radiographs of the thorax and abdomen were obtained, but they revealed no abnormalities apart from the diffuse enlargement of the spleen identified on clinical examination. On the basis of the haematological data, a Coombs test and a serum ANA were performed. The Coombs test was positive (see more on this later) and the ANA test was negative.

22.5.4 Diagnosis and Immunopathology

Springer spaniels have a recognized genetic predisposition to immune-mediated haemolytic anaemia (IMHA). Common underlying trigger factors (drugs, vaccination, infectious disease or neoplasia) are ruled out by historical and diagnostic investigation. The haematological findings here are strongly suggestive of IMHA, in particular the presence of spherocytosis. This suspicion is confirmed by the Coombs test, which demonstrates the presence of a warm-reactive IgG antibody, bound to the red cells in this dog.

This dog has primary idiopathic (autoimmune) IMHA in which IgG antibody attached to antigen on the surface of erythrocytes mediates their removal and partial phagocytosis (leading to the formation of spherocytes) by macrophages within the spleen and liver (extravascular haemolysis). This is an example of a type II hypersensitivity reaction (see Chapter 17). The diffuse splenomegaly in this dog is likely caused by the combination of erythrocyte removal and extramedullary haematopoiesis. The anaemia is strongly regenerative and the dog has a relatively chronic-onset disease for which it compensates physiologically (elevated heart and respiratory rate). Serum hypergammaglobulinaemia likely reflects immune activation and autoantibody production.

22.5.5 Treatment and Prognosis

This dog was treated with an immunosuppressive dose of oral prednisolone. A good response to therapy was observed and the drug was slowly tapered and eventually withdrawn (see Chapter 21). In refractory cases, addition of other immunosuppressive drugs (e.g. azathioprine or ciclosporin) may need to be considered. The owner was warned that the pattern for IMHA is of a relapsing disease, so the dog should be closely monitored by regular PCV checks.

22.6 CASE 2: CANINE IMMUNE-MEDIATED THROMBOCYTOPENIA

22.6.1 Signalment

Five-year-old, neutered female English cocker spaniel (**Figure 22.2a**; photo courtesy S Warman).

ANTISERUM	TITRE AT 37°C	TITRE AT 4°C
Polyvalent Coombs reagent	640	1,280
Anti-dog IgG	2,560	2,560
Anti-dog IgM	0	0
Anti-dog complement C3	0	0

22.2a

PARAMETER	VALUE	NORMAL RANGE
PCV	0.22	0.35–0.55 l/l
Haemoglobin	75.7	120–180 g/l
Red blood cells	2.82	5.4–8.0 × 10^{12}/l
MCV	76.5	65–75 fl
MCHC	350	340–370 g/l
MCH	26.8	22–25 pg
Absolute reticulocytes	Not counted	≤60 × 10^9/l
Platelets	2.0	170–500 × 10^9/l
White blood cells	8.11	5.5–17 × 10^9/l
Neutrophils	4.71	3–11.5 × 10^9/l
Band neutrophils	0.0	0–0.3 × 10^9/l
Lymphocytes	1.30	0.7–3.6 × 10^9/l
Monocytes	1.46	0.1–1.5 × 10^9/l
Eosinophils	0.0	0.2–1.4 × 10^9/l
Basophils	0.0	0–0.1 × 10^9/l
Nucleated red blood cells	0.5 × 10^9/l	

22.6.2 History and Physical Examination

The dog presented with acute onset epistaxis, haematuria, melaena and haematemesis. There have been no recent changes to the lifestyle of the animal and no recent administration of drugs or vaccines. There has been no known contact with anticoagulant rodenticides. The dog has lived all of its life in Northern Europe in an area non-endemic for arthropod-borne infectious diseases such as babesiosis, ehrlichiosis and leishmaniosis. Clinical examination revealed pallor of the oral and conjunctival mucous membranes, mild pyrexia and petechial haemorrhages of the oral mucous membranes and ventral abdominal skin.

22.6.3 Diagnostic Procedures

Routine haematology, serum biochemistry, urinalysis and coagulation profile were performed initially. Haematological examination revealed a moderate regenerative anaemia and severe thrombocytopenia.

Examination of the blood smear revealed polychromasia, anisocytosis and the presence of nucleated erythrocytes. Serum biochemistry indicated mild elevation in ALP and AST. Urinalysis was unremarkable. The prothrombin time (PT) and activated partial thromboplastin time (APTT) were within normal parameters. Survey radiographs of the thorax and abdomen revealed no abnormalities. On the basis of the haematological data, a Coombs test and a serum ANA were performed, but both were negative. The laboratory does not provide an antiplatelet antibody test.

22.6.4 Diagnosis and Immunopathology

Cocker spaniels have a recognized genetic predisposition to developing immune-mediated cytopenias. Common underlying trigger factors (drugs, vaccination, infectious or inflammatory disease or neoplasia) are ruled-out by historical and diagnostic investigation. The haematological findings are strongly suggestive of immune-mediated thrombocytopenia (IMTP) with secondary blood loss anaemia. There is no widely available test for antiplatelet antibodies, so IMTP generally remains a diagnosis of exclusion.

This dog has primary idiopathic (autoimmune) IMTP in which IgG antibody bound to the surface of platelets mediates their removal by macrophages within the spleen and liver or intravascular complement-mediated platelet lysis. This is an example of a type II hypersensitivity reaction (see Chapter 17). The virtual absence of platelets means that the dog is unable to form a primary platelet plug at the site of vascular damage, resulting in the observed bleeding disorder.

22.6.5   Treatment and Prognosis

The dog was treated with an immunosuppressive dose of oral prednisolone and an injection of vincristine. Vincristine enhances bone marrow thrombopoiesis and inhibits macrophage removal of antibody-coated platelets. The dog made no response to therapy and continued to deteriorate. Three days after presentation the owners elected euthanasia. A necropsy examination was performed, which revealed severe gastric haemorrhage with blood clot overlying the gastric mucosa (**Figure 22.2b**). There was active erythropoiesis and plentiful megakaryocytes within the bone marrow, spleen and liver.

IMTP is a severe disease and a number of animals (as in this case) will fail to respond to therapy. Those animals that recover may have relapses of IMTP or develop other autoimmune diseases later in life.

22.7   CASE 3: CANINE SYSTEMIC LUPUS ERYTHEMATOSUS

22.7.1   Signalment

Seven and a half-year-old, male English cocker spaniel (**Figure 22.3a**).

22.7.2   History and Physical Examination

This dog presented with a history of chronic skin disease that had developed progressively over the previous 11 months. Nine months ago the dog had been diagnosed with keratoconjunctivitis sicca (KCS) and was being treated for that condition with topical ciclosporin. Physical examination revealed the presence of severe generalized crusting skin lesions with areas of alopecia. Additionally, there was conjunctival discharge, otitis externa and generalized lymphadenomegaly. The dog had pale ocular and oral mucous membranes. The owner reported removing several ticks from the dog one month before the onset of the skin disease. The dog lives in Northern Europe in an area non-endemic for arthropod-borne infectious diseases such as babesiosis, ehrlichiosis and leishmaniosis.

22.7.3   Diagnostic Procedures

A swab taken from the external ear canal revealed the presence of mixed bacterial species.

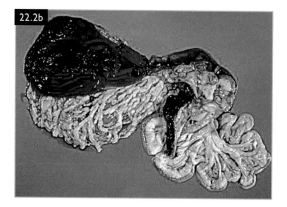

22.2b

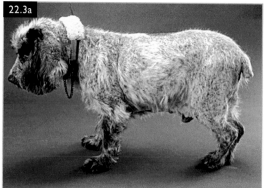

22.3a

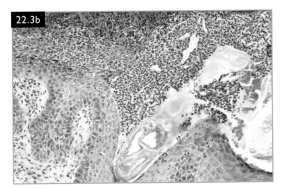

Figure 22.3b From Foster AP, Sturgess CP, Gould DJ et al. [2000] Pemphigus foliaceus in association with systemic lupus erythematosus with subsequent lymphoma in a Cocker Spaniel dog. *Journal of Small Animal Practice* **41**:266–270, with permission.

PARAMETER	VALUE	NORMAL RANGE
PCV	0.25	0.35–0.55 l/l
Haemoglobin	109	120–180 g/l
Red blood cells	2.93	5.4–8.0 × 10^{12}/l
MCV	86.6	65–75 fl
MCHC	431	340–370 g/l
MCH	37.2	22–25 pg
Absolute reticulocytes	83.0	≤60 × 10^9/l
Platelets	9.0	170–500 × 10^9/l
White blood cells	43.7	5.5–17 × 10^9/l
Neutrophils	22.3	3–11.5 × 10^9/l
Band neutrophils	2.19	0–0.3 × 10^9/l
Lymphocytes	6.12	0.7–3.6 × 10^9/l
Monocytes	3.93	0.1–1.5 × 10^9/l
Eosinophils	9.18	0.2–1.4 × 10^9/l
Nucleated red blood cells	Sparse on blood film	

ANTISERUM	TITRE AT 37°C	TITRE AT 4°C
Polyvalent Coombs reagent	40	640
Anti-dog IgG	640	1,280
Anti-dog IgM	0	0
Anti-dog complement C3	0	0

Schirmer tear test readings confirmed the presence of KCS. Skin punch biopsies demonstrated the formation of subcorneal pustules filled by eosinophils and acanthocytes (**Figure 22.3b**). These features are consistent with pemphigus foliaceus and the presence of IgG antibody associated with the desmosomes of the keratinocytes was confirmed by immunohistochemistry.

Haematological examination revealed a mildly regenerative anaemia, marked thrombocytopenia and leucocytosis due to a left shift neutrophilia, monocytosis, lymphocytosis and eosinophilia (see overleaf). Serum biochemical abnormalities included hypoalbuminaemia and hypergammaglobulinaemia. Urinalysis was unremarkable.

On the basis of these findings a Coombs test, serum ANA and rheumatoid factor (RF) were performed. The Coombs test was positive and the ANA titre was 10,240 with homogeneous nuclear staining. RF was negative. Antiplatelet antibody testing was not available. The concentrations of serum IgG, IgM and IgA were measured and the hypergammaglobulinaemia attributed to a selective elevation in IgG.

A bone marrow aspirate was taken, which revealed a normal myeloid:erythroid ratio but with a prominent population of eosinophil precursors. Biopsy of an enlarged lymph node demonstrated reactive and inflammatory changes. A screen for arthropod-borne infectious diseases (serology and PCR) revealed no evidence of exposure to *Bartonella*, *Rickettsia*, several *Ehrlichia* spp. or *Borrelia*. Survey radiographs of the thorax and abdomen showed no abnormalities.

22.7.4 Diagnosis and Immunopathology

This dog has at least four concurrent manifestations of autoimmunity affecting multiple body systems: KCS, pemphigus foliaceus, IMHA and IMTP. Two of these are proven to be associated with the presence of autoantibody (interepithelial IgG within the skin and positive Coombs test). It is likely that antiplatelet antibodies would have been demonstrated had testing been possible. KCS is likely to involve infiltration of lacrimal glandular tissue with pro-inflammatory T lymphocytes. Thus, this autoimmune reaction involves a combination of type II and type IV hypersensitivity mechanisms (see Chapter 17). Additionally, the dog has a very high-titre serum ANA. These findings readily satisfy the criteria for SLE. Cases of true SLE are rare in veterinary medicine; many animals may only partially satisfy the diagnostic criteria and are considered to have an 'SLE overlap syndrome'. At the time of diagnosis this dog was considered to have primary idiopathic multisystemic immune-mediated disease. There was no historical or diagnostic evidence of underlying trigger factors (drug or vaccine administration, neoplasia, infection), although the cocker spaniel breed has a recognized susceptibility to autoimmune disease.

22.7.5 Treatment and Prognosis

The dog was treated with glucocorticoid therapy, medicated baths, antimicrobials and ocular ciclosporin with tear replacement. There was a response to this therapy within three weeks. Seven months after initial presentation the dog re-presented with marked enlargement of peripheral lymph nodes together with enlargement of thoracic and abdominal nodes and diffuse splenomegaly. At this time, lymph node biopsy and immunohistochemistry confirmed a diagnosis of B cell lymphoma. A histological section from the biopsy sample is shown in **Figure 22.3c** and immunohistochemical labelling of the tumour cells with antibody to the B cell

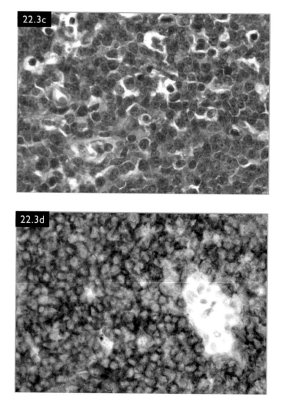

marker CD79a is shown in **Figure 22.3d**. The owner elected for no further treatment and the dog died two months later. This fascinating case demonstrates two distinct 'peaks of the iceberg' (autoimmunity and immune system neoplasia) appearing successively in this dog.

22.8 CASE 4: FELINE BLOOD TRANSFUSION

22.8.1 Signalment

Four-year-old, neutered male British short-hair cat.

22.8.2 History and Physical Examination

This cat was undergoing investigation for a primary problem of chronic non-regenerative anaemia. While hospitalized the PCV decreased from 15% to 10%. The clinician decided that a supportive whole blood transfusion was required.

22.8.3 Diagnostic Procedures

Feline blood transfusion carries a relatively high risk of transfusion reaction. For that reason, all blood transfusions in cats should be based on both blood typing and cross-matching. The major blood group antigen system of the cat consists of A, B or AB types. Another blood group antigen is Mik.

The prevalence of blood type varies by breed and geographical location (and therefore gene pool). The British shorthair is one breed in which there is a recognized higher prevalence of the B blood group. It is therefore very important to ensure that this patient receives compatible blood, as the cat may have high-titre anti-A alloantibody and a mismatched transfusion of type A or AB blood would lead to an adverse reaction.

The blood type of the recipient may be determined by submitting a sample to a specialist diagnostic laboratory or by the use of a rapid in-house test kit. Feline blood typing cards (see Chapter 10) are widely used in practice. This patient was typed using a new tube-based test. The plastic card comprises a set of small columns containing a gel matrix impregnated with typing reagent. A small volume of washed patient erythrocytes is loaded to the top of the column and the card is centrifuged. If the erythrocytes carry a surface antigen reactive with the reagent in the gel, they are unable to pass through the matrix and form a band at the

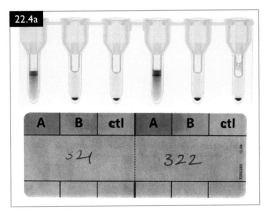

upper surface of the gel. In a negative reaction the cells pass through the gel and button at the base of the column. In this instance it can be seen that the cat (number 322 on the card) is of blood type A (**Figure 22.4a**).

As this cat is of blood group A, it should be able to safely receive whole blood from a type A donor. However, it remains possible that there may be minor incompatibilities (for example related to the Mik antigen) and so a cross-match should still be performed. The aim of the cross-match is to determine whether serum from the recipient cat contains antibody that might react with antigens expressed on the erythrocytes of the donor (major cross-match). The reverse possibility, that donor serum transfused into the recipient might contain antibodies able to react with recipient erythrocytes (minor cross-match), is also tested. Blood from

PHENOTYPE (BLOOD GROUP)	GENOTYPE	INHERITANCE	ALLOANTIBODY
A	AA, Aa^{ab} or Ab	A is dominant over B	Some may have low-titre anti-B
B	bb	B is recessive to A	All have high-titre anti-A
AB	$a^{ab}b$ or $a^{ab}a^{ab}$	Determined by a third allele a^{ab} that is recessive to A and dominant over b	Have no alloantibody
Mik		Some type A cats are Mik⁻, but most are Mik⁺	Mik⁻ cats may have alloantibody to Mik

REACTION NUMBER	COMPONENTS	NATURE OF TEST
1	Recipient serum + donor RBC	Major cross-match
2	Recipient serum + recipient RBC	Control
3	Donor serum + recipient RBC	Minor cross-match
4	Donor serum + donor RBC	Control

several possible donors is normally evaluated. There are a number of methods of performing a cross-match ranging from a simple emergency slide agglutination test to a more complex procedure performed in microtitration plates with the addition of complement. An in-practice card-based cross-match test is also available commercially. The basic procedure for the plate test involves the incubation of serum with a suspension of washed erythrocytes (summarized later in the chapter). In this instance a cross-match was performed with blood from a panel of known type A donor cats and a suitable donor animal identified.

22.8.4   Treatment and Prognosis

Blood was collected from the type A donor cat (**Figure 22.4b**; photo courtesy S Warman) and successfully transfused into the recipient animal, leading to elevation of its PCV. Subsequent diagnostic testing revealed that the non-

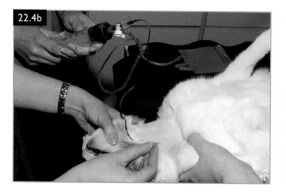

22.4b

regenerative anaemia in this patient was likely to be related to FeLV infection and therefore the long-term prognosis was guarded.

22.9   CASE 5: FELINE MULTIPLE MYELOMA

22.9.1   Signalment

Twelve-year-old, neutered female domestic shorthair cat.

22.9.2   History and Physical Examination

This cat presented with a three-week history of intermittent vomiting and diarrhoea. An endoscopic examination was scheduled to investigate the problem and initial blood samples were taken for haematological and biochemical examination.

22.9.3   Diagnostic Procedures

No haematological abnormalities were detected and the cat was negative for FeLV and FIV by standard serological testing. The main elements of the serum biochemistry data are presented as follows.

PARAMETER	VALUE	NORMAL RANGE
Urea	9.8	6.5–10.5 mmol/l
Creatinine	129	133–175 µmol/l
Total protein	104.1	77–91 g/l
Globulin	77.5	21–51 g/l
Albumin	26.6	24–35 g/l
A:G ratio	0.34	0.4–1.3
Alanine aminotransferase	24	15–45 IU/l
Alkaline phosphatase	33	15–60 IU/l
Total bilirubin	9.9	0–10 µmol/l
Fasting bile acids	15.1	0–15 µmol/l
Calcium	2.69	2.3–2.5 mmol/l

The finding of elevated globulins necessitates further evaluation by serum protein electrophoresis and a densitometric scan of this analysis is presented (**Figure 22.5a**). This reveals the presence of a striking monoclonal gammopathy. Differential diagnoses for this change in cats include multiple myeloma and occasional cases of feline infectious peritonitis (FIP), ehrlichiosis or leishmaniosis. This cat did not have any obvious signs of FIP (although there is currently no reliable antemortem diagnostic test for this infection) and it had not travelled beyond Northern Europe into an area endemic for the relevant arthropod-borne infectious diseases.

Given that multiple myeloma often arises within bone marrow, survey radiography was performed but did not reveal the presence of classical 'punched out' osteolytic foci. A bone marrow aspirate did not reveal any abnormality in haemopoietic precursors and only a low number of plasma cells were noted. The imaging examination did, however, reveal marked enlargement of the spleen and mesenteric lymphadenomegaly. Urinalysis was performed (some cases of multiple myeloma have urine Bence-Jones protein) and although there was a mild proteinuria, the urine protein:creatinine (UPC) ratio was normal. A simple agar gel diffusion test with serum from the patient confirmed that the monoclonal gamma globulin was an IgG antibody.

Exploratory laparotomy confirmed the splenomegaly (**Figure 22.5b**) and lymphadenomegaly.

The liver was noted to have a mottled red–tan parenchyma. A splenectomy was performed and biopsy samples were taken from the mesenteric lymph nodes and liver. Endoscopic biopsy samples of the gastrointestinal mucosa were also collected.

Histopathological examination revealed mild pyogranulomatous enteritis, reactive hyperplasia of the mesenteric lymph nodes, peliosis hepatis (blood-filled spaces in the liver parenchyma) and diffuse infiltration of the spleen by pleomorphic, occasionally multinucleate and mitotic plasma cells (**Figure 22.5c**).

22.9.4 Diagnosis and Immunopathology

The final diagnosis in this case is multiple myeloma with IgG paraproteinaemia. The tumour appears primarily to involve the spleen rather than bone marrow and there was no significant hypercalcaemia of malignancy in this

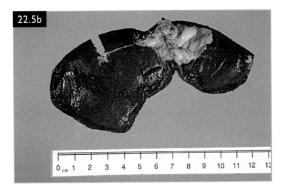

22.5b

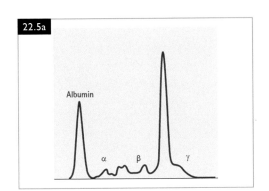

22.5a

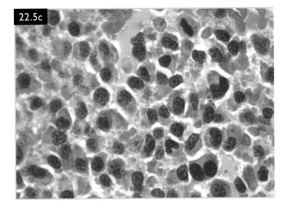

22.5c

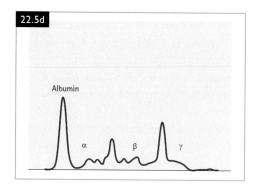

22.5d

Albumin

α   β   γ

case. It is increasingly recognized that feline myeloma more commonly affects abdominal soft tissue than bone.

22.9.5 Treatment and Prognosis

In this cat the monoclonal gammopathy began to resolve soon after splenectomy, suggesting that the primary neoplasm and source of the paraprotein had been removed. A follow-up serum protein electrophoresis is shown (**Figure 22.5d**).

22.10 CASE 6: EQUINE PURPURA HAEMORRHAGICA

22.10.1 Signalment

Eight-month-old Thoroughbred colt.

22.10.2 History and Physical Examination

The animal presented with a four-day history of progressive facial swelling. Clinical examination confirmed that this was because of subcutaneous oedema. The colt was also pyrexic and had an elevated heart rate. Parenteral penicillin was administered and there was an initial reduction in pyrexia and facial oedema. However, two days later the clinical signs returned and the oedema also involved the ventral cervical and thoracic area. The colt was pyrexic, had an elevated heart rate and petechial haemorrhages were present on the oral mucous membranes. Additionally, there was a bilateral sanguineous nasal discharge. The colt died 12 hours later.

22.10.3 Diagnostic Procedures

Necropsy examination confirmed the presence of marked anteroventral subcutaneous oedema (**Figure 22.6a**). There was also pulmonary oedema and an accumulation of fluid within the chest and abdomen. Petechial haemorrhages were present over the mucous membranes and the epicardial and endocardial surfaces of the heart. Blood clots were present within the nasal cavity (**Figure 22.6b**).

There was enlargement of the submandibular and retropharyngeal lymph nodes with cream-coloured pus noted on sectioning these tissues. *Streptococcus equi* was cultured from this material. Histological examination confirmed the presence of lymphadenitis and vasculitis affecting the nasal mucosa.

22.10.4 Diagnosis and Immunopathology

The diagnosis is this case was equine 'strangles' caused by infection with *Streptococcus equi* ssp.

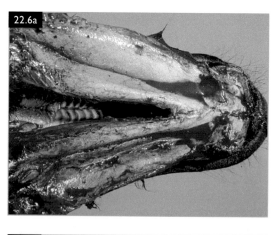

22.6a

22.6b

equi, with secondary purpura haemorrhagica. Purpura haemorrhagica is an uncommon immune-mediated sequela to strangles infection; it arises in 1–2% of cases within weeks of the primary infection. The streptococcal M protein forms immune complexes with IgA or IgM antibody and these immune complexes circulate and deposit within the walls of blood vessels throughout the body (antigen excess type III hypersensitivity; see Chapter 17). This establishes a leucocytoclastic vasculitis leading to fluid loss and haemorrhage. The clinical presentation of this colt followed the classical tissue distribution reported for these lesions. Some animals also develop an immune complex glomerulonephritis, but there was no evidence of this in the present case.

22.11 CASE 7: BOVINE THYMIC LYMPHOMA

22.11.1 Signalment
Twelve-month-old Friesian heifer.

22.11.2 History and Physical Examination
The heifer presented with loss of body condition and marked subcutaneous oedema affecting the ventral cervical and anterior thoracic region. No other animals in the herd were affected. On clinical examination there was an elevated heart rate and dyspnoea.

22.11.3 Diagnostic Procedures
Given the severity of the clinical signs the farmer elected for euthanasia of the animal and a necropsy examination was performed. The main finding was the presence of a very large, solid, cream-coloured mass of 30 cm diameter within the anterior mediastinum. The mass had caused displacement and compression of the heart and lungs within the thoracic cavity. The mass, bisected with the two cut halves lying anterior to the heart and lungs, is shown (**Figure 22.7a**). An enlarged cervical lymph node may be seen just above the mass.

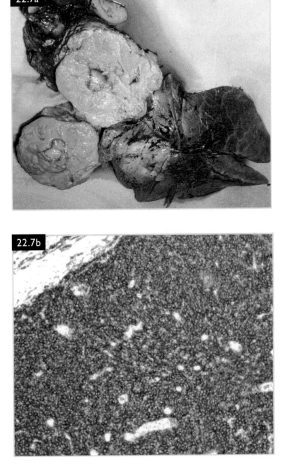

Microscopic examination confirmed that the mass comprised a closely packed sheet of neoplastic lymphocytes. Immunohistochemical examination revealed that these were of T cell lineage (**Figure 22.7b**).

22.11.4 Diagnosis and Immunopathology
The final diagnosis in this case was of thymic T cell lymphoma. The presence of the large mediastinal mass was responsible for the observed clinical signs. The direct compression of heart and lungs led to dyspnoea and elevated heart rate, while occlusion of draining lymphatic vessels was responsible for the observed anteroventral subcutaneous oedema. The ability of the animal to swallow may also have been

compromised, thus contributing to the loss of body condition. This case was notified to the government authority, but it tested negative for bovine leukaemia virus, which may cause multi-centric lymphoma in older cattle.

22.12 CASE 8: CANINE ATOPIC DERMATITIS

22.12.1 Signalment
Three-year-old, neutered male English setter (**Figure 22.8a**).

22.12.2 History and Physical Examination
The dog presented with severe pruritic skin disease that began six months previously and has progressively worsened. The dog continually rubs its face and chews its front paws. Clinical examination revealed erythema and alopecia affecting the face, the inner surface of the pinnae and the axillary and inguinal skin. There was conjunctivitis and cheilitis (inflammation of the lips) and particularly severe lesions with lichenification (thick and fissured skin) of the dorsal surface of the front paws (**Figure 22.8b**). Some of the lesions appeared to have secondary infection and 'tape strip' cytology revealed the presence of coccoid bacteria and yeasts with the appearance of *Malassezia*. *Staphylococcus pseudin-termedius* was subsequently cultured from the lesions.

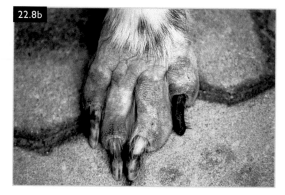

22.12.3 Diagnostic Procedures
A thorough clinical history was obtained. The owner used appropriate and effective flea control and there was no evidence of fleas or flea dirt on the hair coat. The dog lived indoors and was exercised in a suburban environment. The patient's diet was commercial dry kibbles supplemented with occasional canned food and treats. Three months ago the owner switched to a commercially available hydrolyzed protein hypoallergenic diet. After feeding this for three weeks with no change in the skin disease the owner returned the dog to the standard diet.

After discussion with the owner an intra-dermal skin test was performed. This revealed strong positive reactions at 20 minutes to house dust, mattress dust and the two house dust mites *Dermatophagoides pteronyssinus* and *Dermatophagoides farinae*.

22.12.4 Diagnosis and Immunopathology
The final diagnosis is atopic dermatitis with secondary bacterial and yeast infection. The causative allergens are primarily the house dust mites. This is a classical type I hypersensitivity reaction (see Chapters 17 and 18) involving a genetically predisposed dog of young age, with insufficient Treg cell function, becoming sensitized to allergens derived from dust mites. Allergen-specific IgE binds to FcεRI on dermal mast cells, leading to their degranulation on subsequent percutaneous absorption of allergen

through a defective epidermal barrier. With chronicity and secondary infection the disease may extend to involve a CD4+ Th1-mediated inflammatory response.

22.12.5 Treatment and Prognosis

The approach to management of this case is complex and long term, requiring owner compliance. The initial clinical signs are managed with topical antimicrobial baths, systemic antibiotics and a combination of an anti-inflammatory dose of glucocorticoid and essential fatty acid supplementation. Therapeutics targeting IL-31 (lokivetmab) or signaling through the IL-31 receptor (oclacitinib) help to break the 'itch-scratch' cycle (see Chapter 21). Ongoing flea control should be maintained. Once these signs are controlled, long-term management with allergen-specific immunotherapy can be considered (see Chapter 21).

22.13 CASE 9: EQUINE SEVERE COMBINED IMMUNODEFICIENCY

22.13.1 Signalment

Three-month-old Arabian foal.

22.13.2 History and Physical Examination

The foal was born normally and ingested adequate colostrum within the first 24 hours after birth. For the first eight weeks, the foal was clinically normal and appeared to develop appropriately. Over the past two weeks the foal had been lethargic and developed diarrhoea. Physical examination revealed pyrexia, conjunctivitis, laboured breathing and an elevated heart rate.

22.13.3 Diagnostic Procedures

The onset of this collection of clinical signs in an Arabian foal of this age creates a high index of suspicion for severe combined immunodeficiency (SCID). In a stud environment the breeder may be asked whether similar problems have occurred in the past, particularly with the mating of the current dam and sire. Routine haematological examination revealed leucopenia and marked lymphopenia, with virtually no lymphocytes detected on a blood smear. Serum biochemistry and protein electrophoresis revealed hypogammaglobulinaemia. Traces of maternal IgG may still be present and detected by techniques such as SRID, but serum IgM will be undetectable.

22.13.4 Diagnosis and Immunopathology

The clinical and laboratory findings here are consistent with equine SCID and the diagnosis can now be confirmed by a PCR-based genetic test. This detects the 5 base pair deletion in the gene encoding the DNA-dependent protein kinase involved in VDJ recombination that generates lymphocyte antigen receptors. The test allows detection of both homozygous affected foals and heterozygous carrier animals of this autosomal recessive mutation. Affected foals are unable to form functional T and B lymphocytes and therefore have hypoplastic lymphoid tissues and inadequate immune responses. SCID animals readily succumb to multisystemic infection, which is commonly most severe in the respiratory tract, with development of a mixed viral and bacterial pneumonia.

22.13.5 Treatment and Prognosis

There is no practical treatment for this disease. Affected foals may be treated symptomatically, but most will die by five months of age. Experimentally, one animal was kept alive for 12 months with a bone marrow transplant. This foal died and was subject to necropsy examination. There was severe pneumonia with anteroventral consolidation and abscessation of the lung (**Figure 22.9a**). Lymphoid tissue was profoundly hypoplastic, as evidenced by the width of the cross-section of spleen (**Figure 22.9b**).

The molecular diagnostic test should be used to screen Arabian breeding stock for this mutation and controlled breeding programmes

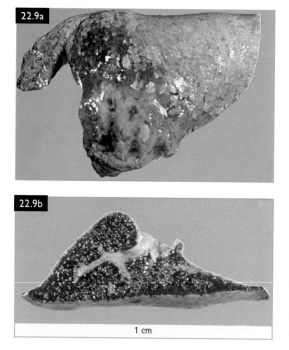

glucocorticoid therapy. The dog has now presented because of clinical signs of increasing lethargy and the breeder was concerned that this may be a relapse of IMHA. Physical examination revealed that the dog was also overweight, has areas of truncal alopecia and scaling and a 'puffy' face with partial closure of the eyelids.

22.14.3   Diagnostic Procedures

Routine haematological and serum biochemical examinations were performed initially and these data are presented here. There were no significant changes to erythrocyte morphology on evaluation of a blood smear.

The main findings of these tests were a mild anaemia, hyperglobulinaemia (determined by protein electrophoresis to be a polyclonal hypergammaglobulinaemia) and hypercholesterolaemia. The clinical history and presenting

instigated. Surveys conducted in 1997 revealed that the mutation was carried by 8% of Arabian horses in the USA and up to 5% in the UK. It is likely that the current prevalence is considerably lower as a result of selective breeding practices.

22.14   CASE 10: CANINE LYMPHOCYTIC THYROIDITIS (HYPOTHYROIDISM)

22.14.1   Signalment

Nine-year-old, neutered female English cocker spaniel (**Figure 22.10a**).

22.14.2   History and Physical Examination

This dog was one of a large breeding kennel of cocker spaniels owned by a successful breeder. Within the colony there was a high prevalence of immune-mediated disease of various types. Two years ago this dog was diagnosed with Coombs-positive IMHA and was successfully treated with a course of immunosuppressive

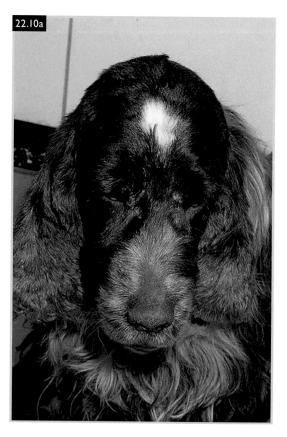

PARAMETER	VALUE	NORMAL RANGE
PCV	0.31	0.35–0.55 l/l
Haemoglobin	50	120–180 g/l
Red blood cells	5.0	5.4–8.0 × 10¹²/l
MCV	67.0	65–75 fl
MCHC	350	340–370 g/l
MCH	23.0	22–25 pg
Platelets	260	170–500 × 10⁹/l
White blood cells	12.0	5.5–17 × 10⁹/l
Neutrophils	8.0	3–11.5 × 10⁹/l
Lymphocytes	2.0	0.7–3.6 × 10⁹/l
Monocytes	1.0	0.1–1.5 × 10⁹/l
Eosinophils	0.5	0.2–1.4 × 10⁹/l

PARAMETER	VALUE	NORMAL RANGE
Urea	6.1	2.0–7.0 mmol/l
Creatinine	115	100–133 pmol/l
Total protein	85	63–71 g/l
Globulin	47.5	20–35 g/l
Albumin	33.8	32–38 g/l
A:G ratio	0.70	0.6–1.5
Alanine aminotransferase	24	20–60 IU/l
Alkaline phosphatase	33	0–110 IU/l
Total bilirubin	2.0	0–10 µmol/l
Glucose	5.2	3.5–5 mmol/l
Cholesterol	14.0	3.5–7 mmol/l
Creatine kinase	114	77–280 IU/l
Calcium	2.3	2.3–2.5 mmol/l

signs, together with these laboratory findings, are suggestive of hypothyroidism.

Further specific tests of thyroid function were performed. There was a low serum T4 concentration and an elevated serum TSH concentration. At this time the dog had a negative Coombs test, but serum autoantibody to thyroglobulin was demonstrated and there was a positive serum ANA (titre of 640 with speckled nuclear staining).

22.14.4 Diagnosis and Immunopathology

This dog is of a breed and familial group highly predisposed to autoimmune disease and it has a previous history of IMHA. The current clinical presentation is consistent with hypothyroidism. This is almost certainly caused by lymphocytic thyroiditis, an infiltration into the thyroid gland of proinflammatory T cells, with destruction of follicular epithelium (**Figure 22.10b**). This will eventually lead to thyroid gland destruction and atrophy.

Thyroglobulin autoantibodies may be detected in the serum in advance of clinical disease in susceptible dogs, although these are likely to represent a 'bystander' immune response to damaged thyroid follicles and release of thyroglobulin into the lymphatic system. The clinical signs and laboratory findings in this dog all relate to depressed metabolism caused by insufficiency of thyroid hormone. The observed lethargy, obesity and hypercholesterolaemia are a direct consequence of this altered metabolism. The cutaneous signs (alopecia and scaling) reflect an atrophic dermatopathy with telogen arrest of hair follicles and hyperkeratosis. The facial puffiness is due to 'myxoedema', an accumulation of mucopolysaccharide matrix

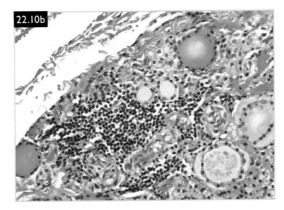
22.10b

within the dermis. The observed mild anaemia is due to bone marrow suppression rather than haemolysis. The autoimmune nature of the disease is confirmed by the presence of serum thyroglobulin and anti-nuclear autoantibodies, but this case does not satisfy the criteria for SLE.

22.14.5   Treatment and Prognosis

Management of lymphocytic thyroiditis involves supplementation with L-thyroxine sodium (a synthetic analogue of T4). Despite the fact that the underlying pathogenesis of thyroid gland destruction involves an autoimmune process, patients are not treated with immunosuppressive drugs.

22.15   CASE 11: CANINE IMMUNE-MEDIATED POLYARTHRITIS

22.15.1   Signalment

Four-year-old, neutered male crossbreed dog.

22.15.2   History and Physical Examination

The dog presented because the owner noticed an increasing reluctance to exercise and a slow and stiff gait when the dog first rises from rest. The problem was described as not continuous, with "some days being better than others". Physical examination revealed mild pyrexia and swelling related to the carpal and tarsal joints. Pain was elicited when these joints were manipulated. There had been no recent history of drug administration or vaccination and the dog had not travelled from Northern Europe. The owner lives in the countryside and the dog is exercised in nearby woodland.

22.15.3   Diagnostic Procedures

Initial diagnostic procedures included radiography and routine haematological and serum biochemical examination. There were similar radiographic changes in both carpal and tarsal joints characterized by increased soft tissue density around the joints with synovial effusion (**Figure 22.11a**). There were no bony changes or damage to articular surfaces. No visceral abnormalities were present on thoracic or abdominal radiographs.

Haematological examination revealed leucocytosis with neutrophilia (19.8×10^9/l; normal range 3–11.5) and monocytosis (3.2×10^9/l; normal range 0.1–1.5) and there was mild polyclonal hypergammaglobulinaemia, recognized on serum biochemical investigation.

A sample of synovial fluid was collected from one carpal and one tarsal joint. The 'mucin clot' was friable, consistent with damage to hyaluronic acid, and there was a high nucleated cell count (82×10^9/l). Cytological examination revealed a dominance of neutrophils with some vacuolated macrophages and a background of blood contamination (**Figure 22.11b**). There

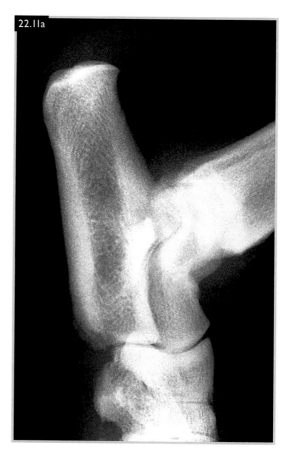

22.11a

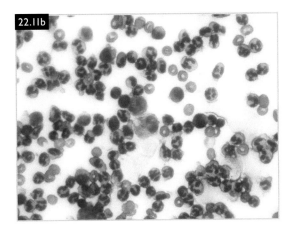

TYPE OF DISEASE	CLINICAL ASSOCIATION
Type I	Joint disease alone
Type II	Joint disease with infectious disease of other body systems ('reactive arthritis')
Type III	Joint disease with gastrointestinal disease ('enteropathic arthritis')
Type IV	Joint disease with underlying neoplasia ('arthritis of malignancy')

were no bacteria associated with these cells or within the fluid.

The dog was negative on serum ANA and RF testing and there was no serological or molecular (PCR) evidence of infection by *Borrelia* (Lyme disease) or *Anaplasma phagocytophilum* (*Ixodes*-transmitted organisms endemic in the local area).

22.15.4 Diagnosis and Immunopathology

The final diagnosis in this case is type I idiopathic polyarthritis. There are four subtypes of canine idiopathic polyarthritis (see later). This dog has no history of exposure to known trigger factors, no evidence of underlying systemic disease or neoplasia, and no evidence of infection by arthropod-borne agents known to lead to joint disease. The pathogenesis of this disease is poorly understood, but is thought to relate to a mixed type III and IV hypersensitivity reaction within the synovium (see Chapter 17).

22.15.5 Treatment and Prognosis

The dog was treated with a course of immunosuppressive glucocorticoid therapy and made an uneventful recovery. Although the prognosis is relatively good, the owner was warned about the relapsing/remitting nature of immune-mediated diseases and the potential for recurrence.

22.16 CASE 12: BOVINE ALLERGIC RHINITIS

22.16.1 Signalment

A herd of 50 milking Jersey cows established six years ago with calves bought in from three different farms.

22.16.2 History and Physical Examination

Fifteen of the cows have recently shown signs of snorting, coughing and nasal discharge. The herd is up-to-date with its vaccination against IBR.

22.16.3 Diagnostic Procedures

One of the affected animals (a five-year-old female) has been culled and a postmortem examination conducted. The nasal passages were oedematous, with mucosal nodular swellings and a mucopurulent exudate (**Figure 22.12a, b**).

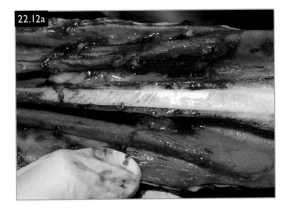

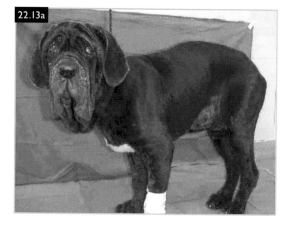

Samples of the nasal mucosa were examined histologically. There was hyperplasia, ulceration and inflammation of the mucosal epithelium with oedema and inflammation of the underlying lamina propria. There was marked infiltration of eosinophils into the affected mucosa.

22.16.4   Diagnosis and Immunopathology

The clinical presentation and pathological findings are consistent with the condition known as 'allergic nasal granuloma' or 'allergic rhinitis'. The aetiopathogenesis is unconfirmed, but the disease is believed to have an allergic basis (most likely type I hypersensitivity), as it occurs when animals are at grass and generally resolves over winter. Bovine allergic rhinitis has been documented most often in Australia and New Zealand as sporadic or herd outbreaks.

22.17   CASE 13: CANINE DRUG REACTION

22.17.1   Signalment

Ten-month-old, neutered female Neapolitan mastiff (**Figure 22.13a**; photo courtesy P. Mellor).

22.17.2   History and Physical Examination

This dog had surgical correction of bilateral entropion (inversion of the eyelid margin) 15 days ago. Before the surgery, the dog had been on a course of cephalexin and immediately after surgery had been placed on a course of co-amoxiclav and the non-steroidal anti-inflammatory drug carprofen. The dog had received both antimicrobial drugs at different times earlier in its life. Seven days ago there was a sudden onset of extensive generalized skin eruptions and since that time the dog had become increasingly weak and lethargic with periods of collapse.

Physical examination at presentation revealed that the lesions were alopecic, erythematous and ulcerated and several had a distinctive 'target ring' appearance (**Figure 22.13b**). The dog also showed generalized peripheral lymphadenomegaly, pale mucous membranes and an elevated heart rate with a systolic heart murmur.

22.17.3   Diagnostic Procedures

Initial diagnostic procedures included cardiac imaging, routine haematology and serum biochemistry, urinalysis and skin biopsy. Haematological data are presented here.

There is a marked non-regenerative anaemia, thrombocytopenia (confirmed on evaluation of the blood smear) and monocytosis.

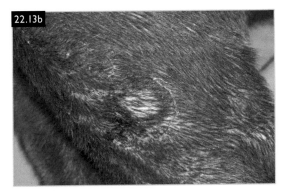

Figure 22.13b From Mellor PJ et al. [2005] Neutrophilic dermatitis and immune-mediated haematological disorders in a dog: suspected adverse reaction to carprofen. *Journal of Small Animal Practice* **46**:237–242, with permission.

PARAMETER	VALUE	NORMAL RANGE
PCV	0.08	0.37–0.55 l/l
Haemoglobin	36.5	120–180 g/l
Red blood cells	1.19	$5.5–8.5 \times 10^{12}$/l
MCV	73.6	60–77 fl
MCHC	305	340–370 g/l
MCH	41.5	32–37 pg
Absolute reticulocytes	0	$\leq 60 \times 10^9$/l
Platelets	63	$150–500 \times 10^9$/l
White blood cells	8.97	$6–17.5 \times 10^9$/l
Neutrophils	5.1	$3–11.5 \times 10^9$/l
Lymphocytes	1.5	$1–4.8 \times 10^9$/l
Monocytes	2.3	$0.2–1.5 \times 10^9$/l
Eosinophils	0.04	$0.05–1.3 \times 10^9$/l
Basophils	0.0	$0–0.5 \times 10^9$/l

There were no abnormalities on serum biochemistry, but urinalysis revealed the presence of haemoglobinuria. No cardiac abnormalities were detected on imaging examination and the heart murmur was attributed to being of haemic origin. Skin biopsy revealed the presence of a neutrophilic vasculitis.

These observations were suggestive of a multisystemic immune-mediated process requiring further specific diagnostic procedures. An EDTA blood sample showed autoagglutination of red cells, which was not dispersed by the addition of saline (positive in-saline agglutination test). This was considered presumptive evidence for IMHA, but a Coombs test was not performed. A direct immunofluoresence test demonstrated antibody associated with the patient's platelets. Serum ANA was negative and immunohistochemistry did not demonstrate immunoglobulin or complement associated with the cutaneous vascular lesions.

22.17.4 Diagnosis and Immunopathology

The history, clinical presentation and laboratory findings in this dog were consistent with an immune-mediated drug reaction. There was evidence for IMHA, IMTP (both of which are type II hypersensitivities) and cutaneous vasculitis (type III hypersensitivity). The pathogenesis of this disease may involve the causative drug acting as a hapten and binding to host carrier protein. The dog had received three drugs in the period immediately before the onset of disease and both antimicrobials had been given in the past, thus providing an opportunity for sensitization.

22.17.5 Treatment and Prognosis

The dog was given an infusion of polymerized bovine haemoglobin (as oxygen-carrying support) and started on a course of immunosuppressive prednisolone and azathioprine. A different antimicrobial (enrofloxacin) was administered and the dog was given topical medicated baths and supplemented with essential fatty acids. There was rapid clinical improvement with erythroid regeneration, normalization of platelet count, resolution of the heart murmur and reduced severity of skin lesions at discharge, nine days after presentation. The dog was healthy at a recheck examination, seven months later.

22.18 CASE 14: CANINE TRAPPED NEUTROPHIL SYNDROME

22.18.1 Signalment

Eight-month-old, female border collie (**Figure 22.14a**). The dog lives on a farm and has been bred to be a working sheepdog.

22.18.2 History and Physical Examination

The owner reported that this young dog has always been 'small for its breed' and since a young age has had episodes of inappetence, vomiting, diarrhoea, lethargy and apparent pain on walking. These episodes appeared to begin at around the time of the second vaccination in the puppy primary course. Although not currently diarrhoeic, the dog showed signs of pyrexia, conjunctivitis and cheilitis and apparent joint pain on manipulation of the limbs.

22.18.3 Diagnostic Procedures

Initial investigations included routine haematology, serum biochemistry, urinalysis and radiographs of major joints. Haematological data are presented here.

The most significant haematological findings are leucopenia with profound neutropenia and a mild anaemia. There were no abnormalities on serum biochemistry or urinalysis. Radiographs did not reveal active joint pathology and survey radiographs of the thorax and abdomen were

PARAMETER	VALUE	NORMAL RANGE
PCV	0.32	0.35–0.55 l/l
Haemoglobin	109	120–180 g/l
Red blood cells	4.64	5.4–8.0 × 10^{12}/l
MCV	69.4	65–75 fl
MCHC	337	340–370 g/l
MCH	23.4	22–25 pg
Platelets	295	170–500 × 10^9/l
White blood cells	2.57	5.5–17 × 10^9/l
Neutrophils	0.05	3–11.5 × 10^9/l
Lymphocytes	0.77	0.7–3.6 × 10^9/l
Monocytes	0.72	0.1–1.5 × 10^9/l
Eosinophils	0.0	0.2–1.4 × 10^9/l
Basophils	0.0	0–0.1 × 10^9/l

considered normal. Samples of joint fluid from the carpal and tarsal joints had low cellularity and normal mucin clot formation. Bacteriological culture of these fluids led to no significant growth. PCR failed to identify infection with either *Borrelia* or *Anaplasma* and serum ANA and RF were negative.

Further blood samples confirmed that the neutropenia was persistent and a bone marrow core biopsy was collected. This was markedly cellular with profound elevation in the myeloid:erythroid ratio. The majority of cells were of the granulocytic lineage (**Figure 22.14b**).

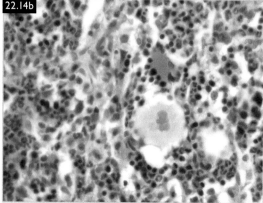

22.18.4 Diagnosis and Immunopathology

The main features of the disease in this young dog are a severe neutropenia (predisposing to multisystemic infectious/inflammatory disease), but plentiful granulocytic precursors within the bone marrow. Given the breed of dog and clinical history, the most likely diagnosis is 'trapped neutrophil syndrome'. This inherited primary immunodeficiency disorder is also known as myelokathexis and is essentially a failure of the bone marrow to release neutrophils into the circulation.

22.18.5 Treatment and Prognosis

There is no effective treatment for this disease. A PCR molecular diagnostic test for the causative mutation in the VPS13B gene is available to confirm the diagnosis and identify carriers. This autosomal recessive mutation appears to be widespread in this breed throughout the world.

22.19 CASE 15: FELINE LYMPHOID LEUKAEMIA

22.19.1 Signalment

Six-year-old, neutered female domestic shorthair cat.

22.19.2 History and Physical Examination

The cat presented with a five-day history of anorexia and lethargy. On physical examination the cat was seen to be in thin bodily condition, lethargic, with pale mucous membranes and increased heart and respiratory rates (**Figure 22.15a**; photo courtesy Feline Centre, University of Bristol).

22.19.3 Diagnostic Procedures

Initial diagnostic procedures included routine haematology, serum biochemistry, urinalysis and evaluation of retroviral status. Haematological data are presented here.

PARAMETER	VALUE	NORMAL RANGE
PCV	0.04	0.27–0.50 l/l
Haemoglobin	14	80–150 g/l
Red blood cells	0.85	$5.5–10.0 \times 10^{12}$/l
MCV	47.1	40–55 fl
MCHC	350	310–340 g/l
MCH	16.5	13–17 pg
Platelets	<5 per × 100 field	
Neutrophils	1.5	$2.5–12.5 \times 10^9$/l
Lymphocytes	3.9	$1.5–7.0 \times 10^9$/l
Monocytes	0.1	$0.0–0.8 \times 10^9$/l
Eosinophils	0.0	$0.0–1.5 \times 10^9$/l

Examination of the blood smear revealed the presence of a population of atypical lymphoblastic cells. The only abnormality on serum biochemistry was elevation of alkaline phosphatase (71 IU/l; normal range 0–20) and urinalysis was normal. The cat was negative for FeLV and FIV.

The major findings are, therefore, a severe anaemia, thrombocytopenia and neutropenia (pancytopenia) associated with the presence of atypical lymphoid cells within the circulation. These observations are consistent with bone marrow disease and presumptive leukaemia. Survey radiographs of the thorax and abdomen did not reveal visceral disease. A bone marrow core biopsy was taken. The marrow was hypercellular with haemopoietic tissue almost

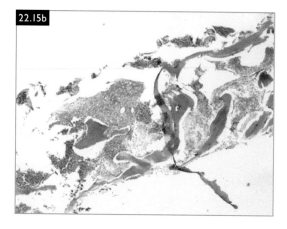

22.15b

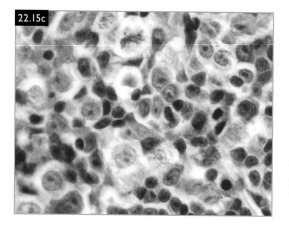

22.15c

entirely replaced by a monomorphic population of pleomorphic and mitotic lymphoblastic cells (**Figures 22.15b, 22.15c**).

22.19.4 Diagnosis and Immunopathology

The diagnosis is chronic lymphocytic leukaemia of bone marrow origin with neoplastic cells seeding to the blood. It is also possible that these cells had started to colonize the liver and spleen. The tumour does not appear to be FeLV related in this case. The severity of disease in this cat might seem at odds with the relatively late clinical presentation, but it is a good example of the propensity of this species to compensate and present with relatively subtle clinical signs despite severe disease.

22.19.5 Treatment and Prognosis

The cat was placed onto a combination chemotherapeutic protocol, but the long-term prognosis will be guarded.

22.20 CASE 16: EQUINE NEONATAL ISOERYTHROLYSIS

22.20.1 Signalment

One-day-old Thoroughbred foal.

22.20.2 History and Physical Examination

The foal was born normally in the afternoon and was seen to take in adequate colostrum. The following morning the foal was collapsed. Physical examination revealed pallor (but not jaundice) of mucous membranes (**Figure 22.16**) and a urine sample had red discolouration. The foal was not pyrexic and there was no clinical evidence of sepsis. There was an elevated heart rate, a weak pulse and occasional 'yawning' was observed. This is the second foal born to this mare by the same stallion, with the first foaling being uneventful.

22.20.3 Diagnostic Procedures

A blood sample revealed severe anaemia with a PCV of 0.8 l/l (18%) (normal range for adult horse 0.32–0.53 [32–53%]). Platelet and leucocyte

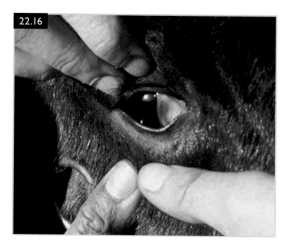

22.16

counts were normal. Analysis of the urine sample confirmed that the discolouration was due to the presence of haemoglobin. A Coombs test performed with equine polyvalent Coombs reagent was positive (titre 320 at 37°C) and when serum from the mare was mixed with a suspension of washed foal erythrocytes, there was also a strong agglutination reaction, with a titre of 1,280.

22.20.4 Diagnosis and Immunopathology

This foal has neonatal isoerythrolysis (see Chapter 14). The mare had become sensitized to foreign blood group antigens from the first foal and developed antibodies to these. These antibodies concentrated in the colostrum that was ingested and absorbed by the foal. The maternally derived antibody in the bloodstream rapidly attached to the foal's red cells, resulting in peracute onset, severe IMHA. The most likely blood group antigen incompatibility in this case is the Aa or Qa antigen (expressed by the stallion but not the mare). The disease may have an acute (two to four days old) or subacute (four to five days old) onset, with progressively less severe clinical signs and greater likelihood of icterus at presentation.

22.20.5 Treatment and Prognosis

This foal must immediately be prevented from further access to milk from the mare for the remainder of that day. A foster mother or milk replacement should be used, but after 48 hours it should be safe to permit the foal to suckle the mare. The severity of anaemia in this foal means that the prognosis will be guarded and the foal requires a blood transfusion. A whole blood transfusion must be cross-matched and this can be done by testing serum from the mare against potential donor erythrocytes. A second transfusion may be required depending on the degree of elevation of the foal's PCV. The stud should be warned not to mate this mare in the future. For future pregnancies, it would be wise to test the mare's serum with the stallion's erythrocytes during the final two weeks of pregnancy. Any reaction would indicate that the foal should be fostered or hand-reared using alternative source of colostrum.

22.21 CASE 17: A HERD OUTBREAK OF BOVINE VIRAL DIARRHOEA VIRUS INFECTION

22.21.1 Signalment

Full-term calf born dead with diffuse cataracts and cerebellar hypoplasia.

22.21.2 History and Physical Examination

This 30-cow dairy herd has a history of multiple abortions (especially in first calf heifers), weak calves at birth and calves developing pneumonia during the first few months of life.

22.21.3 Diagnostic Procedures

Blood and tissue samples were collected and tested from an aborted fetus and a two-week-old calf by IFA testing and virus isolation. Tissue samples were positive for BVDV by IFA testing. At necropsy examination, the fetus and calf both had multiple gross abnormalities. The fetus had cerebellar hypoplasia (**Figures 22.17a, 22.17b**) and thymic aplasia. The calf also had thymic aplasia and hypoplasia of all lymphoid tissues.

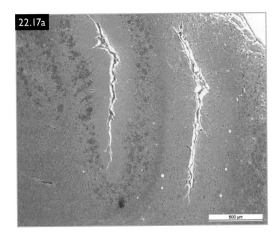
22.17a

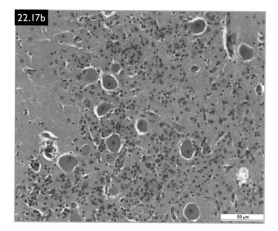

All cattle were ear notched and tested for the presence of BVDV to detect animals that are persistently infected with the virus. Persistently infected (PI) animals develop when fetuses from conception to 120 days are infected *in utero* with a non-cytopathic strain of BVDV. These animals often fail to develop an immune response (cellular and/or humoral) because the immune system becomes tolerant towards the BVDV antigens that are expressed in primary lymphoid tissues during this period of development.

22.21.4 Diagnosis and Immunopathology

This herd did not have a history of vaccination against BVDV, IBR, parainfluenza-3 and bovine respiratory syncytial virus. It was also not a closed herd, and three four- to nine-month-old heifers had been purchased during the past year. It is very likely that one or more of those heifers were persistently infected with BVDV. The PI animal is infected for life and are a source of infection to the rest of the herd. Although many will die by one to three years of age, some PI cattle have been known to survive for more than ten years.

22.21.5 Treatment and Prognosis

Acutely infected animals often require antibiotic treatment for secondary bacterial infections due to the immunosuppressive effects of BVDV causing the animal to be at greater risk for bacterial diseases, especially gastrointestinal and respiratory diseases. Additional supportive therapy (e.g. fluids) is often necessary since diarrhoea is common. An animal found to be persistently infected must be culled from the herd as it cannot be permitted to remain as a source of BVDV infection. All new animals entering the herd should be tested for BVDV before they are brought in and all herd animals need to be vaccinated as calves and then revaccinated with the cattle core vaccines every three or more years to maintain herd immunity. If these recommendations are followed, the herd should not develop similar problems in the future.

Active immunization Injection of antigen in an immunogenic form to stimulate an immune response and induce immunological memory.

Acute phase proteins Proteins in the blood whose concentration increases during the early stages of an infection or inflammation.

Adaptive immunity A form of immunity mediated by antigen-specific lymphocytes that undergo clonal selection, proliferation, and differentiation. The adaptive immune response generates immunological memory. Also 'acquired immunity'.

Adhesion molecule Cell surface molecule that mediates the binding to other cells or to extracellular matrix components.

Adjuvant Vaccine component that enhances the immune response by inducing the accumulation and activation of antigen-presenting cells at the injection site and regional lymph nodes.

Afferent lymphatic Lymph vessel draining lymph from tissue to the regional lymph node.

Affinity Strength of binding of a single binding site of a molecule to another; specifically the binding of a single epitope of an antigen to a Fab fragment of an immunoglobulin.

Affinity maturation Process that leads to increased affinity of antibodies. It consists of acquiring somatic mutations in the genes that encode for the Ig variable domains followed by selection of B cells that produce high affinity immunoglobulins.

Agammaglobulinaemia Absence of immunoglobulins in the blood.

Agglutination Aggregation of particulate antigens by antibodies; multimeric antibodies such as IgA and IgM are particularly effective.

Allele Two or more alternative forms of a gene with different nucleotide sequences.

Allergen Antigen that induces an allergic or hypersensitivity reaction.

Allergen-specific immunotherapy (ASIT) Repeated subcutaneous injection or sublingual administration of gradually increasing quantities of antigen to which an individual is sensitized to diminish the immune response on subsequent natural exposure to antigen. Used in the management of atopic dermatitis. Also 'hyposensitization'

Allergy Clinical syndrome caused by a type I hypersensitivity reaction mediated by IgE and Th2 cytokines (IL-4, IL-5, IL-9 and IL-13)

Alloantigen Tissue antigen derived from a genetically dissimilar individual of the same species. Typically stimulates an immune response following incompatible grafting of tissue or cells.

Allograft Transplant of tissue or organ between genetically dissimilar individuals of the same species.

Anaphylaxis Systemic type I hypersensitivity reaction.

Anergy Failure of an immune response to an antigen due to lack of co-stimulation by antigen-presenting cells. It is one of the mechanisms that contributes to immunological tolerance. Clinically, patients are 'anergic' when they fail to mount a DTH response to a cutaneous antigen administration.

Anaphylatoxin Complement fragments C5a and C3a generated during complement activation that induce acute inflammation by recruiting neutrophils and inducing mast cell degranulation.

Anchor residues   Specific amino acid residues of an antigenic peptide that bind to the peptide-binding groove of MHC molecules.

Antibody   Glycoproteins that serve as antigen-specific receptors on B cells and are secreted by plasma cells. Secreted antibodies bind to pathogens to target them for killing or removal. Also referred to as immunoglobulins.

Antibody-dependent cell-mediated cytotoxicity (ADCC)   Destruction of a target cell coated with antigen-specific IgG by NK cells. The NK cells bind to IgG via their Fc-receptors.

Antigen   Substance that binds to an antibody or T cell receptor.

Antigen presentation   Display of peptides bound to the binding groove of MHC I or MHC II molecules on the surface of cells for recognition by T cell receptors.

Antigen processing   Breakdown of proteins into peptides within the cytoplasm (endogenous pathway) or endosomes (exogenous pathway) and binding of the peptides to MHC molecules.

Antinuclear antibody   Autoantibody with specificity for one of the components of the nucleus of a cell.

Antiserum   Serum collected from an individual following exposure to an antigen. Contains antibody that can be utilized in serological assays. Antiserum is polyclonal with different antibodies that recognize different epitopes of antigens with varying avidities.

Apoptosis   Cell death resulting from activation of an 'internal death programme' within the cell by activation of intracellular caspases leading to DNA degradation, nuclear condensation and fragmentation, and phagocytosis of the cellular remains. Also 'programmed cell death'.

Arthus reaction   Local form of a type III hypersensitivity reaction involving immune complex formation in tissue of a sensitized individual following exposure to antigen.

Asthma   A chronic respiratory disease, characterized by intermittent bronchoconstriction, that has several causes, one of which is type I hypersensitivity to inhaled allergens.

Atopic dermatitis   Skin disease, characterized by pruritus (itching), that may have several causes, one of which is type I hypersensitivity to allergens that penetrate the skin barrier.

Attenuation   Reduction in virulence of an infectious agent so that it retains antigenic structure and limited replicative capacity, but cannot produce disease (attenuated vaccine; modified live vaccine).

Autoantigen   Antigen derived from self-tissue that is recognized by self-reactive T cells or autoantibodies.

Autogenous vaccination   Culture of a pathogen from a patient and injection of a killed preparation of that pathogen with an adjuvant to promote the immune response to the organism.

Autoimmune Regulator (AIRE)   Transcription factor expressed in medullary epithelial cells that stimulates the expression of peripheral antigens. This is important for central tolerance through negative selection of self-reactive T cells.

Autoimmunity   Generation of an immune response to self-antigen caused by failure of tolerance mechanisms that may result in tissue pathology and autoimmune disease.

Autophagy   Physiologic process of degradation of dysfunctional or unnecessary cellular components via the lysosomal pathway. Such components are contained within autophagosomes, which then fuse with lysosomes for degradation of their content.

Avidity   Strength of overall binding between two molecules that contact at multiple sites; specifically the binding of multiple epitopes of an antigen by multiple binding sites of an immunoglobulin.

B cell receptor (BCR)   Immunoglobulin receptor for antigen on the surface of a B lymphocyte.

Bence-Jones proteins   Free immunoglobulin light chains that may be produced in multiple myeloma and pass through the glomerular filter into the urine (Bence-Jones proteinuria).

Bursa of Fabricius Primary lymphoid organ associated with the cloaca that is the main site of B-cell development in birds.

Bystander suppression Non-specific suppression of a local immune response not directly related to that being controlled by a suppressor lymphocyte.

Caninized monoclonal antibody Genetically engineered monoclonal antibody with canine constant regions and mouse antigen-binding sites. Reduces the induction of immune responses against the monoclonal antibody when injected therapeutically in dogs.

Carrier protein Protein that when linked to a hapten can initiate an immune response to the hapten in addition to responses to its own 'native' epitopes.

CD molecules Nomenclature used to define immune cell surface molecules.

Chediak–Higashi Syndrome Immunodeficiency associated with abnormal formation of cytoplasmic granules in the cytoplasm of neutrophils, macrophages, cytotoxic T cells and NK cells.

Chemokines Soluble proteins released from a cell to form a chemotactic gradient responsible for the migration of specific leucocytes (e.g. from blood into tissue).

Chemotaxis Migration of cells along a chemotactic gradient of increasing concentration of a chemoattractant.

Cell-mediated immunity (CMI) Type of immune response mediated by T lymphocytes involving macrophage activation by IFN-γ and cytotoxic destruction of target cells (e.g. infected cells, neoplastic cells or incompatible graft cells).

Clonal deletion Removal by apoptosis of autoreactive lymphocytes bearing receptors specific for self-antigens. May occur within the thymus or bone marrow for T and B cells, respectively (central tolerance), or within secondary lymphoid tissues (peripheral tolerance).

Clonal expansion Exponential proliferation of antigen-specific lymphocytes following activation by antigen to generate large numbers of B and T cells from rare naïve antigen-specific lymphocytes.

Clonal selection Selection of the few antigen-specific lymphocytes from the entire repertoire on exposure to antigen for activation and proliferation.

Complement Series of plasma and cell surface proteins which, when activated, interact sequentially, forming a self-assembling enzymatic cascade that generates inflammatory mediators, opsonins and a terminal membrane attack complex.

Complementarity determining regions (CDRs) Regions within the variable domains of TCRs and BCRs with the most amino acid variability. Six CDRs fold together to form an antigen-binding site. Also known as hypervariable regions.

Constant region Portion of an antibody or TCR that is not involved in binding of an epitope or peptide and is conserved among different B and T cells.

Contact allergy Type IV hypersensitivity reaction to chemicals that penetrate the skin barrier. They modify proteins which then elicit an immune response from CD4+ and CD8+ T cells.

Coombs test Agglutination test used to detect the presence of immunoglobulin and/or complement on the surface of erythrocytes in immune-mediated hemolytic anemia (IMHA).

Co-stimulatory molecules Molecules expressed on antigen-presenting cells that provide a second signal required for the activation of naïve lymphocytes following recognition of antigen.

CpG motif Sequence of unmethylated cytosine and guanidine nucleotides present in bacterial DNA that is a potent stimulus for innate immune responses. CpG motifs are recognized by TLR9.

CR1–4 Cell surface receptors that bind complement components.

Cross-linking Aggregation of receptor molecules by binding to a ligand which induces signaling through the receptors. E.g., joining together of two or more mast cell surface Fcε receptors by allergen binding to individual receptor-bound IgE molecules triggers mast cell activation and degranulation.

Cross-presentation Mechanism that allows dendritic cells to activate naïve CD8 T cells specific for extracellular antigens or antigens expressed by other cells (e.g. virus-infected or tumor cell). It involves the uptake of antigens and processing of antigenic peptides for presentation by MHC I molecules.

Cross-reactive Two antigens share common or similar epitopes, enabling recognition of different antigens by the same antigen receptor (immunoglobulin or TCR).

C-type lectin receptors (CLRs) Family of pattern recognition receptors that recognize carbohydrate groups on PAMPs and DAMPs. Binding of CLRs induces phagocytosis and activation of the innate immune system.

Cyclic GMP-AMP synthase (cGAS) Pattern recognition receptor that senses the presence of DNA in cytoplasm. Binding of DNA induces the synthesis of cyclic GMP-AMP which activates STING.

Cyclic haematopoiesis Immunodeficiency caused by a genetic mutation in grey collie dogs resulting in cyclic cytopenia.

Cyclosporin An immunosuppressive drug that inhibits the activation of NFAT in T cells.

Cytokine Proteins secreted by cells that act as intercellular messengers of the immune system and are important mediators of immune and inflammatory reactions.

Cytotoxic T cell (CTL) Type of T cell usually characterized by the expression of CD8 that can recognize and kill virus-infected cells and tumor cells by inducing apoptosis.

Damage-associated molecular pattern (DAMP) Molecules released by damaged cells that induce inflammation and enhance immune responses by binding to pattern recognition receptors (PPRs).

Defensins Antimicrobial peptides secreted by epithelial cells and present in neutrophil granules that can kill a broad spectrum of bacteria and fungi.

Degranulation Release of biological mediators from cytoplasmic granules following activation of immune and inflammatory cells.

Delayed-type hypersensitivity Type IV hypersensitivity reaction involving activation of macrophages by T cells in a sensitized individual re-exposed to antigen. Takes 24–72 hours to become manifest clinically.

Dendritic cell Bone marrow-derived antigen-presenting cell located in lymphoid and nonlymphoid tissues. Plays a critical role in the initiation of the adaptive immune response by presenting antigens and costimulatory molecules to naïve T cells.

Determinant See 'epitope'.

Diapedesis Process by which leucocytes squeeze between endothelial cells to enter tissue from the blood.

Dietary hypersensitivity Hypersensitivity to food-derived allergen (see also 'food allergy'). Presents as cutaneous or alimentary disease.

Diversity (D) gene segment Short nucleotide sequences located between the variable gene segments and constant gene segments in the immunoglobulin heavy chain and β and γ TCR chain loci. Random selection of one D gene segment contributes to the diversity of the antigen receptors, in particular the third hypervariable region.

Duration of immunity Period after vaccination or infection in which a protective immune response to a pathogen can be detected (by experimental challenge or immune assays).

Effector cell Cell that actively participates in an immune response (e.g. by producing antibody, cytokines, or killing infected cells).

Efferent lymphatic Lymphatic vessel draining lymph from secondary lymphoid tissue (lymph node, spleen, tonsil, Peyer's patches) or the thymus to the common thoracic duct.

Eicosanoids Family of inflammatory mediators, including prostaglandins and leukotrienes, derived from arachidonic acid

ELISA Enzyme-linked immunosorbent assay. Used to determine the concentration of either antigen or antibody. Based on detection by an enzyme-conjugated antibody and an appropriate substrate that leads to a colour change, which can be measured spectrophotometrically.

Endogenous antigens Antigens produced within the cytoplasm of a cell (i.e. self-antigens or molecules produced by intracellular pathogens).

Endosome A cytoplasmic vesicle formed by invagination of the cell membrane around extracellular proteins during the process of antigen uptake by an APC. See also 'phagosome'.

Endotoxin Toxin released from the cell wall of dying gram-negative bacteria (e.g. *E. coli*), that stimulate innate immune cells by binding to TLR4.

Enterocytes Epithelial cells lining the intestinal tract.

Enterotoxin Toxin secreted by enteric pathogenic bacteria that binds to receptors on enterocytes and mediates secretory diarrhoea.

Eotaxin Chemokine that attracts eosinophils into tissue.

Epitope The specific part of an antigen that is bound by an antibody or TCR. Also called an antigenic 'determinant'.

Epitope spreading Plays a role in the progression of autoimmune disease. Spreading of the immune response to other epitopes within a self-antigen (intramolecular) or other self-antigens (intermolecular) after the response is initiated by presentation of one epitope.

Erythema Redness of the skin due to increased blood flow. One of the cardinal signs of inflammation.

Exogenous antigen External antigen taken up by an APC.

Fab Fragment antigen binding. Produced by proteolysis of antibody molecules and consists of the entire light chain and the variable domain and first constant domain of the heavy chain. The antigen binding site is present at the N terminal end of Fab.

Fc Fragment crystallizable. Produced by proteolysis of antibody molecules and consists of two or three heavy chain domains of the immunoglobulin molecule connected by one or more disulfide bridges.

Fc receptor Cell surface receptor that binds the constant region of an immunoglobulin.

FcεRI High affinity receptor that binds to the constant region of IgE. Expressed on mast cells and basophils. Antigen-induced cross-linking of IgE bound to FcεRI induces mast cell and basophil degranulation.

FcγR Receptor for subclasses of IgG. FcγR on neutrophils and macrophages bind IgG immune complexes facilitating phagocytosis (opsonization). FcγR on NK cells mediates antibody-dependent cell-mediated cytotoxicity.

Felinized monoclonal antibody Genetically engineered monoclonal antibody with feline constant regions and mouse antigen-binding sites. Reduces the initiation of immune responses against the monoclonal antibody when injected therapeutically in cats.

Flea allergy dermatitis Type I hypersensitivity to antigens within flea saliva.

Follicle See Lymphoid follicle.

Food allergy An immunological reaction to dietary antigens. Most often considered a type I hypersensitivity reaction. (See also 'dietary hypersensitivity').

FOXP3 Transcription factor expressed by CD4+ regulatory T cells.

Frustrated phagocytosis May occur when a phagocytic cell cannot ingest a large antigen, but is able to release cytoplasmic granule contents adjacent to the particle to produce surface damage to that target.

Gene conversion Generation of BCR diversity in certain mammals and birds by incorporation of V region pseudogenes into the expressed V region segment.

Germinal centres   Central region of secondary lymphoid follicles where antigen-activated B lymphocytes divide, undergo somatic mutation, and selection into plasma cells or memory B cells.

Germline repertoire   Theoretical number of different immunoglobulin or TCR molecules that can be generated from the multiple gene segments encoding variable domains inherited by an individual. The repertoire can be further expanded by imprecise joining of fragments, random addition of nucleotides and somatic mutation.

Giant cells   Cells that may be formed during chronic granulomatous inflammation by fusion of activated macrophages.

Granuloma   Localized accumulation of activated macrophages surrounded by lymphocytes in response to a persistent stimulus. May have a necrotic centre.

Gut-Associated Lymphoid Tissues (GALT) Organized lymphoid tissues associated with the alimentary mucosa. Includes Peyer's patches in the small intestine and solitary lymphoid nodules in the stomach and intestine.

Haematopoiesis   Generation of blood cells, in the fetal liver and postnatal bone marrow.

Haplotype   Linked set of genes associated with one haploid genome. Used to describe a set of linked MHC genes inherited from one parent.

Hapten   Small chemical compound that can bind an antibody molecule, but which by itself cannot elicit an immune response. It is only capable of generating an antibody or T-cell response when bound to a 'carrier protein'.

Hassall's corpuscles   Epithelial structures in the medulla of a thymic lobule.

Herd immunity   Protection of a population from an infection when a large proportion of individuals in that population is vaccinated.

Heterologous vaccine   Use of an antigenically similar, related infectious agent that lacks virulence to cross-protect against another more virulent pathogen.

High Endothelial Venules (HEVs)   Specialized blood vessels in secondary lymphoid tissues in which venular endothelial cells take on a cuboidal morphology and express adhesion molecules that allow binding and extravasation of naïve lymphocytes.

Hinge region   Portion of an immunoglobulin molecule between Fab and Fc that permits movement of the 'arms' (Fab) of the Y-shaped structure.

Histocompatibility   Immunological similarity between individuals. Determined by the cellular expression of histocompatibility molecules and forms the basis of tissue matching before transplantation.

Histiocytic sarcoma   Malignant proliferation of dendritic cells in dogs that may be either localized or disseminated.

Histiocytoma   Benign cutaneous neoplasm of epidermal Langerhans cells in the dog.

HLA-DM   Molecule in the endosomal compartment that removes the CLIP peptide, a remnant of the invariant chain, from the peptide-binding groove of MHC II molecules allowing the binding of exogenous peptides.

Homing receptors   Cell surface molecules expressed by subpopulations of lymphocytes that enable interaction with specific endothelial vascular addressins to allow those lymphocytes to enter tissue at particular anatomical locations.

Humoral immunity   Immunity mediated by antibodies.

Hybridoma   A cell arising from the fusion of an antigen-specific plasma cell with an immortal myeloma cell. Hybridoma cells secrete monoclonal antibodies.

Hypercalcaemia of malignancy   Elevated blood calcium concentration due to the ability of some tumours (e.g. lymphoma, myeloma) to produce parathyroid hormone (PTH)-like peptides that mediate osteoclast resorption of calcium from bone.

Hypergammaglobulinaemia Elevation in serum gamma globulin concentration.

Hyperplasia Increase in size of a tissue or organ due to an increase in the number of constituent cells (e.g. reactive hyperplasia of a lymph node due to antigenic stimulation of the lymphoid cells in that node).

Hypersensitivity State of immunological sensitization to an innocuous antigen that leads to an excessive (symptomatic) immune response on re-exposure to the antigen. See also 'allergy'.

Hypervariable regions Portions of the variable domains of antibody molecules and T cell receptors with the greatest variability in amino acid sequence. Typically forms the contact residues with antigen. See also 'complementarity determining region'.

Hypogammaglobulinaemia Decrease in the concentration of serum gammaglobulin.

Hypoplasia (e.g. thymic) Reduced size of an organ or tissue from birth (i.e. a congenital defect due to incomplete development).

Hyposensitization See 'Allergen-specific immunotherapy'.

Iatrogenic hyperadrenocorticism Induction of hyperadrenocorticism by prolonged medical administration of glucocorticoids.

IgA deficiency Lack of IgA in serum and at mucosal surfaces with resulting predisposition to infectious and immune-mediated disease. In man and dogs is a relative, rather than absolute, deficiency and does not involve genetic mutation of genes encoding the IgA molecule.

Immediate hypersensitivity Immune reaction in a sensitized individual within minutes of re-exposure to antigen caused by degranulation of mast cells.

Immune complex Complex of antigen and antibody molecules.

Immune deviation Polarization of the immune response to a specific antigen from one type of Th cells such as Th1 cells to another set such as Th2 cells resulting in a suppression of the inflammatory response.

Immune surveillance Surveillance of the body by the immune system to detect and destroy infected or transformed cells before they can cause disease.

Immunity gap Period of time when there is insufficient maternally derived immunoglobulin to protect a young animal from infection, but sufficient maternal immunoglobulin to prevent onset of the endogenous immune response in that animal. Also termed the 'window of susceptibility'.

Immunodeficiency Failure of part/s of the immune system caused by inherited gene mutations (primary immunodeficiency) or acquired suppression of immune function (secondary immunodeficiency).

Immunodominant epitopes Those epitopes within an antigen that are most likely to engender an immune response.

Immunofluorescence Technique in which fluorochromes such as fluorescein isothiocyanate (FITC) may be coupled to antibodies to detect molecules within tissue (immunofluorescence microscopy) or on the surface of cells (by flow cytometry).

Immunogen An antigenic substance that induces an adaptive immune response following administration to an individual.

Immunoglobulin A general term for all antibody molecules. In monomeric form, immunoglobulins consist of two identical heavy chains and two identical light chains.

Immunoglobulin class switch Commitment of an antigen-activated B lymphocyte to change to the expression of a single immunoglobulin class (IgA, IgG or IgE) instead of the combination of IgM and IgD that characterizes the naïve B cell.

Immunoglobulin superfamily Family of molecules that contain a globular domain first detected in antibody molecules and that likely arose through gene duplication during evolution. Includes immunoglobulins, TCRs, MHC molecules and certain adhesion molecules.

Immunohistochemistry Method to probe a tissue section for the presence of a molecule with antigen-specific antibodies conjugated with an enzyme. The enzyme converts a colorless soluble substrate into a colored insoluble product at the location of the antigen that can be detected by light microscopy.

Immunological ignorance Tolerance of a lymphocyte due to failure to interact with its cognate antigen.

Immunological synapse Interface between anantigen-presenting cell and T cell formed by multiple intercellular molecular interactions with TCR–MHC–peptide interactions at the core.

Immunopathology Tissue pathology caused by an immune response.

Immunoregulation Control of the immune response; may be a positive effect ('upregulation') or suppressive effect ('downregulation').

Immunosenescence Decline in immune function with advancing age.

Immunosuppression Inhibition of innate and/or adaptive immune responses by a disease process or drug treatment.

Inflammasome Multiprotein complex in the cytoplasm of cells induced by microbial and danger signals. The complex is composed of NLR receptors, adaptor proteins, and caspase-1. The inflammasome drives inflammation and a form of cell death called pyroptosis.

Inflammation A localized protective response to a stimulus (e.g. infection or trauma) aimed at containing and removing the injurious agent and restoring homeostasis. The cardinal signs of acute inflammation are heat, redness, swelling, pain and loss of function. Inflammation may be acute or chronic.

Inflammatory Bowel Disease Chronic diarrhea caused by dysregulation of the immune response to components of the intestinal microbiome.

Innate immunity A form of immunity that is evolutionarily older and provides immediate protection against infections.

Integrins Cell surface molecules consisting of an α and a β chain that serve as adhesion molecules with other cells and with the extracellular matrix.

Interferons A family of cytokines. The type I and type III interferons are antiviral molecules, whereas type II interferon (IFN-γ) has a key immunoregulatory function.

Interleukins A heterogeneous group of cytokines with autocrine (e.g., IL-2), paracrine (e.g., IL-4), and endocrine (e.g., IL-1) activity.

Intradermal (skin) test Intradermal injection of a small quantity of antigen to determine causative allergens in allergic disease. Formation of a local area of erythema and oedema (wheal) within minutes after injection is consistent with a type I hypersensitivity response.

Intravenous Immunoglobulin (IVIG) Immunoglobulins pooled from many human donors that is sometimes used in the management of immune-mediated diseases in companion animals.

Invariant chain Protein that binds to MHC II molecules in the endoplasmic reticulum. It prevents the binding of peptides in the ER, stabilizes the MHC II, and directs it to endosomal compartments.

J chain Joining chain that links together monomers of IgA or IgM into multimeric forms via their Fc regions.

Joining (J) gene segment Short nucleotide sequence between the variable and constant regions of an immunoglobulin or TCR locus. Random selection of a J gene segment contributes to the diversity of the antibody and TCR repertoire.

Keratoconjunctivitis sicca Lymphocyte-mediated destruction of the lacrimal glands leading to reduced production of tears and 'dry eye'.

Killed vaccine Vaccine containing an organism that has been killed in a way that retains its antigenic structure. Also 'inactivated vaccine'.

Langerhans cell A type of bone marrow-derived dendritic cell located within the epidermis.

Late-phase response Occurs several hours after an immediate hypersensitivity reaction and involves the infiltration of eosinophils, lymphocytes and basophils into affected tissue.

Lectin A protein that binds specific carbohydrate groups.

Lethal acrodermatitis Immunodeficiency disease affecting bull terrier dogs.

Ligand Molecule that binds specifically to a receptor.

Lipopolysaccharide See 'endotoxin'.

Lymph nodes Encapsulated secondary lymphoid organs present throughout the body that filter lymph and trap lymph-derived antigens. The structure of lymph nodes enables the efficient induction of antigen-specific adaptive immune responses.

Lymphadenomegaly Enlargement of lymph nodes.

Lymphadenopathy Pathological change in lymph nodes; often used to describe lymph node enlargement in clinical setting.

Lymphoblast Antigen-activated lymphocyte.

Lymphocyte recirculation The migration of naïve lymphocytes from blood into secondary lymphoid organs and via lymphatic vessels back into the blood circulation.

Lymphoid follicle Structures in secondary lymphoid organs within which B lymphocytes reside. Primary follicles contain circulating naïve B cells, whereas secondary follicles with a germinal centre and mantle zone reflect activation of the immune system by antigen.

Lymphoid leukaemia Neoplasia of lymphocytes that begins within the bone marrow with subsequent release of neoplastic cells into the circulating blood.

Lymphoma Tumour of lymphocytes presenting as diffuse infiltration or mass lesions of the viscera.

Lysosome Organelle in the cytoplasm of a phagocytic cell with an acidic pH and a collection of proteolytic enzymes.

Lysozyme An enzyme found in mucosal secretions and in macrophages and neutrophils that breaks down the cell wall of bacteria.

M cells Specialized epithelial cells found within the epithelial layer overlying mucosal lymphoid tissues, including Peyer's patches. Transport antigens from the lumen into the underlying lymphoid tissue.

Macrophages Bone marrow-derived phagocytic cells derived from circulating monocytes. Classical activation of macrophages by IFN-γ leads to M1 macrophages that are efficient in killing intracellular organisms and alternative activation generates M2 macrophages involved in tissue repair.

Major histocompatibility complex (MHC) Genetic locus that encodes for highly polymorphic molecules that present peptides to T cells. Also encodes other proteins involved in antigen processing, some complement proteins and cytokines.

Mantle zone Outer region of a secondary lymphoid follicle comprised of recirculating naïve B lymphocytes.

Marginal zone Region at the periphery of splenic white pulp that populated by macrophages and B cells.

Marker vaccine Vaccine containing a genetically modified organism that induces an immune response that can be distinguished from the immune response made to natural infections with field strains of the organism.

Membrane Attack Complex (MAC) Collection of terminal pathway complement components (C5-C9) that form a pore in the membrane of a target cell and lead to osmotic lysis.

Memory Ability of the adaptive immune system to make a more rapid and more potent immune response on re-exposure to an antigen in comparison with the primary response to that antigen. Underlies the process of vaccination.

β2-Microglobulin Small non-polymorphic protein that is part of MHC I molecules and the neonatal Fc receptor (FcRn).

Modified live vaccine Vaccine composed of pathogens with reduced virulence that can replicate, but do not cause disease. See also 'attenuation'.

Molecular mimicry Shared structure or epitope expressed by a pathogen and a self-molecule. Infection by such pathogens is a possible mechanism underlying autoimmune diseases.

Monoclonal antibody Antibody of a single specificity and avidity produced *in vitro* for research, diagnostic or therapeutic purposes.

Monoclonal gammopathy Single species of immunoglobulin secreted by neoplastic plasma cells in multiple myeloma that presents as a sharply defined peak in the gamma globulin region on serum protein electrophoresis. May occasionally occur in some infectious diseases (e.g. monocytic ehrlichiosis or leishmaniasis).

Mononuclear cells Leucocytes with a large round to oval nucleus that includes primarily lymphocytes and monocytes. The term peripheral blood mononuclear cells (PBMCs) is often used to refer to these populations in the blood.

Mucosa-associated lymphoid tissues (MALT) Non-encapsulated organized lymphoid tissues beneath the mucosal epithelium, including tonsils and Peyer's patches.

Multiple myeloma Malignant tumour of plasma cells that often secrete antibodies (monoclonal gammopathy).

Myasthenia gravis Autoimmune disease in which autoantibody binds the acetylcholine receptor at the neuromuscular junction. Inhibition of acetylcholine binding leads to muscle weakness.

Naïve lymphocyte Mature lymphocyte that has not previously encountered the antigen for which it has a specific receptor.

Natural killer cell Innate lymphoid cell with granular cytoplasm that mediates lysis of infected cells and tumor cells and secrete IFN-γ. Activation is determined by the balance between activating and inhibitory receptors. Can also kill via antibody-dependent cell-mediated cytotoxicity.

Negative selection Selection of those developing T lymphocytes within the thymus that bear a TCR able to recognize self-antigen with high affinity. These cells are deleted by apoptosis within the thymus. This process contributes to immunological tolerance.

Neonatal Fc-receptor (FcRn) Receptor expressed on neonatal enterocytes that transports IgG across the intestinal epithelium. Also expressed by endothelial cells throughout life which protects IgG from degradation and accounts for its long half-life.

Neonatal isoerythrolysis Disease caused by incompatability between blood group antigens between the dam and her offspring. Absorption of antibodies with specificity for the neonatal red blood cells from colostrum leads to haemolytic anaemia in the newborn animal.

Neonatal tolerance Failure of immune response to an antigen in adult life caused by exposure to that antigen *in utero* or during the early neonatal period.

NOD-like receptors (NLRs) Family of pattern recognition receptors located in the cytoplasm of cells that recognize pathogen-associated and danger-associated molecular patterns (PAMPs and DAMPs).

Nuclear factor of activated T cells (NFAT) Transcription factor in T cells required for the expression and secretion of IL-2. Inhibited by cyclosporin and tacrolimus.

Opsonization Coating of a particle with molecules such as IgG and C3b that enhance the effectiveness of phagocytosis.

Oral tolerance Suppression of the systemic immune reponse following oral administration of an antigen. The same effect can be induced by primary antigen exposure at other mucosal surfaces (e.g. nasal tolerance).

Paracortex Intermediate area of a lymph node that envelopes the cortical follicles. Area of the lymph node in which T lymphocytes reside.

Passive immunization Transfer of antibody from an immune donor to a recipient for treatment or to provide immediate protection.

Pathogen-associated molecular pattern (PAMP) A conserved molecule or molecular structure uniquely expressed by broad classes of microorganisms. PAMPs are recognized by pattern-recognition receptors (PRRs) and activate the innate immune system.

Pathogen Organism that causes disease or tissue damage when it infects a host.

Pattern Recognition Receptors (PRRs) Receptors expressed by cells of the innate immune system that recognize PAMPs and DAMPs.

Pelger-Huët anomaly Genetically mediated hyposegmentation of neutrophil nuclei of no apparent clinical significance.

Pemphigus A group of autoimmune blistering skin diseases characterized by the formation of vesicles/pustules within or below the epidermis.

Perforin A molecule released from a cytotoxic T cell or NK cell that polymerizes to form a pore within the cell membrane of the target cell, allowing the influx of granzymes that induce apoptosis.

Periarteriolar Lymphocyte Sheath (PALS) Area of white pulp of the spleen in which T lymphocytes reside. The T cells form a cylindrical sheath around the central arteriole.

Peyer's patches Unencapsulated organized lymphoid tissue within the small intestinal mucosa and submucosa.

Phagocytosis Internalization of a particle by a phagocytic cell (e.g. neutrophil or macrophage).

Phagolysosome Digestive vacuole within the cytoplasm of a phagocytic cell formed by the fusion of the phagosome with the enzyme-rich lysosome.

Phagosome Membrane-bound vesicle within the cytoplasm of a phagocytic cell containing phagocytosed particles.

Plasma cells Terminal differentiation stage of B lymphocytes. Plasma cells synthesize and secrete immunoglobulin.

Plasmacytoma A localized, benign tumour of plasma cells usually in the skin or mucous membranes.

Polyclonal An immune response by activating multiple clones of lymphocytes with different antigen specificities.

Polymeric Ig receptor (poly-IgR) Receptor expressed at the basolateral surface of mucosal epithelial cells that captures IgA or IgM and transfers these immunoglobulins across the mucosal barrier.

Polymorphism The presence of two or more variants of a gene (alleles) that occupy a genetic locus within a population. The different alleles encode for structurally different proteins.

Positive selection Selection of developing T lymphocytes within the thymus that bear a functional TCR able to recognize a peptide presented by a self MHC molecule.

Precipitation Interaction of soluble antigen and antibody to form a visible precipitate in solution or a line of precipitation in an agar gel.

Primary immune response Immune response made by an individual on first encounter with an antigen.

Primary lymphoid organs Organs in which lymphocytes develop and mature in an antigen-independent manner.

Programmed cell death A form of cell death ("cellular suicide") carried out by a series of regulated intracellular steps. Includes apoptosis and pyroptosis.

Proteasome A large multiprotein complex in the cytoplasm of cells that degrades proteins that are misfolded or no longer needed into peptides by proteolysis.

Pyrexia Fever.

Pyrogen Substance that induces fever.

Pyroptosis Form of programmed cell death induced by inflammasome-mediated activation of caspase-1 and associated with the release of IL-1β.

RAG1 and RAG2 Recombination-activating genes that encode for the V(D)J recombinase expressed exclusively in developing B and T cells.

Receptor editing Process by which immature B cells that recognize self antigens in the bone marrow can create an immunoglobulin with different specificity.

Recombinant vaccine Pure source of protein produced *in vitro* by inserting a gene into a vector (e.g. bacterial, insect or mammalian cell). The gene is expressed and the protein released from the cells into the culture medium. Such recombinant proteins can form the basis for vaccines.

Regulatory T cell (Treg) Population of T cells that inhibit immune responses. Tregs are CD4+ and express the transcription factor FOXP3.

Repertoire Immunologically, the complete set of antigen-specific receptors of B and T cells within an individual.

Rheumatoid factor Typically an IgM auto-antibody with specificity for IgG. Found in the serum and synovial fluid of patients with auto-immune polyarthropathies.

RIG-like receptors (RLR) Family of pattern recognition receptors in the cytoplasm of cells that are activated by viral RNA and induce the secretion of type I IFN.

Secondary immune response Immune response made upon secondexposure to an antigen. The secondary immune response occurs more rapidly, is more powerful and persists for a longer period.

Secondary lymphoid organs Organized lymphoid tissues in which adaptive immune responses to antigens are generated.

Secretory component Portion of the polymeric Ig receptor that remains bound to the Fc portion of an IgA (or IgM) molecule when it is released from the mucosal epithelial cell and that protects the molecule from enzymatic degradation.

Self-tolerance Failure of the normal animal to respond immunologically to self-antigens.

Sensitivity The probability that a diagnostic test will correctly identify those animals in a population that are affected by a particular disease (true positives).

Sensitization Exposure to an antigen resulting in an immune response.

Sepsis Dysregulated immune response to systemic infection resulting in life-threatening organ dysfunction. See also 'Systemic Inflammatory Response Syndrome'.

Serology The study of antigen-antibody interactions *in vitro*.

Seropositive Having serum antibodies against a specific antigen as detected by a serological test.

Severe Combined Immunodeficiency (SCID) Genetic mutation resulting in lack of functional T and B lymphocytes and severe impairment of the immune sytem.

Single radial immunodiffusion Serological technique based on the principle of precipitation of antigen and antibody. Often used to measure the concentration of immunoglobulin in serum.

Somatic mutation Mutations that spontaneously arise in genes in somatic cells (e.g. nucleotide substitutions or deletions) giving rise to diversity in the encoded proteins.

Specificity The probability that a diagnostic test will correctly identify those animals in a population that do not have the disease under consideration (true negatives).

Spleen Abdominal secondary lymphoid organ that is the major site for the induction of adaptive immune responses to blood-borne pathogens. Divided into red and white pulp.

Splenomegaly Enlargement of the spleen.

STING Stimulator of Interferon Genes. Adaptor molecule that binds cyclic dinucleotides generated by the cytosolic DNA sensor cyclic GMP-AMP synthase (cGAS) or by intracellular bacterial infections, resulting in secretion of type I IFN.

Subunit vaccine Vaccine containing an antigenic fragment of an organism rather than the entire organism itself.

Superantigen Molecules, typically secreted by bacteria, that activate a large number of T cells that express certain Vβ genes by crosslinking the TCR with MHC II molecules.

Systemic Inflammatory Response Syndrome (SIRS) Life-threatening systemic disease caused by excessive production of cytokines in response to disseminated infections and severe tissue damage resulting in the release of abundant PAMPs and DAMPs.

Systemic lupus erythematosus (SLE) Systemic autoimmune disease caused by excessive formation of immune complexes which often contain DNA and other nuclear antigens. The clinical manifestations are caused by deposition of immune complexes (type III hypersensitivity).

T cell receptors (TCR) Dimeric antigen receptors on the surface of T lymphocytes. The TCRαβ recognizes antigenic peptides bound to MHC molecules, whereas the TCRγδ recognizes other antigens without the MHC requirement.

T follicular helper (Tfh) cells A subpopulation of CD4+ T cells in germinal centers of secondary lymphoid follicles that provide signals for B cells to undergo isotype switching and differentiation into plasma cells or memory B cells.

T helper 1 (Th1) cells A subpopulation of CD4+ helper T lymphocytes that preferentially produces IFN-γ and stimulates cell-mediated immune responses.

T helper 2 (Th2) cells A subpopulation of CD4+ helper T lymphocyte that preferentially produces IL-4, IL-5, IL-9 and IL-13. Th2 cells-direct allergic inflammatory reactions and tissue repair.

T helper17 (Th17) cells A subpopulation of CD4+ T lymphocyte that preferentially produces IL-17and IL-22. Th17 cells play a protective role in mucosal bacterial and fungal infections and are involved in the pathogenesis of certain immune-mediated diseases.

Tacrolimus An immunosuppressive drug that inhibits the activation of NFAT in T cells. Also known as FK506.

T-dependent antigens Protein antigens that induce the generation of Tfh cells which provide help for B cell activation and differentiation.

Tertiary lymphoid organs Organized lymphoid tissues that develop at sites of chronic inflammation or in association with tumors.

Thoracic duct Major lymphatic vessel of the body into which all lymphatics flow. The thoracic duct in turn empties into the bloodstream to permit the recirculation of leucocytes within the body.

Thymic aplasia Congenital defect resulting in absence of the thymus. Often associated with hairlessness and results in impairment of T-cell-mediated immunity.

Thymus Bilobed primary lymphoid organ in the cranial mediastinum that is the site of T cell development and maturation.

T-independent antigens Antigens such as polysaccharides and lipids that can activate B cells without help from T cells.

Titre The inverse of the last serum dilution giving an unequivocally positive reaction in a serological test. Measure of the level of antibodies in a sample.

Tolerance A state of unresponsiveness of the immune system to a specific antigen following exposure to that antigen.

Toll-like receptors (TLRs) A family of pattern-recognition receptors present on the cell surface or in endosomes that recognizes PAMPs and DAMPs.

Tonsils Organized secondary lymphoid tissues in the pharynx that are sites for induction of adaptive immune responses to inhaled and ingested antigens.

Toxoids Inactive forms of toxins created through chemical treatment or genetic manipulation that are used as vaccine antigens.

Transporter associated with antigen processing (TAP) Dimer composed of TAP-1 and TAP-2 molecules that acts as an ATP-dependent transporter of peptides from the cytoplasm into the endoplasmic reticulum.

Tumor necrosis factor Cytokine that can induce apoptosis in cells and has a variety of regulatory and inflammatory effects.

Ubiquitin Polypeptide molecule that 'tags' a cytoplasmic protein for proteolytic degradation by the proteasome.

Urticaria Cutaneous reaction characterized by the sudden appearance of localized erythema and oedema (wheals). Most often secondary to a type I hypersensitivity reaction.

Vaccination Administration of a vaccine to induce immune-mediated protection against an infectious disease.

Vaccine An immunogenic preparation of an antigen used to induce immunity against an infectious disease. May contain inactivated or attenuated pathogens, subunits, or nucleic acids that encode for an antigen.

Vaccine efficacy Measure of how well a vaccine works, by preventing infection and/or the tissue pathology and clinical signs of disease. Determined experimentally by challenge of vaccinated versus unvaccinated animals with virulent organism or by field studies through natural exposure to the organism.

Valence The number of antigen binding sites carried by an Ig (e.g. IgG has a valence of two).

Variable (V) gene segments DNA sequences that encode for most of the variable domains of immunoglobulins and T cell receptors. Immunoglobulin light- and heavy chain loci and TCR loci contain many V gene segments. Selection of one of the V gene segments and recombination with D and J gene segments contributes to the specificity of the antigen receptors.

Vasculitis Inflammation of the wall of a blood vessel leading to increased permeability of the vessel.

Vasodilation Dilation of a blood vessel with increased permeability between endothelial cells allowing egress of fluid, protein and cells into the surrounding tissue.

Vector vaccine Vaccine containing a carrier organism (bacteria or virus) genetically modified with a gene that encodes for a microbial protein. Infection of cells by the carrier allows protein expression and induction of an immune response to that protein.

Virulence The ability of an infectious agent to produce tissue pathology and clinical disease.

Western blotting Serological technique used to detect a protein in a sample. Involves separation of the constituent parts of an antigen by polyacrylamide gel electrophoresis and transfer to a membrane, which is then incubated with enzyme- or fluorochrome-labeled antibodies. Also called 'immunoblotting'.

Wheal A small raised, erythematous and oedematous focus within the skin. Most often associated with type I hypersensitivity reactions.

Xenograft A tissue or organ transplanted from one species into another.

Note: Page numbers in *italic* indicate a figure and page numbers in **bold** indicate a table on the corresponding page.